Methods in Molecular Biology

Series Editor
John M. Walker
School of Life and Medical Sciences
University of Hertfordshire
Hatfield, Hertfordshire, UK

For further volumes:
http://www.springer.com/series/7651

For over 35 years, biological scientists have come to rely on the research protocols and methodologies in the critically acclaimed *Methods in Molecular Biology* series. The series was the first to introduce the step-by-step protocols approach that has become the standard in all biomedical protocol publishing. Each protocol is provided in readily-reproducible step-by-step fashion, opening with an introductory overview, a list of the materials and reagents needed to complete the experiment, and followed by a detailed procedure that is supported with a helpful notes section offering tips and tricks of the trade as well as troubleshooting advice. These hallmark features were introduced by series editor Dr. John Walker and constitute the key ingredient in each and every volume of the *Methods in Molecular Biology* series. Tested and trusted, comprehensive and reliable, all protocols from the series are indexed in PubMed.

Basophils and Mast Cells

Methods and Protocols

Second Edition

Edited by

Bernhard F. Gibbs

Division of Experimental Allergology and Immunodermatology, Department of Human Medicine, School of Medicine and Health Sciences, Carl von Ossietzky University of Oldenburg, Oldenburg, Germany

Franco H. Falcone

Institute for Parasitology, Justus Liebig University Giessen, Biomedizinisches Forschungszentrum Seltersberg (BFS), Giessen, Germany

Humana Press

Editors
Bernhard F. Gibbs
Division of Experimental Allergology
and Immunodermatology
Department of Human Medicine
School of Medicine and Health Sciences
Carl von Ossietzky
University of Oldenburg
Oldenburg, Germany

Franco H. Falcone
Institute for Parasitology
Justus Liebig University Giessen, Biomedizinisches
Forschungszentrum Seltersberg (BFS)
Giessen, Germany

ISSN 1064-3745 ISSN 1940-6029 (electronic)
Methods in Molecular Biology
ISBN 978-1-0716-0695-7 ISBN 978-1-0716-0696-4 (eBook)
https://doi.org/10.1007/978-1-0716-0696-4

This Humana imprint is published by the registered company Springer Science+Business Media, LLC, part of Springer Nature.
The registered company address is: 1 New York Plaza, New York, NY 10004, U.S.A.

Acknowledgements

The cover image shows a peripheral blood basophil immunostained with BB1 antibody, showing the presence of basogranulin (green) in the secretory granules (copyright Andrew F Walls and Akinori Mochizuki).

Preface to the Second Edition

The well-received first edition of our book, *Basophils and Mast Cells: Methods and Protocols*, highlights the need for detailed methods that focus on basophils and mast cells, especially the use of primary cells. This second edition builds on and extends the strengths of the previous one. The large majority of chapters have been revisited and, where necessary, updated by their original authors. In addition to refining existing methods, the new edition also contains newly developed methods provided by cutting-edge researchers in the field.

To accommodate the new chapters, we have expanded the structure of the book from originally four into six parts: after an introduction to the basic biology of basophils and mast cells in Part I, Part II focuses on obtaining the cells by purification or culture of stem cells progenitors. Here, a new chapter has been added by Siebenhaar and colleagues, describing the generation of peripheral $CD34^+$ stem cell-derived mast cells (Chapter 5). Similarly, MacGlashan describes two protocols for the generation of basophils from $CD34^+$ progenitors, starting with two different materials: the convenient (but more expensive) commercial sources of $CD34^+$ progenitors, or leukapheresis pack. As an interesting aside, MacGlashan reminds us in his introduction to Chapter 6 that although basophils are normally considered short-lived cells, their residence time in circulation (with estimates ranging from 12 h to a few days) is more likely to be just the tip of the iceberg, as a much longer time is spent during the process leading to maturation in the bone marrow, before they are released into circulation. Our understanding here is very patchy.

Part III of the book is now dedicated to basophil and mast cell lines. After the updated summary of the most commonly used mast cell and basophil lines by Passante (Chapter 10), three new chapters describe the use of humanized reporter systems. As a possible alternative to BATs and MATs, described in the next part, which require speedy processing of blood samples and flow cytometry equipment, IgE reporter systems are also available. These can be combined with allergen arrays, as described by Kalli and coauthors in Chapter 12. Barwary et al. (Chapter 13) describe a new, easy to measure humanized IgE reporter system. Here, the cells are preloaded with a red fluorescent protein targeted to the granules, which is then released upon activation, and can be conveniently measured without the need for expensive reagents or long incubation times. Of course, other research arenas have also advanced since the publication of our first edition, such as biomedical nanotechnology, and we therefore felt that it was important to include protocols tailored to the study of the interactions between nano- or submicron-scale materials and basophils or mast cells, as in Chapter 14 by Alelwani and her coauthors.

Part IV describes diagnostic applications. This part has also been expanded and updated. The new Chapter 16 by Ebo and coauthors now introduces BATs in more depth, in particular with respect to their use in understanding drug hypersensitivity reactions. Two useful flow cytometry techniques applied to basophils are presented by Elst and coauthors: in Chapter 17, a technique which, rather than detecting changes in surface expression of activation markers such as CD63 or CD203c, allows staining of exteriorized granular materials. The process takes advantage of the ability of fluorescently labeled avidin to bind to negatively charged proteoglycans contained in the granule matrix, which are exteriorized during degranulation; the method correlates well with CD63 upregulation [1]. In Chapter 18, the same authors remind us that not all activation pathways for mast cells and basophils are mediated via crosslinking of the high affinity IgE receptor and describe a protocol for the assessment of Mas-related G protein-coupled receptor (MRGPRX2) expression on the two cell types. The MRGPRX2 receptor is increasingly recognized as being involved in many systemic pseudo-allergic (or anaphylactoid) reactions to small molecules, such as basic secretagogues and several drugs [2]. Measuring cellular activation by flow cytometry is by no means limited to basophils. Similar techniques can be also applied to mast cells: the new Chapter 19 by Bahri et al. describes a novel method for mast cell activation tests (MATs), extending the instrumentarium available to researchers in our field.

Part V focuses on functional studies. The new Chapter 20 by Bechmann et al. should be of great interest to the reader planning to study gene expression patterns in basophils; as shown in this chapter, the method chosen to purify RNA has a strong bearing on the results (i.e., the quality and integrity of RNA) obtained—an important lesson in times where many labs rely on "convenient" commercial kits for this kind of work.

A new chapter now enables to assess the role of basophils in lung disease. Although basophils have been known for several decades to play an important role in asthma, and were described in sputum by Kimura and coworkers in 1975 [3], studying their role in different forms of asthmatic disease has been difficult due to the lack of appropriate protocols. The technique described by Fux et al. in Chapter 23 will help laboratories interested in examining the roles of basophils in different asthma phenotypes. Part VI closes with protocols for knockout and disease models.

In total, 11 new chapters have been added in this edition. We are confident that the second edition of our book will provide you with even more essential cutting-edge tools for your research into the functions of mast cells and basophils. Given that these cells appear to be involved in numerous diseases as well as regulating many physiological functions, there is a considerable unmet need to shed more light on these cell types. We hope that these protocols will be of great benefit and wish you every success in your research.

Oldenburg, Germany — ***Bernhard F. Gibbs***
Giessen, Germany — ***Franco H. Falcone***

References

1. Joulia R, Mailhol C, Valitutti S, et al (2017) Direct monitoring of basophil degranulation by using avidin-based probes. J Allergy Clin Immunol 140:1159–1162.e6

2. McNeil BD, Pundir P, Meeker S, et al (2015) Identification of a mast-cell-specific receptor crucial for pseudo-allergic drug reactions. Nature 519:237–241

3. Kimura I, Tanizaki Y, Saito K, et al (1975) Appearance of basophils in the sputum of patients with bronchial asthma. Clin Allergy 5:95–8

Preface to the First Edition

According to the latest epidemiological findings, it appears that allergic diseases will soon be affecting over 40% of the population in both Europe and elsewhere. There is, therefore, an overwhelming unmet need to unravel the immunological causes of allergic inflammation and to improve anti-allergic therapies. In recent years, it has come to light that mast cells and their blood-borne basophil counterparts not only contribute significantly to the symptoms of allergy but also play a role in supporting the underlying tendency of the immune system to respond in a pro-allergic manner. This immunomodulatory function has particularly led to a resurgence into basophil research and may one day finally uncover the biological role for this enigmatic and rare cell type which so readily releases a number of major inflammatory and immunomodulatory chemical mediators within very short periods of time. Moreover, the discovery that both mast cells and basophils potentially contribute to a variety of diseases, such as urticaria, cancer, and autoimmune diseases, as well as physiological functions, has increased the need for improved tools for researching these cells (*see* Chapter 1).

One of the most problematic features of mast cells and basophils is their striking differences in behavior and response to various stimulating agents and drugs depending on their location and the species from which they are studied. This limits, to some extent, our ability to extrapolate findings from, for example, mouse to human, but also tissue to tissue and the use of tumor mast cell and basophil lines. It is therefore imperative that better tools for the isolation of these cells from primary tissues are forthcoming. This indeed has been the case in recent years and we now present a book which, we feel, will significantly help anyone conducting research into these cells from the most practical point of view: definitive lab-based protocols targeted to those who actually do the work!

Whether you are planning to purify basophils or mast cells from peripheral blood (Chapter 2) or tissues (Chapter 3), or want to culture them from precursors from buffy coat blood (Chapter 4) or bone marrow (Chapter 5), you will find well-tested protocols in this book. Other chapters discuss the use of mast cells in organotypic skin models (Chapter 6), the difficulty of counting basophils (Chapter 7), how to stain mast cells and basophils in tissues (Chapter 8), how to assess mediator release (Chapter 10), and signal transduction (Chapter 15). Also discussed is the suitability of using mast cell or basophil-like cell lines as surrogates for primary mast cells and basophils (Chapter 9) and the use of reporter cell lines (Chapter 13) for the detection of allergen-specific IgE in serum samples. Chapter 12 presents protocols which can be used to measure mast cell migration using microscopy. Chapter 14 describes protocols that can be used for gene silencing of mast cells and basophils, two cell types that are notoriously difficult to transfect.

Researchers more interested in ex vivo or in vivo analysis will find protocols for human basophil activation tests (BATs) in Chapter 11, for mast cell phenotyping (Chapter 16), or the study of murine basophils by flow cytometry in Chapter 17. Chapter 18 closes with in vivo models that can be used for the analysis of mast cell functions.

Our book is written by leading research scientists in the field of basophil and mast cell biology. The impetus for this consortium of authors was given by the recently established EU/ESF-BMBS COST Action BM1007 "Mast Cells and Basophils—Targets for Innovative Therapies" which we would like to acknowledge for providing the ideal platform for scholarly collaboration in this field.

We sincerely hope that this methodology book will provide you with all the necessary tools for your research into mast cells and basophils, and we wish you every success in the quest of shedding more light on these fascinating cell types.

Oldenburg, Germany *Bernhard F. Gibbs*
Nottingham, UK *Franco H. Falcone*

Contents

Part III Basophil/Mast Cell Lines

Part IV Diagnostic Applications

Part V Whole Cells: Functional Studies

Part VI Whole Organism: Disease and Knockout Models

Contributors

MARCOS J. C. ALCOCER • *School of Biosciences, University of Nottingham, Nottingham, UK*
WALLA ALELWANI • *Department of Biochemistry, Faculty of Science, University of Jeddah, Jeddah, Kingdom of Saudi Arabia*
RAED A. ALHARBI • *Department of Laboratory Medicine, Faculty of Applied Medical Sciences, Albaha University, Albaha, Kingdom of Saudi Arabia*
CORNELIA AMALINEI • *"Grigore T. Popa" University of Medicine and Pharmacy, Iasi, Romania; Institute of Legal Medicine, Iasi, Romania*
METIN ARTUC • *Department of Dermatology and Allergy, Universitätsmedizin Berlin, Berlin, Germany*
MAGDA BABINA • *Department of Dermatology and Allergy, Universitätsmedizin Berlin, Berlin, Germany*
RAJIA BAHRI • *Lydia Becker Institute of Immunology and Inflammation, Division of Musculoskeletal and Dermatological Sciences & Manchester Collaborative Centre for Inflammation Research (MCCIR), School of Biological Sciences, Faculty of Biology, Medicine and Health, University of Manchester, Manchester, UK*
MONIKA BAMBOUSKOVÁ • *Laboratory of Signal Transduction, Institute of Molecular Genetics of the Czech Academy of Sciences, Prague, Czech Republic; Department of Pathology and Immunology, Washington University in St. Louis, St. Louis, MO, USA*
EWA A. BARTKO • *Laboratory for Medical Allergology, Herlev-Gentofte Hospital, 2900 Hellerup, Denmark*
NAFAL J. S. BARWARY • *Department of Biology, School of Science, University of Duhok, Duhok, Iraq*
KATRINE BAUMANN • *RefLab ApS, COBIS building, Copenhagen N, Denmark*
VICTOR G. BECHMANN • *Laboratory for Medical Allergology, Herlev-Gentofte Hospital, Hellerup, Denmark*
ANDREW J. BLOK • *Division of Molecular Therapeutics and Formulation, School of Pharmacy, University of Nottingham, Nottingham, UK*
ELENA BORZOVA • *Department of Dermatology and Venereology, I.M. Sechenov First Moscow State Medical University, Moscow, Russian Federation*
CHRIS H. BRIDTS • *Department of Immunology, Allergology, Rheumatology, University of Antwerp, Antwerp University Hospital, Antwerp, Belgium; Faculty of Medicine and Health Science, Department of Immunology, Allergology, Rheumatology and the Infla-Med Centre of Excellence, University of Antwerp, Antwerp University Hospital, Antwerp, Belgium*
SILVIA BULFONE-PAUS • *Lydia Becker Institute of Immunology and Inflammation, Division of Musculoskeletal and Dermatological Sciences & Manchester Collaborative Centre for Inflammation Research (MCCIR), School of Biological Sciences, Faculty of Biology, Medicine and Health, University of Manchester, Manchester, UK*
LUC S. DE CLERCK • *Department of Immunology, Allergology, Rheumatology, University of Antwerp, Antwerp University Hospital, Antwerp, Belgium; Faculty of Medicine and Health Science, Department of Immunology, Allergology, Rheumatology and the Infla-Med Centre of Excellence, University of Antwerp, Antwerp University Hospital, Antwerp, Belgium*

LEANDER DE PUYSSELEYR • *Faculty of Medicine and Health Science, Department of Immunology, Allergology, Rheumatology and the Infla-Med Centre of Excellence, University of Antwerp, Antwerp University Hospital, Antwerp, Belgium*

LUBICA DRÁBEROVÁ • *Laboratory of Signal Transduction, Institute of Molecular Genetics of the Czech Academy of Sciences, Prague, Czech Republic*

PAVEL DRÁBER • *Laboratory of Biology of Cytoskeleton, Institute of Molecular Genetics of the Czech Academy of Sciences, Prague, Czech Republic*

PETR DRÁBER • *Laboratory of Signal Transduction, Institute of Molecular Genetics of the Czech Academy of Sciences, Prague, Czech Republic*

DIDIER G. EBO • *Department of Immunology and Allergology, AZ Jan Palfijn Gent, Ghent, Belgium; Faculty of Medicine and Health Science, Department of Immunology, Allergology, Rheumatology and the Infla-Med Centre of Excellence, University of Antwerp, Antwerp University Hospital, Antwerp, Belgium*

JESSY ELST • *Faculty of Medicine and Health Science, Department of Immunology, Allergology, Rheumatology and the Infla-Med Centre of Excellence, University of Antwerp, Antwerp University Hospital, Antwerp, Belgium*

MARGARETHA A. FABER • *Faculty of Medicine and Health Science, Department of Immunology, Allergology, Rheumatology and the Infla-Med Centre of Excellence, University of Antwerp, Antwerp University Hospital, Antwerp, Belgium*

FRANCO H. FALCONE • *Institute for Parasitology, Justus Liebig University Giessen, Biomedizinisches Forschungszentrum Seltersberg (BFS), Giessen, Germany*

JASMINE FARRINGTON • *Department of Biomedical Science, University of Sheffield, Western Bank, Sheffield, UK*

STEFAN FRISCHBUTTER • *Department of Dermatology and Allergy, Charité—Universitätsmedizin Berlin, Berlin, Germany*

MICHAELA FUX • *University Institute of Clinical Chemistry, University Hospital Bern, Inselspital, Bern, Switzerland*

BERNHARD F. GIBBS • *Division of Experimental Allergology and Immunodermatology, Department of Human Medicine, School of Medicine and Health Sciences, Carl von Ossietzky University of Oldenburg, Oldenburg, Germany*

SVEN GUHL • *Department of Dermatology and Allergy, Universitätsmedizin Berlin, Berlin, Germany*

MARGO M. HAGENDORENS • *Faculty of Medicine and Health Science, Department of Immunology, Allergology, Rheumatology and the Infla-Med Centre of Excellence, University of Antwerp, Antwerp University Hospital, Antwerp, Belgium; Department of Pediatrics, University of Antwerp, Antwerp University Hospital, Antwerp, Belgium*

ANETTE TEO HANSEN • *Laboratory for Medical Allergology, Herlev-Gentofte Hospital, Hellerup, Denmark*

HANS JÜRGEN HOFFMANN • *Department of Respiratory Diseases and Allergy, Aarhus University Hospital and Department of Clinical Medicine, Aarhus University, Aarhus, Denmark*

BETTINA M. JENSEN • *Laboratory for Medical Allergology, Herlev-Gentofte Hospital, 2900 Hellerup, Denmark*

MARINA KALLI • *Division of Molecular Therapeutics and Formulation, School of Pharmacy, University of Nottingham, Nottingham, UK; School of Biosciences, University of Nottingham, Nottingham, UK*

JONGHUI KIM • *Department of Dermatology and Allergy, Universitätsmedizin Berlin, Berlin, Germany*

Edward F. Knol • *Department of Immunology, University Medical Center Utrecht, Utrecht, The Netherlands*
Yanyan Luo • *Department of Dermatology and Allergy, Charité—Universitätsmedizin Berlin, Berlin, Germany*
Donald MacGlashan Jr • *Asthma and Allergy Center, Johns Hopkins University, Baltimore, MD, USA*
Lina Mayorga • *Allergy Research Group, Instituto de Investigación Biomédica de Málaga-IBIMA and Allergy Unit, Hospital Regional Universitario de Málaga, ARADyAL, Malaga, Spain*
Christel Mertens • *Department of Immunology, Allergology, Rheumatology, University of Antwerp, Antwerp University Hospital, Antwerp, Belgium; Faculty of Medicine and Health Science, Department of Immunology, Allergology, Rheumatology and the Infla-Med Centre of Excellence, University of Antwerp, Antwerp University Hospital, Antwerp, Belgium*
Luis María Escribano Mora • *Instituto de Estudios de Mastocitosis de Castilla La Mancha, Hospital Virgen del Valle, Toledo, Spain; Spanish Mastocytosis Network (REMA), Salamanca, Spain; Centro de Investigación del Cáncer, Salamanca, Spain*
José Mário Morgado • *Instituto de Estudios de Mastocitosis de Castilla La Mancha, Hospital Virgen del Valle, Toledo, Spain; Spanish Mastocytosis Network (REMA), Salamanca, Spain; Cytognos S.L. Biotechnology, Salamanca, Spain*
Ryosuke Nakamura • *Japan Agency for Medical Research and Development (AMED), Tokyo, Japan*
Ana Olivera • *Mast Cell Biology Section, Laboratory of Allergic Diseases, National Institute of Allergy and Infectious Diseases, National Institutes of Health (NIH), Bethesda, MD, USA; Molecular Immunology Section, Laboratory of Molecular Immunogenetics, National Institute of Arthritis and Musculoskeletal and Skin Diseases, National Institutes of Health, Bethesda, MD, USA*
Egle Passante • *Sanofi Genzyme, Berkshire, UK*
Peter T. Peachell • *Department of Infection, Immunity and Cardiovascular Disease, The Medical School, University of Sheffield, Sheffield, UK*
Lars K. Poulsen • *Laboratory for Medical Allergology, Herlev-Gentofte Hospital, Hellerup, Denmark*
Juan Rivera • *Molecular Immunology Section, Laboratory of Molecular Immunogenetics, National Institute of Arthritis and Musculoskeletal and Skin Diseases, National Institutes of Health, Bethesda, MD, USA*
Lionel Rohner • *University Institute of Clinical Chemistry, University Hospital Bern, Inselspital, Bern, Switzerland*
Zuzana Rubíková • *Laboratory of Biology of Cytoskeleton, Institute of Molecular Genetics of the Czech Academy of Sciences, Prague, Czech Republic*
Vito Sabato • *Department of Immunology and Allergology, AZ Jan Palfijn Gent, Ghent, Belgium; Faculty of Medicine and Health Science, Department of Immunology, Allergology, Rheumatology and the Infla-Med Centre of Excellence, University of Antwerp, Antwerp University Hospital, Antwerp, Belgium; Department of Immunology and Allergology, AZ Jan Palfijn Gent, Ghent, Belgium*
Laura Sánchez-Muñoz • *Instituto de Estudios de Mastocitosis de Castilla La Mancha, Hospital Virgen del Valle, Toledo, Spain; Spanish Mastocytosis Network (REMA), Salamanca, Spain*

JÖRG SCHEFFEL • *Department of Dermatology and Allergy, Charité—Universitätsmedizin Berlin, Berlin, Germany*
CHRISTIAN SCHWARTZ • *Institute of Clinical Microbiology, Immunology and Hygiene, University Hospital Erlangen, Erlangen, Germany*
ELIZABETH P. SEWARD • *Department of Biomedical Science, University of Sheffield, Western Bank, Sheffield, UK*
FRANK SIEBENHAAR • *Department of Dermatology and Allergy, Charité—Universitätsmedizin Berlin, Berlin, Germany*
PER S. SKOV • *RefLab ApS, COBIS building, Copenhagen N, Denmark*
SNOW STOLNIK • *Division of Molecular Therapeutics and Formulation, School of Pharmacy, University of Nottingham, Nottingham, UK*
VADIM V. SUMBAYEV • *Medway School of Pharmacy, University of Kent, Chatham Maritime, UK*
EMILY J. SWINDLE • *School of Clinical and Experimental Sciences (MPT 810, Rm LF73, SAB, Level F), Faculty of Medicine, University of Southampton, University Hospital Southampton, Southampton, UK*
CRISTINA TEODÓSIO • *Spanish Mastocytosis Network (REMA), Salamanca, Spain; Centro de Investigación del Cáncer, Salamanca, Spain*
ATHINA VAN GASSE • *Faculty of Medicine and Health Science, Department of Immunology, Allergology, Rheumatology and the Infla-Med Centre of Excellence, University of Antwerp, Antwerp University Hospital, Antwerp, Belgium; Faculty of Medicine and Health Science, Department of Pediatrics, University of Antwerp, Antwerp University Hospital, Antwerp, Belgium*
MICHEL VAN HOUDT • *Faculty of Medicine and Health Science, Department of Immunology, Allergology, Rheumatology and the Infla-Med Centre of Excellence, University of Antwerp, Antwerp University Hospital, Antwerp, Belgium*
DRITON VLLASALIU • *Institute of Pharmaceutical Science, King's College London, London, UK*
DAVID VOEHRINGER • *Infektionsbiologische Abteilung, Universitätsklinikum Erlangen, Erlangen, Germany*
ANDREW F. WALLS • *Southampton General Hospital, University of Southampton, Southampton, UK*
DANIEL WAN • *University of Nottingham, School of Pharmacy, Nottingham, UK; Merck Group, Glasgow, Scotland, UK*
XIAOWEI WANG • *School of Pharmacy, University of Nottingham, Nottingham, UK; School of Biosciences, University of Nottingham, Nottingham, UK*
TORSTEN ZUBERBIER • *Department of Dermatology and Allergy, Universitätsmedizin Berlin, Berlin, Germany*

Part I

Introduction and Historical Background

Chapter 1

Paradigm Shifts in Mast Cell and Basophil Biology and Function: An Emerging View of Immune Regulation in Health and Disease

Ana Olivera and Juan Rivera

Abstract

The physiological role of the mast cell and basophil has for many years remained enigmatic. In this chapter, we briefly summarize some of the more recent studies that shed new light on the role of mast cells and basophils in health and disease. What we gain from these studies is a new appreciation for mast cells and basophils as sentinels in host defense and a further understanding that dysregulation of mast cell and basophil function can be a component of various diseases other than allergies. Perhaps the most important insight reaped from this work is the increasing awareness that mast cells and basophils can function as immunoregulatory cells that modulate the immune response in health and disease. Collectively, the recent knowledge provides new challenges and opportunities toward the development of novel therapeutic strategies to augment host protection and modify disease through manipulation of mast cell and basophil function.

Key words Allergy, Atherosclerosis, Autoimmunity, Basophils, Cancer, Dermatitis, Immunoglobulin E, Infection, Mast cells, Proteases

1 Introduction

It is more than a century since mast cells and subsequently basophils were described by Paul Ehrlich based on the staining of their "metachromatic" granules with basic aniline dyes [1–4]. For many years, thereafter, the biological role of mast cells and basophils has remained enigmatic. Mast cells are known to be localized around blood vessels in connective tissues, but they are not part of the perivascular system [5]. These cells are long-term tissue residents with slow turnover; they fully differentiate in tissues but still retain the ability to proliferate and replenish themselves [6]. On the other hand, the basophil was for many years viewed as a blood "mast cell" and because of its rarity (~1% of blood leukocytes) had long been considered as a redundant cell type or possibly a circulating precursor of tissue mast cells [7]. Features common to both cell types

Bernhard F. Gibbs and Franco H. Falcone (eds.), *Basophils and Mast Cells: Methods and Protocols*, Methods in Molecular Biology, vol. 2163, https://doi.org/10.1007/978-1-0716-0696-4_1,

include the presence of preformed metachromatic granules, as described by Paul Ehrlich [1–5, 8], that contain a vast array of releasable allergic mediators [9], the surface expression of the high-affinity receptor for IgE [5, 10], and the ability to similarly synthesize T helper 2-polarizing cytokines and chemokines de novo [5, 10, 11]—characteristics that have served to cement the view of a common biological function. Nonetheless, beyond the difference in where the cells are localized in vivo, there are distinct features of both cell types: basophil maturation occurs in the bone marrow and not in peripheral tissues as seen for mast cells; unlike mast cells, basophils do not proliferate after maturation; and the life span of a mast cell can be weeks (perhaps months), whereas that of a basophil is normally measured in days (however, *see* Chapter 6 for discussion of maturation of basophils in the bone marrow). Moreover, both murine and human basophils express intracellular and cell surface molecules that are distinct from those expressed on a mast cell [7, 12]. These differences suggest discrete in vivo functions for these cells, but only recently have we begun to appreciate their specific roles. This new awareness arises in large part as a result of novel discoveries on previously unrecognized roles for both mast cells and basophils in health and disease [7, 13].

In this introductory chapter, we can only capture a few of these recent discoveries. For the sake of brevity, we cannot summarize the vast recent literature that has contributed to all the new insights in this field, but we hope the examples presented herein set the tenor for the following chapters, which will provide additional details. The rapidly expanding awareness of a role for mast cells and basophils in innate immunity and as immunomodulatory cells [14] has driven much of this new investigation. While there is still some controversy on some of the specific roles assigned to these cells [15, 16], there is an increasing appreciation that mast cells and basophils play a role in immune protection and, as important players in the immune system, can also become dysregulated and contribute to diseases, other than allergic disease. Such discoveries underlie a shifting paradigm in mast cell and basophil biology and function.

2 Novel Roles for Mast Cells and Basophils: Protecting the Host

2.1 An Unexpected Protective Role of Mast Cells in Atopic Dermatitis

Contact dermatitis is a common form of skin inflammation that may appear after exposure to an irritant or, usually, an allergen. In mice, studies on allergic contact dermatitis are normally conducted using contact hypersensitivity models (CHS) (*see* Chapter 28), also known as delayed-type hypersensitivity reactions (DTH) [17]. In these models, mice are sensitized and consequently challenged epicutaneously with a hapten such as dinitrofluorobenzene (DNFB), oxazolone, fluorescein isothiocyanate (FITC), or trinitrochlorobenzene (TNCB). This elicits a local immune reaction with characteristics of allergic contact dermatitis. In contrast to these models that normally employ a single challenge, other models use periodic challenges with these haptens, producing a chronic inflammatory response that is more akin to human atopic dermatitis [18]. The role of mast cells in CHS or delayed-type IV hypersensitivity reaction is still controversial because depending on the type of mast cell-deficient mice or the model of CHS, the elicitation of the inflammatory response was reduced, enhanced, or unchanged [15, 16, 19–21]. This may not be surprising considering the immunomodulatory functions of mast cells (discussed in more detail in a following section) and the variety of protocols and models used to elicit CHS (with different types of haptens, concentrations of hapten, solvent, sites of sensitization, and times during the course of CHS measuring inflammation), which may affect some aspects of the immune response. In fact, Norman et al. showed that the mast cell-deficient $Kit^{w/w\text{-}v}$ mice had a more severe response than their wild-type counterparts (mast cell sufficient) after elicitation with high doses of oxazolone, but when low doses of this irritant were used, the $Kit^{w/w\text{-}v}$ mice had reduced responses [22]. These experiments are important as they reflect how the strength of the stimulus elicits an immune response that may require that mast cells serve a dual role (protective versus exacerbating). Similar to the strength of stimulus, all other factors mentioned above may differentially promote the type of immune response, what particular immune cells are involved, and which pro- or anti-inflammatory components are induced in the local environment and, thus, which mast cell responses are elicited and their specific role in the inflammatory process.

In models of chronic skin inflammation that show hallmarks of human atopic dermatitis (including barrier abnormalities, infiltration of mast cells and eosinophils, elevated IL-4 and IgE, and a predominant Th2-type of inflammation [23] with a late-phase Th-1 component (Hershko and Rivera, unpublished observation)), mast cells have a predominant role dampening inflammation [21]. Instead of a single challenge, oxazolone applied repeatedly over the course of 20–25 days caused chronic skin inflammation that was markedly exacerbated in mast cell-deficient $Kit^{W\text{-}sh/W\text{-}sh}$ mice [21]. Interestingly, the suppression of inflammation by mast cells was dependent on their production of IL-2 (a cytokine not normally associated with the biological function of mast cells in vivo), since the enhanced disease in $Kit^{W\text{-}sh/W\text{-}sh}$ mice was reduced when the mice were engrafted with wild-type mast cells but not with IL-2-deficient mast cells. The view of IL-2-mediated suppression in atopic dermatitis was further supported by a similarly exacerbated skin inflammation in IL-2-deficient mice or in wild-type mice treated with anti-CD25 antibody (which binds and transiently inactivates the IL-2 receptor). Of particular interest, however, was that the rescue of the suppressive role of mast cells in *Kit*$^{W\text{-}sh/W\text{-}sh}$ mice after engraftment of wild-type mast cells was observed even when these cells were not engrafted at the site of inflammation (by intradermal reconstitution). Engraftment distal to the local site (by i.v. reconstitution, which results in engraftment in the spleen, lymph nodes, and other tissues other than the skin) was sufficient to promote the protective effect and dampen the inflammation. Several lines of evidence suggested that suppression of skin inflammation by mast cells occurred in the spleen: in *Kit*$^{W\text{-}sh/W\text{-}sh}$ mice engrafted with wild-type mast cells, production of IL-2 by these cells was observed in the spleen. In addition, the numbers of mast cells increased in this organ (together with the skin and lymph nodes) after induction of chronic inflammation, and the increase of mast cells in the spleen showed a marked association with increased disease intensity. Thus, as the inflammation in the skin worsened, more mast cells were recruited to the spleen, and this recruitment plateaued at a point where disease intensity entered a steady state. At the local site of inflammation, expansion of regulatory T cells (Tregs) was observed in wild-type mice. In contrast, in *Kit*$^{W\text{-}sh/W\text{-}sh}$ mice reconstituted with IL-2-deficient mast cells, Treg expansion was reduced but could be rescued by reconstitution of *Kit*$^{W\text{-}sh/W\text{-}sh}$ with wild-type mast cells. These findings suggested that the recruitment of mast cells to the spleen during oxazolone -induced chronic allergic dermatitis played an important role in regulating Treg expansion and increased the numbers of these cells at the local site of inflammation. It is interesting that systemic rather than local mast cell production of IL-2 (in the skin) led to Treg expansion, but one might speculate that such circumstances might allow mast cells to play a dual role in both promoting (at the local site) and preventing (by distal regulation) inflammation.

The protective role for mast cells in DTH reactions was also manifested when DTH was elicited secondarily to the inflammatory response induced by either UVB irradiation [24] or by mosquito bites [25]. In the latter study, downregulation of the skin inflammatory response by mast cells was attributed to their production of IL-10 at a distal site, likely in the lymph nodes since *Kit*$^{W\text{-}sh/W\text{-}sh}$ mice showed reduced IL-10 production in the lymph node compared to wild-type mice or mast cell-reconstituted *Kit*$^{W\text{-}sh/W\text{-}sh}$ mice [25]. IL-10-mediated mast cell-dependent suppression was also found in long-term DTH responses (up to 15 days after a single epicutaneous challenge with a hapten) or after UVB chronic irradiation. Reconstitution experiments using IL-10-deficient mast cells in *Kit*$^{w/wv}$ mice demonstrated a requirement for mast cell produced IL-10 in these models [26]. Thus, although the cytokine involved in suppressing inflammation differs depending on the type of model studied, overall, the findings demonstrate an anti-inflammatory role of mast cells in long-term DTH [26], late stages of chronic allergic dermatitis [21], or short-term DTH responses secondary to another inflammatory stimulus like UVB irradiation [24] or mosquito bites [25]. These findings reveal an unsuspected role for mast cells in controlling the inflammatory response, and thus one must consider if manipulation of the ability of mast cells to produce such cytokines may be of potential interest in controlling skin diseases in humans.

2.2 Protection by Basophils to Reinfection (Hookworms, Ticks)

The innate immune contribution of mast cells to protection from infection by various species of parasites (reviewed in [13, 16, 27]) has long been recognized. Interestingly, however, even though peripheral basophilia during parasite infections has long been recognized [28], the actual involvement of basophils in parasite-elicited T-helper type 2 responses, and their role in parasite expulsion had not been fully appreciated. This was, in large part, due to the erroneous belief that mice did not have basophils, and thus studies in mouse models were hampered by this notion. Subsequent recognition that indeed basophils were present in mice [29–31] (*see* Chapter 29 for identification of murine basophils), and the discovery of antibodies that depleted basophils in vivo [7] as well as the development of basophil-null [32] and basophil ablation mouse models [33] ushered in a new era of investigation on the role of these cells in health and disease. While basophils comprise only about 1% of circulating leukocytes, they are the major innate immune cell-producing IL-4 after helminth infection [34–36]. At the moment, it is unclear under what circumstances the IL-4-producing role of basophils is essential for host protection versus situations where IL-4 production by basophils may play a contributory role in Th2-mediated immunity. Evidence of the latter role is provided by models of basophil deficiency [32, 33], or transient depletion of basophils by injection of anti-FcεRI or anti-CD200R3 mAbs [37], which demonstrated no effect on the immune response and the expulsion of the *Nippostrongylus brasiliensis* (*N. brasiliensis*) following initial infection but showed a contributory role in worm expulsion when Th2 cells were absent [38]. These findings suggested that during the initial infection, the role of basophils is redundant. During a secondary infection to *N. brasiliensis*, however, the presence of basophils seemed critical for the expulsion of this helminth, regardless of the presence or absence of mast cells or CD4 T cells, which were both required for protection during the initial infection [32]. The mechanisms governing basophil-mediated protection against reinfection are not well understood. Nonetheless, because IL-4 and IL-13 are important for Th2 cell polarization, effector cell recruitment, and tissue responses to worm infection [39], and basophils can abundantly secrete both cytokines during these infections [34, 35, 38, 40], part of the protection may be mediated through the production of IL-4/IL-13 [32]. This concept is partly supported by the lack of recruitment of eosinophils to mediastinal lymph nodes in the absence of basophils [32] and by the requirement of STAT6 (which is activated by IL-4Rα) for worm expulsion during repeated infections [37]. A less well-appreciated function of basophils (most clearly demonstrated in the mouse) is their capacity to express MHC class II and other surface molecules, like the B-cell survival and activation factor (BAFF) [41–43], which may confer the ability to prime naïve T cells into Th2 cells and also increase humoral responses through support of B-cell differentiation and survival. The expression of such molecules in human basophils is less well defined, yet some evidence for basophil MHC class II expression (HLA-DR) has emerged from studies of the role of basophils in systemic lupus erythematosus (discussed below) [44–46]. While the role of such molecules on basophils is not completely understood, they may play a role during parasite infections and help contribute to basophil-induced protection [37, 47].

A role for basophils in the protection against tick reinfestation was also recently demonstrated. These studies used antibody-mediated depletion of basophils and an elegantly engineered mouse that allows for diphtheria toxin-induced, basophil-specific ablation [33]. After repeated infestations, many animals develop resistance to ticks, but the nature of this anti-tick immunity was unclear. Wada and colleagues showed that upon repeated infestation, engorgement of these ectoparasites is diminished and is associated with recruitment of basophils to the tick-feeding site. This resistance to reinfection was transferred by sera of previously tick-infected animals, indicating the importance of a component of the humoral immune response. Furthermore, they showed that the presence of immunoglobulin Fc receptors on the basophils was required [33]. While the specific mechanism for the basophil-mediated protective effect and which basophil-effector molecules involved has not been investigated, these findings also demonstrate that basophils function in protection to reinfection by parasites. Further studies will be needed to investigate how basophils mediate this protection and to determine if these findings can be extended to other species of parasites and to parasite infections of humans.

2.3 The Protective Effect of Mast Cell and Basophil Proteases

An area of considerable interest that has emerged from studies of the protective role of mast cells and basophils is that of the role of mast cell proteases in host defense. While the release of mast cell proteases (MCP) from activated mast cells or basophils is known to promote inflammation in certain pathological conditions, such as asthma or arthritis, their role is not restricted to only these pro-inflammatory functions [48]. In many circumstances their actions have been shown to be protective instead of harmful, contributing to the yin-yang of protective and harmful consequences of mast cell- and basophil-derived products.

Proteases cleave a variety of functionally diverse protein substrates through recognition of specific peptide sequences. Cleavage can result in protein activation or inhibition, and thus it is not surprising that the overall physiopathological impact of mast cell proteases depends on the environmental circumstances in which they are released. When mice are exposed to venoms from certain honeybees, snakes, lizards, and scorpions, mast cell carboxypeptidase 3 (CPA3) and other mast cell proteases are essential for limiting the toxicity, morbidity, and mortality. These venoms contain toxins (i.e., sarafotoxins and helodermin in snake and lizard venoms, respectively) that share structural similarities with endogenous mammalian peptides (endothelin-1 and VIP, respectively) and bind their corresponding receptors in mast cells, thus causing them to degranulate [49, 50]. While mast cell degranulation causes changes in vascular permeability, local inflammation, and anaphylaxis, the symptoms to intradermal injection of these venoms were more severe and even lethal in mice lacking mast cells, CPA3, or the corresponding peptide receptors, indicating that mast cell CPA3-mediated proteolytic degradation of these toxins benefits the host [49, 51].

Importantly, the control of toxicity by mast cells is not limited to animal venoms. These cells can protect against hypothermia, diarrhea, and death induced by an injection of endothelin-1 [50] or VIP [49], findings that may have implications in multiple pathologies, such as sepsis, where any of these or other vasoactive peptides are elevated. Certain models of moderate to moderately severe sepsis have revealed a protective function of mast cells using kit-dependent mast cell-deficient mouse models [52, 53], and the findings suggest that these protective effects may be mediated by mast cell-derived cytokines and/or mast cell proteases. For example, an interesting study in a model of severe septic peritonitis by cecal ligation and puncture [54] showed that the peptide neurotensin, which is increased and contributes to hypotension and mortality in this model, can be degraded by activated mast cells [54, 55] and can cause a reduction in the levels of neurotensin diminishing its negative effects during sepsis [54]. These authors identified neurolysin as the main mast cell protease involved in the degradation of neurotensin [54]. Similarly, other studies described a protective role for mast cell-derived mMCP4 in eliminating TNF, which can promote inflammation [53, 56], during moderately severe septic peritonitis [56]. Others have shown an essential role for mMCP6 in the clearance of *K. pneumoniae* injected intraperitoneally [57]. In addition to the models of peritonitis, mast cell-derived mMCP4 was shown to play a key role in restricting inflammation in the brain after trauma [58], and mMCP4 [59] and CPA [60], presumably derived from mast cells or basophils, were found to reduce lung inflammation during allergic asthma by degrading, respectively, IL-33 and IL-13 [59]. Thus, an emerging view from such studies is that mast cell and basophil proteases can serve not only to promote inflammation but may be essential in limiting toxicity and infection.

3 Unexpected Physiopathological Roles of Mast Cells and Basophils

3.1 Adverse Effects of Mast Cell Protection in Atopic Dermatitis: The Atopic March

Recent studies in our lab indicated that mice subjected to chronic allergic dermatitis induced by repeated exposure to a hapten were more susceptible to airway inflammation induced by a different allergen [61], a phenomenon that resembles the increased incidence and severity of asthma after AD in humans commonly known as the atopic march [62]. Thus, as postulated, occurrence of atopic dermatitis in early life increases susceptibility and/or severity of asthma or other allergies in later life. As described in the previous section, our findings indicated that mast cells suppressed skin inflammation in a chronic allergic dermatitis model [21], and thus it was of interest to determine if in such a model, one might observe a relationship between atopic dermatitis and allergic asthma. Interestingly, the findings indicated an enhanced sensitivity to the inhaled allergen following the resolution of dermatitis, which was dependent on mast cells since an enhanced asthmatic response was not seen in mice lacking mast cells ($Kit^{W\text{-}sh/W\text{-}sh}$) but was recovered when such mice were reconstituted with bone marrow-derived mast cells prior to the induction of dermatitis [61]. In fact, the findings showed that the increased airway reactivity was associated with a persistent increase in the overall number of mast cells after the induction and resolution of atopic dermatitis. However, how an increase in mast cell numbers results in enhanced sensitivity to an oral challenge is unclear. Similar to their role in suppressing allergic chronic dermatitis, their potentiating effect in airway sensitivity did not appear to be consequence of a direct effector function in the local environment because mast cell numbers were not increased in the lung but only in distal sites such as the skin, lymph nodes, and spleen. This was particularly obvious in $Kit^{W\text{-}sh/W\text{-}sh}$ mice that were intradermally reconstituted with mast cells, since they showed increased airway hypersensitivity following atopic dermatitis even though no mast cells could be detected in the lungs [61].

The mechanism(s) underlying this distal effect of mast cells on airway reactivity is unclear. Nonetheless, it was previously shown that activation of mast cells in the skin at the time of immunization was effective in enhancing an antibody response to vaccine antigens [63] presumably by recruiting or activating dendritic cells and/or lymphocytes to the lymph nodes [64, 65]. Similarly, mast cells that are elevated in numbers after AD may potentially promote interactions with dendritic cells or T cells (discussed further below) in the lymph nodes to facilitate an immune response to an inhaled challenge. Regardless, the idea that mast cells can suppress allergic dermatitis while promoting airway hypersensitivity highlights the plasticity of the mast cell [66], which is likely influenced by the overall pathological environment of the disease. Consistent with this view, while mast cells produced IL-2 in the spleen during chronic dermatitis [21], after resolution of the dermatitis, they no longer produced IL-2, and induction of this cytokine was not detected after provocation of asthma [61]. This change in phenotype suggests that such transformations ultimately shape the type of response by the mast cell and thus its role in health and disease.

3.2 An Unexpected Role for Mast Cells in Obesity

Obesity is a condition characterized by excess adiposity, metabolic dysregulation in white adipose tissue, and a chronic low-grade inflammation in various organs, thereby increasing the incidence of insulin resistance, diabetes, cardiovascular disease, and other diseases [67]. Although the exact mechanisms involved in the progression of this condition and the specific contribution of immune cell types is not clear, it is generally accepted that nutrient excess causes vascular changes in the tissue, infiltration of immune cells, and chronic overload of macrophages, which then causes metabolic dysfunction in the adipocytes, as well as induction of inflammatory cytokine production by immune cells. Mast cells, which are present in adipose tissue, as well as other innate and adaptive immune cells, have been implicated in the initiation and progression of obesity [67–69]. Interestingly, increase in the epididymal fat tissue and other fat depots in mice with diet-induced obesity correlates with increased mast cell numbers in these tissues [70–72]. In humans, the white adipose tissue of obese subjects contains higher numbers of mast cells than that from lean donors, and these cells are localized in the proximity of blood vessels [71], particularly in diabetic individuals [73]. Furthermore, these cells were also found preferentially localized in fibrosis bundles and their presence correlated, as in mice [71], with macrophage accumulation and endothelial cell inflammation [73]. In obesity-related glomerulopathy, mast cells infiltrate renal tubulointerstitial regions, and their presence is associated with tubular atrophy and interstitial fibrosis [74]. These mast cells had a higher tryptase or chymase content [71, 73] than those in normal individuals and might be activated in this disease, since subjects with body mass index $\geq$32 kg mm^{-2} had significantly higher serum tryptase concentrations than lean individuals in two independent studies [71, 75]. In contrast, this correlation was not found in a smaller cohort study of children [76], and thus, whether the absence of correlation may be due to a smaller cohort or to age-dependent factors or other unknown variables is not clear. Regardless, adipose tissue has been described as a reservoir of committed mast cell progenitors with the endowed potential to be home to peripheral tissues [77]. Thus, the possibility exists that nutrient overload may directly or indirectly affect the differentiation of these progenitors into mature mast cells and promote their activation or homing to other tissues.

A mouse study that demonstrates a role for mast cells in obesity employed two models of Kit-dependent mast cell-deficient mice, $Kit^{W\text{-}sh/W\text{-}sh}$ and $Kit^{W/Wv}$, fed with a Western-type diet consisting of high fats and sugar [71]. These mice gained significantly less body weight than wild-type congenic controls on the same diet [71, 72]. The lower weight gain was accompanied by significantly lower concentrations of serum leptin, insulin, and various cytokines, and these mice had a higher resting metabolic rate and higher glucose tolerance and insulin sensitivity than WT mice [71]. The adoptive transfer of wild-type bone marrow-derived mast cells in such mast cell-deficient mice caused a gain in body weight as well as recovery of most of the aforementioned parameters. While reconstitution of the mast cell-deficient mice with mast cells from TNF-deficient mice resulted in a similar weight gain as mice reconstituted with wild-type mast cells, mice reconstituted with mast cells from IL-6 or INF γ-deficient mice had poor weight gain, indicating that mast cell-derived IL-6 and INF γ play a role in obesity progression but not TNF. Mast cell-stabilizing drugs, such as cromolyn and ketotifen, also prevented diet-induced obesity and improved preestablished obesity and diabetes in mice that had consumed a Western diet for 12 weeks, suggesting a role for mast cells not only in the onset but also in the progression of obesity [71]. A caveat of this work is that the discovery of a role for mast cells in diet-induced obesity was almost entirely based on Kit-dependent mast cell-deficient mouse models, which clearly have mast cell-unrelated defects [15, 16]. Moreover, it should be noted that the efficacy and selectivity of cromolyn at inhibiting mast cell functions have been questioned [78]. Yet, taken together with the correlative studies in other mouse models and in humans, this study makes a strong argument for the role of mast cells in the promotion of obesity.

As mentioned above, there is evidence of a requirement for mast cell cytokines (IL-6 and IFN-γ) in promoting obesity, but the exact mechanism by which these products might regulate obesity is not understood. Given the involvement of mast-derived products in immune cell recruitment, activation, and in the production of cytokines that can affect the surrounding endothelial cells, fibroblasts, and smooth muscle cells, the consequences of accumulating mast cells in adipocyte tissues may be manyfold [79]. While the production of cytokines by mast cells may contribute to the pro-inflammatory nature of obesity, it should be noted that the production of prostanoids by mast cells may also promote adipocyte dysfunction not only by inhibiting lipolysis but also by promoting fat deposition. Thus, palmitic acid, a common and abundant fatty acid in Western-style diets, was shown to induce PGE2 production by mast cells [80], a lipid product that has anti-lipolytic responses in adipocytes [81]. In addition, another mast cell-derived prostanoid product, PGJ2, a metabolite of PGD2, was found to activate PPAR-γ, which mediates differentiation of adipocytes [72]. Thus, collectively, these mast cell products may all contribute to enhance fat deposition.

3.3 Mast Cells in Atherosclerosis

Atherosclerosis is a progressive disease defined by the appearance of lesions or plaques in the walls of large and midsize arteries and often is the underlying cause of cardiovascular disease. The pathogenesis of atherosclerosis involves the accumulation of lipids from blood in the subendothelial region of the arterial wall and a complex interplay between endothelial, vascular, and inflammatory cells that leads to the progression of the atherosclerotic plaque into a rupture-prone unstable plaque that is the main cause for the complications of atherosclerosis. A number of inflammatory cells including macrophages, T cells, dendritic cells, and, relatively recently, mast cells have been suggested to be involved in the initiation and/or progression of the disease [82]. Mast cells are usually found in the perivascular area of vessel walls and are more abundant in the intima and adventitia of atherosclerotic plaques in mice and humans and in intraplaque neovessels in humans. These MCs contain higher tryptase and chymase [83–86], and their presence and content correlate with disease progression and plaque hemorrhage, destabilization, and lipidosis [69, 85]. While cause and effect have not been clearly demonstrated, there is considerable correlation between mast cell activation or the presence of mast cell-derived products and atherosclerotic plaque progression [82, 87]. Moreover, the functional importance of mast cells in the progression of plaque formation was demonstrated by the use of the mast cell-deficient mouse, $Kit^{w/wv}$. Crossing of these mice with two atherosclerotic mouse models, the $LDLr^{-/-}$ mouse [88, 89] and the $ApoE^{-/-}$ mice [90], which develops atherosclerosis upon feeding with a high-fat diet or spontaneously, respectively, attenuated the progression of atherosclerosis. This manifested as a decrease in lipid deposition, lesion size, and immune cell infiltration in arterial walls [88, 89], a phenotype that could be reversed by adoptive transfer of wild-type bone marrow-derived mast cells, strongly suggesting a cause-and-effect relationship between mast cells and atherosclerosis progression in this model [89]. However, when mast cells from IL-6 or IFN γ-deficient mice were used in the adoptive transfer experiments, the atherosclerotic phenotype was not restored, thus implicating these mast cell-produced cytokines in the development of the disease [89]. Similar conclusions were drawn from studies in $ApoE^{-/-}/Kit^{w/wv}$ mice fed with a high-fat diet for 6 months to accelerate the disease [90]. These mice showed a diminished development of atherosclerotic lesions and reduced hepatic steatosis but also reduced serum levels of LDL, HDL, and considerable elevation of the cytokines IL-6 and IL-10 in the earlier phase of the disease [90].

Potentially mast cells could contribute to the development of atherosclerosis at many points in this process [82, 91]. Some studies have suggested that mast cell-derived products can promote the recruitment of inflammatory cells to the developing atherosclerotic plaque by inducing vascular permeability [92] and the expression of adhesion molecules by the endothelium [89, 93]. Mast cell-derived proteases may also degrade the various components of the extracellular matrices in the plaque, destabilizing the plaque [87]. Others have shown that mast cell-derived IL-6 or IFN γ promotes apoptosis of smooth muscle cells, which together with microvessel formation triggered by mast cell products promotes the development of arterial aneurysms [94].

The triggers for mast cell activity in this disease are not well understood. It is possible that the lipid components, such as oxidized LDL molecules, induce mast cell degranulation and leukocyte adhesion [64]. Oxidized LDL-IgG complexes have been detected in atherogenic lesions, and they were found to induce pro-inflammatory cytokine production by mast cells [95]. Recently, IgE was reported to be required for atherosclerosis development since ApoE$^{-/-}$-/FcεRIα-deficient mice had reduced atherosclerosis and IgE was found to be elevated in patients with myocardial infarction or unstable angina pectoris. However, the contribution of IgE was primarily due to effects on macrophage apoptosis, and the possible role of mast cells in these mice was not investigated [96]. Nonetheless, the collective studies provide evidence of a contributory role for mast cells in this disease.

3.4 Mast Cells in Tumor Growth and Development

The connection between inflammation and the development of cancer is widely appreciated [97]. Although lymphocytes have taken the center stage, and the presence of high numbers of Tregs in tumors has been associated with a dampening of the immune response in this environment [98], a contributory role for mast cells in tumor development is also beginning to emerge. A number of studies in human cancers and mouse tumor models established an association between the location and numbers of mast cells (as well as mast cell protease content in the tumors) and the stage of malignancy, microvessel density, or prognosis and recurrence of tumors. This, together with the findings of mast cell-derived cytokines in the developing tumors, has suggested a role for mast cells in promoting the development of tumors [99–111]. Tumor-derived SCF [112] and tumor-induced cytokines [113], for example, TNF [102], are thought to be involved in the recruitment and responsiveness of mast cells to particular types of tumors, although it is not known whether such factors have a general role in mast cell biology and function in most tumors. An evolving view, based on numerous studies, is that at early stages of malignancy, mast cells are recruited to the tumor where they proliferate and are activated to produce mediators that promote tumor vessel formation, recruitment of immune cells, and immunosuppression, all of which would favor tumor progression. As the tumor evolves and becomes more aggressive, the tumor environment does not support mast cell proliferation/recruitment within the tumor, and tumor growth becomes more independent of mast cells. However, not all tumors show a positive correlation with mast cells, but instead some show no correlation or a negative correlation, which would be more consistent with an immunomodulatory or immunoregulatory role of mast cells on the progression of the tumor [105, 106, 108, 114]. Given the plasticity of mast cells to differentiate and respond differently depending on environmental cues, it is not surprising that the role of mast cells may vary depending on the type of tumor and the stage of tumor development.

The role of mast cells as promoters of tumor growth has primarily been demonstrated in Kit-dependent mast cell-deficient mouse models, including $Kit^{w/wv}$ [115] and $Kit^{W\text{-}sh/W\text{-}sh}$ mice [106, 113, 116–119]. Implantation of a variety of tumor types in these mice showed reduced numbers of tumors and reduced size of tumors for orthotopically implanted B16 melanoma cells [115, 118], pancreatic ductal adenocarcinoma [116], Lewis lung carcinoma [119], prostate adenocarcinomas [106], and urothelial carcinoma cells expressing HY antigens [117]. That mast cells were important for tumor development was demonstrated by injection of mast cells derived from wild-type syngeneic mice into the mast cell-deficient mice, which reversed this phenotype, enhancing tumor growth and numbers. A similar reduction in tumor growth and vascularization was observed when an inducible mouse model of Myc-promoted β-cell and skin tumorigenesis was crossed with the mast cell-deficient $Kit^{W\text{-}sh/W\text{-}sh}$ mouse [113] as well as in a mouse model of polyposis lethally irradiated and reconstituted with marrows of $Kit^{W\text{-}sh/W\text{-}sh}$, CD34-, or CD43-deficient mice that are unable to differentiate mast cells or recruit their precursors to peripheral sites [102]. Studies in these models have also revealed some clues about how mast cells may regulate tumor development, although the exact mechanisms are unclear. Nonetheless, reconstitution of mast cell-deficient mice with mMCP9 deficient mast cells, unlike reconstitution with wild-type mast cells, failed to rescue the development of well-differentiated, slow-growing prostate cell adenocarcinomas in these mice, suggesting that mMCP9 plays a role in advancing the development of this type of tumor probably by promoting cell migration and invasion [106]. Others studies have suggested a role for mast cell-derived TNF in the proliferation of adenomatous polyps [102], and a possible immunoregulatory effect of mast cells in enhancing immunosuppression, thus reducing protective antitumor immunity [117]. It is likely that mast cells may employ multiple mechanisms depending on the environment and tumor type.

It is important to stress, however, that the role of mast cells, even in mouse models, is not unidirectional, since they may influence tumorigenesis in many other ways. For example, it was shown that even though the presence of mast cells promoted early-stage prostate cancer in a mouse model, paradoxically, targeting mast cells in a tumor-prone transgenic adenocarcinoma of the mouse prostate resulted in a more aggressive tumor, indicating a suppressive function for mast cells in this latter circumstance [106]. Another recent study suggests that reduction of mast cell (and possibly other cells) TNF production by mast cell-derived PGD2 restricts vascular permeability and inflammation in the tumor environment [119]. In this study, the authors reconstituted $Kit^{W\text{-}sh/W\text{-}sh}$ mice with PGD2-deficient mast cells and found increased size and number of lung carcinoma cell tumors. This effect was counteracted by reconstitution of $Kit^{W\text{-}sh/W\text{-}sh}$ mice with mast cells deficient in both PGD2 and TNF production [119]. Other studies unraveling a suppressive role for mast cells in the evolution into aggressive tumors include a genetic model of early stage of intestinal neoplasia (*APC-Min* mouse) that when crossed with a mast cell-deficient mouse showed more numerous and larger tumors [120]. These findings exemplify the complexity of the potential role of mast cells in tumor growth and development. While one might consider the potential targeting of mast cells to treat cancer, this may be restricted to certain types and stages of tumors. Of considerable concern is that targeting of mast cells may worsen the disease given the dual and opposing roles for these cells in some tumors [106]. Thus, it is essential to understand the underlying mechanisms by which mast cells can promote versus inhibit tumor development. An important cautionary note on the aforementioned studies is the consideration that Kit itself is a factor that may be dysregulated within the tumor cells in some types of cancer [121], and thus the results used in Kit-dependent mast cell deficient models should be further validated for a better understanding of the role of mast cells in cancer.

3.5 Mast Cells and Basophils in Autoimmune Diseases

The ability of mast cells to respond to IgE and IgG immune complexes and inflammatory stimuli, and to secrete mediators that affect immune and nonimmune cells [9], positions them as likely contributors in the development of autoimmune disease. Although the potential contribution of mast cells in autoimmune diseases has been long recognized, their specific role in disease progression is still not well defined [122]. In animal models, this is in part due to nonuniformity in the experimental approaches to induce autoimmune diseases or differences intrinsic to the mouse models used. While many of such models share a somewhat similar clinical manifestation of disease, they may still differ in the cell types involved, the exact sequence of events required for disease development, and the underlying mechanisms allowing for disease development. In addition, the interpretation of the contribution of mast cells using mast cell-deficient models has, in instances, been hampered by the known defects in these models.

For more than two decades, there have been reports on the contributory role of mast cells in rheumatoid arthritis (RA) and multiple sclerosis (MS). For RA, the overall picture, using different approaches, generally supports the concept that mast cell activity exacerbates the inflammatory process and severity of this disease, although this effect may not always manifest or be predominant due to the redundancy of the immune system, depending on the specific characteristics of the human disease or the animal model under study. The findings are somewhat more ambiguous in MS as there is considerable divergence in the experimental approaches, the mouse models used, and the influence of dietary and other environmental and genetic factors which may be key in this disease. Thus, studies on the role of mast cells in mouse models of MS (experimental autoimmune encephalomyelitis, EAE) have resulted in a differing opinion on the role of these cells in this disease (reviewed in [15, 16, 122, 123]). Importantly, it should be noted that the baseline environment and the type or strength of the stimulus used to induce MS in mouse models may reveal or mask a role for mast cells in this disease [122]. This view is highly consistent with the aforementioned plasticity of mast cells as influenced by the environment in which they reside.

Rheumatoid arthritis (RA) is an autoimmune disorder characterized by chronic inflammation of the joints, with synovial hyperplasia, cartilage destruction, bone erosion, and changes in joint function. An association between the inflamed joint and the presence of mast cells has long been recognized. MCs and MC-derived products are elevated in the synovial fluid and synovial tissue of RA patients [124–127] and in the arthritic lesions of mice [128–130], suggesting their participation in joint inflammation. In addition, mast cell stabilizers such as nedocromil, tranilast, cromolyn, or salbutamol have been shown effective in reducing neutrophil infiltration and clinical scores of arthritis in collagen-induced and K/BxN serum-induced arthritis in mice [128, 130, 131], and treatment of some arthritis patients with antihistaminic or antiserotoninergic drugs was reported as effective in reducing arthritic attacks [124]. Such studies support the concept of a contributory role of mast cells in this inflammatory condition. In addition, the inhibitor of the tyrosine kinase KIT, imatinib mesylate (Gleevec), which induces apoptosis in rheumatoid synovial mast cells [132], was reported to be efficacious in reducing arthritic swelling and pain and other markers of disease in two out of three patients with severe RA [133]. While further studies are required to determine if the effect is indeed mast cell-specific (given the off target effects of such inhibitors), collectively the studies to date provide some level of confidence in a role for mast cells in RA. Unfortunately, the results from mast cell-deficient mouse models used to address the specific role of mast cells in arthritis have not been clear-cut. While the $Kit^{W/W\text{-}v}$ mast cell-deficient mouse is resistant to arthritis induction by K/BxN serum transfer [131, 134–137] and to anti-collagen monoclonal antibody and LPS-induced arthritis [138], the $Kit^{W\text{-}sh/W\text{-}sh}$ mast cell-deficient mice or mice in which mast cells are eradicated by insertion of Cre recombinase in the carboxypeptidase A3 locus (Cre-Master) were fully susceptible to collagen- [139], anticollagen- and LPS- [138], and K/BxN-induced arthritis [134, 140]. The defect in the development of arthritis in $Kit^{W/W\text{-}v}$ mice can be restored by reconstitution of the mice with wild-type bone marrow-derived mast cells. It has been reasoned that the results obtained in the $Kit^{W/W\text{-}v}$ may be attributed to a more prominent role for mast cells in the context of differences in the numbers of neutrophils in these mice relative to $Kit^{W\text{-}sh/W\text{-}sh}$ mice [15, 134]. Nevertheless, these studies have been helpful as they not only clarify some aspects of the mechanisms of action of mast cells in this model but also serve to unravel the harmful role of mast cells that may be obscured in other models by either the redundancy with other cell immune cells or by the potential opposing effects of mast cells to promote or protect against disease progression. This may be of relevance when considering the targeting of mast cells in disease.

Several mast cell products have been implicated in the pathogenesis of arthritis, although some are not necessarily produced only by mast cells. Thus, mice lacking histidine decarboxylase, the enzyme-producing histamine [141] and mice deficient in the tryptases mMCP7 [142]; mMCP6 [142, 143], which appears to be expressed in all synovial mast cells as demonstrated in the K/BxN serum transfer-induced arthritic synovium [143]; or mMCP4 [144], showed markedly reduced scores of arthritis in various models. In addition, mice deficient in the cytokine IL-1, which is produced by mast cells via the activation of FcγRIIIA in autoantibody-induced arthritis, also had reduced arthritis [135]. Many stimuli have been proposed to activate mast cell degranulation or cytokine production in the environment of the arthritic synovium [135]. Autoantibody-induced arthritis was shown to require FcγRIIIA [134, 135, 145] and the complement receptor C5aR, which may be needed for the release of neutrophil chemoattractants into the synovium [136], while in collagen-induced arthritis models, the IL33 receptor in mast cells [146] has been implicated. Interestingly, in RA patients the presence of high IgE that is complexed with anti-IgE has been reported [147]. Thus, in some cases of RA, mast cell or basophil contribution may be a consequence of FcεRI activation.

The role of basophil in autoimmunity has not been as extensively investigated. Nonetheless, one classic example where these cells can promote autoimmunity is in the well-described case of autoimmune chronic urticaria where patients generate IgG autoantibodies directed to the FcεRIα chain [148]. More recently, we uncovered a previously unrecognized role for basophils in the pathogenesis of systemic lupus erythematosus (SLE), a complex autoimmune disease affecting multiple target organs and characterized by the presence of autoreactive antibodies [148]. Using Lyn kinase-deficient mice, which develop a lupus-like disease in late life, we demonstrated the presence of elevated levels of autoreactive IgE. The high levels of autoreactive IgE were required for the development of the lupus phenotype in the Lyn-deficient mice as demonstrated by the reduced kidney disease when the IgE locus was also genetically deleted in these mice [44]. The link between autoreactive IgE and progression of autoimmune disease is not restricted to this animal model of lupus, since it is observed in other lupus models in which deletion of the IgE locus reduces mortality and onset of nephritis (Dema and Rivera, unpublished observations). More importantly, studies in human SLE cohorts of more than 200 individuals showed that a high percentage (>50%) of patients have high levels of autoreactive IgEs, and this correlates with increased disease activity and lupus nephritis ([44]; Dema, Charles, Rivera, unpublished observations). The presence of autoreactive IgE and the deposition of immunocomplexes in the kidney of Lyn-deficient mice were dependent on IL-4, since these were absent in Lyn-deficient mice also carrying a deletion in the IL-4 gene. Notably, investigation of the underlying mechanism leading to increased IL-4 and to the presence of circulating autoreactive IgE led us to uncover that basophils (and not mast cells) were activated in these mice and produced large amounts of IL-4. In fact, depletion of basophils prevented the increased IL-4 levels, markedly reduced the presence of autoreactive IgE, and reversed a Th2 bias found in the Lyn-deficient mice as well as the glomerulonephritis seen at late stages of the disease. This, together with the finding that autoreactive IgE-containing IgG immune complexes can activate basophils to produce IL-4, demonstrated that basophils can serve as amplifiers of autoimmunity, whereby their activation by IgE-containing immune complexes results in increased B-cell-mediated autoantibody production (mechanistic details provided in the next section). These findings clearly link the basophil to the promotion of autoimmune disease, such as SLE, and raise the question of whether basophils may play a role in the immunomodulation of other autoimmune diseases.

4 The Immunomodulatory Role of Mast Cells and Basophils

Given the contributory, rather than essential, nature of mast cells and basophils in many of the aforementioned protective and exacerbating roles, perhaps it is not surprising that these cells exert immunoregulatory function. A regulatory role for mast cells was inferred by Paul Ehrlich when he named them "Mastzellen" in the belief that they provided nutritional function because of their high content of granules. The role of mast cells in modulating the immune response has long been recognized due, in part, to the effects of mast cell granule mediators on other immune cells [149]. However, until recently the molecular basis by which mast cells or basophils exerted such influence on the immune system was inadequately investigated. In this section, we briefly summarize some of the recent work with a primary focus on how mast cells influence and are in turn influenced by T cells. In addition, we briefly summarize what we are learning about the role of basophils in modulating B-cell function and the humoral response.

4.1 Recent Advances in Mast Cell Immune Modulation

The idea that mast cells and T cells can communicate is supported by the presence of a diverse array of molecules on their cell surface that can mediate their cross-talk [150]. Mast cells can also be found close to T cells in inflamed tissues as well as in lymphoid organs [19, 151–153]. Recent studies have shown that mast cells and activated T cells can interact via ICAM-1 (CD54) and LFA-1 (CD11a) expressed, respectively, on their cell surface, an interaction that enhances FcεRI-mediated mast cell degranulation [154]. Mast cells can also influence T-effector cells as they are able to act as antigen presenting cells [155]. Expression of MHC class II molecules (MHC-II) by mast cells can be induced by treatment with LPS and IFN-γ [156]. While such MHC-II bearing mast cells failed to stimulate naïve T cells, they effectively supported activated T-effectors and caused expansion of regulatory T cells (Tregs). This antigen-presenting role of mast cells is further supported by the demonstration that treatment of peritoneal cell-derived mast cells (PCMCs) with IFN-γ and IL-4 also induced expression of MHC-II molecules. In this case, the authors found that these MHC-II-competent PCMCs were able to present antigen to effector T cells causing their activation, proliferation, and formation of an immunological synapse between the PCMC and the T cell [157]. Antigen presentation by mast cells is not limited to MHC class II, as it was recently reported that MHC-I-dependent cross presentation of mast cells to $CD8^+$ T cells caused increased $CD8^+$ T-cell proliferation and effector functions [158]. While it is clear that such interactions can have reciprocal effects on both mast cells and T-cell effector functions, a mechanistic understanding of the processes underlying these events is currently limited.

Perhaps, where a considerable advance has been made toward understanding mechanisms associated with Treg, mast cell interactions is in the discovery of OX40 and OX40L interactions, respectively, between these cells [159–165]. It was first reported that this interaction results in the suppression of MC degranulation [161]. A consequence of this interaction was impaired influx of extracellular Ca^{2+} after FcεRI stimulation of mast cells. It was demonstrated that this suppression of Ca^{2+} influx was a result of increased cAMP generation in the mast cells after OX40L engagement. The use of reagents that inhibited cAMP generation in mast cells restored normal Ca^{2+} influx and normal degranulation even in the presence of Tregs. In addition, OX40-deficient Tregs failed to induce cAMP production in mast cells or suppress Ca^{2+} influx and degranulation. Further mechanistic studies demonstrated that soluble OX40 could effectively block Treg-mediated suppression of mast cells and that in mast cells the suppressive effects of OX40L engagement were mediated through dampening of Fyn-dependent signals [163, 164].

However, the interaction of Tregs with mast cells via the OX40-OX40L axis also has consequences on Treg function. Piconese and colleagues demonstrated [162] that mast cells can abrogate Treg-suppressive activity upon OX40/OX40L interaction. This was dependent, at least in part, on IL-6 production and resulted in an increase in IL-17-producing T cells. These results suggest that in some inflammatory conditions, one might find the presence of mast cells in close proximity to Tregs and Th-17 cells. Consistent with this view, in the EAE mouse model of multiple sclerosis, mast cells were found to localize with Tregs and Th-17 cells. Similarly, these cell types could be identified in the BAL fluid of OTII transgenic mice challenged with ovalbumin intranasally. While there are numerous other examples where mast cells have been shown to play a role in regulating Treg and T-effector function, the findings briefly highlighted herein begin to unravel the mechanisms involved and promote the view that such interactions may play a key role in regulating the immune response in health and disease.

4.2 Basophils and B Cells: Boosting the Humoral Response

While mast cell interactions with B cells were recently described to enhance B-cell proliferation and differentiation [166], in the past few years, there has been considerable focus on the role of basophils in influencing B-cell biology and function, and we will spotlight these studies herein. What is emerging from this recent work is that basophils can deliver help to B cells via both direct and indirect mechanisms.

As mentioned earlier in this chapter, under certain circumstances, basophils appear to migrate to the draining lymph nodes where they release IL-4 and induce Th2 cell differentiation [41, 42, 47, 167]. Th2 cell differentiation is essential for host defense to various pathogens and allergens through induction of protective IgG1 and IgE production by B cells. It is important to note, however, that basophils may not necessarily mediate Th2 cell differentiation alone but are likely to cooperate with dendritic cells [168, 169] and thus should not be considered as a replacement of dendritic cell function but may be more appropriately thought to augment dendritic cells in promoting Th2 cell differentiation.

Thus, in this manner, basophils are likely to indirectly promote B-cell function through cooperating with dendritic cells to increase T helper functions that amplify B-cell responses.

Perhaps a good example of more direct interdependence of basophils and B cells, in regulating humoral responses, is provided by the observation that immune surveillance is enhanced by IgD-mediated activation of B-cell-stimulating programs in basophils [43]. IgD is secreted by plasmablasts in the mucosa where it can afford protection by binding to pathogens and other virulence factors. It also binds to basophils, via an unknown receptor, and cross-linking of basophil-bound IgD was shown to cause the release of BAFF, IL-4, and IL-13 [43], which stimulates B cells to undergo IgM production as well as class switch recombination to IgG and IgA. In addition to IgD, other molecules like CD40L, which is expressed on basophils, provide further help for class switch recombination by engaging CD40 on B cells. Other examples of the direct interdependence of basophils and B cells are provided by in vivo and in vitro studies where depletion of basophils caused loss of plasmablasts in the spleen [44, 170]. These studies showed that basophil activation via IgE/Ag interactions causes a marked enhancement of plasma cell numbers with this increase being dependent on IL-4 and IL-6 [44, 170]. This is consistent with our recent finding (T. Ricks and J. Rivera, unpublished observation) that basophils localize to the plasmablast-enriched regions of the spleen in the *lyn*$^{-/-}$ mouse model, which has a basophilia that plays a role in amplification of autoantibody production and development of lupus [44, 171].

Finally, while the data suggests that both mouse and human basophils can augment memory responses of B cells and T cells [172–174], it is less clear if there are circumstances in which basophils alone can promote differentiation of naïve T- and B cells. Nonetheless, it appears certain that basophils can now be viewed as important immunomodulatory cells with an apparent amplifying role in humoral responses.

5 Concluding Remarks

A number of points should be considered as the new era in mast cell and basophil biology and function further develops. A guiding principle can be garnered from the accumulating literature that not all mast cells are the same [66]. The intrinsic property of mast cells to develop in tissues provides a remarkable level of genomic and proteomic plasticity that makes a skin mast cell quite different from one residing in the spleen or in the joints (even though they are all considered connective tissue-type mast cells). While basophils mature in the bone marrow, it is also apparent that their phenotype may be distinct when circulating in the blood versus recruited to the tissues. In mouse studies, minimal expression of MHC-II or BAFF was detected on blood basophils, whereas upon recruitment to the lymph nodes, these cells express considerable amounts of these molecules [44]. Thus, there is considerable danger in generalizing the findings on mast cell and basophil functions when studies are conducted with a limited sampling of mast cell or basophil types (or tissues) or by the use of one or another mouse model.

Importantly, one should keep in mind that the immune system is highly regulated and finely tuned and, as such, even minor perturbations (like differences in stimuli strength or change in commensal flora) can result in markedly different responses [175]. Moreover, given that we accept a regulatory role for mast cells and basophils in immunity, the contributory (rather than essential) role of these cells positions them in a niche where their mode of activation, or the particular effector responses elicited, may be a determinant of the type or extent of inflammatory response initiated. Thus, arguments for or against a particular (patho)physiological role of mast cells and basophils should always be viewed in the context of the model or experiment. Of note is that even minor differences in the genetics (or alterations made to genes) of a presumed homogeneous population (such as seen in congenic mice) can cause marked variations in responses [176]. In our opinion, there is little uniformity in individual mouse responses (similar to what is seen in the study of humans), and thus only by large cohort samplings can one obtain the statistical power to provide a more general view of the (patho)physiological role of a given cell type or gene. However, such studies are both resource- and labor-intensive and, frankly, not practical given the current fiscal constraints encompassing the worldwide scientific community. With this in mind, it becomes important to translate basic laboratory discoveries to human health and disease. Finding correlates of human (patho)physiology provide the degree of confidence that justifies further dissection of a particular role for a cell, gene, or molecule in health and disease.

Regardless of these caveats, the enthusiasm for the expanding importance of mast cells and basophils in immunity is at an all-time high. It is a most heartening time to be a scientist in this field, as the rapidly evolving technologies provide promise of future discoveries that advance a better understanding of mast cell and basophil biology and function. We envision that this new knowledge will translate into novel treatments for allergies and other diseases in which mast cells and basophils serve as effectors and modulators of the immune system.

References

1. Ehrlich P (1877) Beitrage zur Kenntnis der Anilinfarbungen und ihrer Verwendung in der mikroskopischen Technik. Arch mikr Anat 13:263–277
2. Ehrlich, P. (1878) Beitrage zur Theorie und Praxis der histologischen Farbung. 6-17-1878. Thesis, Leipzig University

3. Ehrlich P (1891) Farbenanalytische Untersuchungen zur Histologie und Klinik des Blutes. Hirschwald, Berlin
4. Ehrlich P, Lazarus A (1898) Die Anaemie, 1. Normale und pathologische Histologie des Blutes, Holder
5. Galli SJ (1993) New concepts about the mast cell. N Engl J Med 328:257–265
6. Metcalfe DD, Baram D, Mekori YA (1997) Mast cells. Physiol Rev 77:1033–1079
7. Karasuyama H, Mukai K, Obata K, Tsujimura Y, Wada T (2011) Nonredundant roles of basophils in immunity. Annu Rev Immunol 29:45–69
8. Dvorak AM, Nabel G, Pyne K, Cantor H, Dvorak HF, Galli SJ (1982) Ultrastructural identification of the mouse basophil. Blood 59:1279–1285
9. Galli SJ, Kalesnikoff J, Grimbaldeston MA, Piliponsky AM, Williams CM, Tsai M (2005) Mast cells as "tunable" effector and immunoregulatory cells: recent advances. Annu Rev Immunol 23:749–786
10. Kinet JP (1989) The high-affinity receptor for IgE. Curr Opin Immunol 2:499–505
11. Min B, Paul WE (2008) Basophils and type 2 immunity. Curr Opin Hematol 15:59–63
12. Falcone FH, Haas H, Gibbs BF (2000) The human basophil: a new appreciation of its role in immune responses. Blood 96:4028–4038
13. Galli SJ, Tsai M (2010) Mast cells in allergy and infection: versatile effector and regulatory cells in innate and adaptive immunity. Eur J Immunol 40:1843–1851
14. Tsai M, Grimbaldeston M, Galli SJ (2011) Mast cells and immunoregulation/immunomodulation. Adv Exp Med Biol 716:186–211
15. Reber LL, Marichal T, Galli SJ (2012) New models for analyzing mast cell functions in vivo. Trends Immunol 33:613–625
16. Rodewald HR, Feyerabend TB (2012) Widespread immunological functions of mast cells: fact or fiction? Immunity 37:13–24
17. Honda T, Egawa G, Grabbe S, Kabashima K (2013) Update of immune events in the murine contact hypersensitivity model: toward the understanding of allergic contact dermatitis. J Invest Dermatol 133:303–315
18. Matsumoto K, Mizukoshi K, Oyobikawa M, Ohshima H, Tagami H (2004) Establishment of an atopic dermatitis-like skin model in a hairless mouse by repeated elicitation of contact hypersensitivity that enables to conduct functional analyses of the stratum corneum with various non-invasive biophysical instruments. Skin Res Technol 10:122–129
19. Dudeck A, Dudeck J, Scholten J, Petzold A, Surianarayanan S, Köhler A, Peschke K, Vöhringer D, Waskow C, Krieg T, Müller W, Waisman A, Hartmann K, Gunzer M, Roers A (2011) Mast cells are key promoters of contact allergy that mediate the adjuvant effects of haptens. Immunity 34:973–984
20. Galli SJ, Nakae S, Tsai M (2005) Mast cells in the development of adaptive immune responses. Nat Immunol 6:135–142
21. Hershko AY, Suzuki R, Charles N, Alvarez-Errico D, Sargent JL, Laurence A, Rivera J (2011) Mast cell interleukin-2 production contributes to suppression of chronic allergic dermatitis. Immunity 35:562–571
22. Norman MU, Hwang J, Hulliger S, Bonder CS, Yamanouchi J, Santamaria P, Kubes P (2008) Mast cells regulate the magnitude and the cytokine microenvironment of the contact hypersensitivity response. Am J Pathol 172:1638–1649
23. Man MQ, Hatano Y, Lee SH, Man M, Chang S, Feingold KR, Leung DY, Holleran W, Uchida Y, Elias PM (2008) Characterization of a hapten-induced, murine model with multiple features of atopic dermatitis: structural, immunologic, and biochemical changes following single versus multiple oxazolone challenges. J Invest Dermatol 128:79–86
24. Hart PH, Grimbaldeston MA, Swift GJ, Jaksic A, Noonan FP, Finlay-Jones JJ (1998) Dermal mast cells determine susceptibility to ultraviolet B-induced systemic suppression of contact hypersensitivity responses in mice. J Exp Med 187:2045–2053
25. Depinay N, Hacini F, Beghdadi W, Peronet R, Mecheri S (2006) Mast cell-dependent downregulation of antigen-specific immune responses by mosquito bites. J Immunol 176:4141–4146
26. Grimbaldeston MA, Nakae S, Kalesnikoff J, Tsai M, Galli SJ (2007) Mast cell-derived interleukin 10 limits skin pathology in contact dermatitis and chronic irradiation with ultraviolet B. Nat Immunol 8:1095–1104
27. Abraham SN, St John AL (2010) Mast cell-orchestrated immunity to pathogens. Nat Rev Immunol 10:440–452
28. Costa JJ, Weller PF, Galli SJ (1997) The cells of the allergic response: mast cells, basophils, and eosinophils. JAMA 278:1815–1822
29. Dvorak AM, Seder RA, Paul WE, Kissell-Rainville S, Plaut M, Galli SJ (1993) Ultrastructural characteristics of Fc epsilon R-positive basophils in the spleen and bone marrow of mice immunized with goat anti-

mouse IgD antibody. Lab Investig 68:708–715

30. Seder RA, Paul WE, Dvorak AM, Sharkis SJ, Kagey-Sobotka A, Niv Y, Finkelman FD, Barbieri SA, Galli SJ, Plaut M (1991) Mouse splenic and bone marrow cell populations that express high-affinity Fc epsilon receptors and produce interleukin 4 are highly enriched in basophils. Proc Natl Acad Sci U S A 88:2835–2839
31. Seder RA, Plaut M, Barbieri S, Urban J Jr, Finkelman FD, Paul WE (1991) Purified Fc epsilon R+ bone marrow and splenic non-B, non-T cells are highly enriched in the capacity to produce IL-4 in response to immobilized IgE, IgG2a, or ionomycin. J Immunol 147:903–909
32. Ohnmacht C, Schwartz C, Panzer M, Schiedewitz I, Naumann R, Voehringer D (2010) Basophils orchestrate chronic allergic dermatitis and protective immunity against helminths. Immunity 33:364–374
33. Wada T, Ishiwata K, Koseki H, Ishikura T, Ugajin T, Ohnuma N, Obata K, Ishikawa R, Yoshikawa S, Mukai K, Kawano Y, Minegishi Y, Yokozeki H, Watanabe N, Karasuyama H (2010) Selective ablation of basophils in mice reveals their nonredundant role in acquired immunity against ticks. J Clin Invest 120:2867–2875
34. Min B, Prout M, Hu-Li J, Zhu J, Jankovic D, Morgan ES, Urban JF Jr, Dvorak AM, Finkelman FD, LeGros G, Paul WE (2004) Basophils produce IL-4 and accumulate in tissues after infection with a Th2-inducing parasite. J Exp Med 200:507–517
35. van Panhuys N, Prout M, Forbes E, Min B, Paul WE, Le Gros G (2011) Basophils are the major producers of IL-4 during primary helminth infection. J Immunol 186:2719–2728
36. Voehringer D, Shinkai K, Locksley RM (2004) Type 2 immunity reflects orchestrated recruitment of cells committed to IL-4 production. Immunity 20:267–277
37. Ohnmacht C, Voehringer D (2010) Basophils protect against reinfection with hookworms independently of mast cells and memory Th2 cells. J Immunol 184:344–350
38. Ohnmacht C, Voehringer D (2009) Basophil effector function and homeostasis during helminth infection. Blood 113:2816–2825
39. Voehringer D, Reese TA, Huang X, Shinkai K, Locksley RM (2006) Type 2 immunity is controlled by IL-4/IL-13 expression in hematopoietic non-eosinophil cells of the innate immune system. J Exp Med 203:1435–1446
40. Lantz CS, Min B, Tsai M, Chatterjea D, Dranoff G, Galli SJ (2008) IL-3 is required for increases in blood basophils in nematode infection in mice and can enhance IgE-dependent IL-4 production by basophils in vitro. Lab Investig 88:1134–1142
41. Perrigoue JG, Saenz SA, Siracusa MC, Allenspach EJ, Taylor BC, Giacomin PR, Nair MG, Du Y, Zaph C, van Rooijen N, Comeau MR, Pearce EJ, Laufer TM, Artis D (2009) MHC class II-dependent basophil-CD4+ T cell interactions promote T(H)2 cytokine-dependent immunity. Nat Immunol 10:697–705
42. Yoshimoto T, Yasuda K, Tanaka H, Nakahira M, Imai Y, Fujimori Y, Nakanishi K (2009) Basophils contribute to T(H)2-IgE responses in vivo via IL-4 production and presentation of peptide-MHC class II complexes to CD4+ T cells. Nat Immunol 10:706–712
43. Chen K, Xu W, Wilson M, He B, Miller NW, Bengtén E, Edholm E-S, Santini PA, Rath P, Chiu A, Cattalini M, Litzman J, Bussel JB, Huang B, Meini A, Riesbeck K, Cunningham-Rundles C, Plebani A, Cerutti A (2009) Immunoglobulin D enhances immune surveillance by activating antimicrobial, proinflammatory and B cell-stimulating programs in basophils. Nat Immunol 10:889–898
44. Charles N, Hardwick D, Daugas E, Illei GG, Rivera J (2010) Basophils and the T helper 2 environment can promote the development of lupus nephritis. Nat Med 16:701–707
45. Charles N, Dema B, Rivera J (2012) Reply to: basophils from humans with systemic lupus erythematosus do not express MHC-II. Nat Med 18:489–490
46. Charles N, Rivera J (2011) Basophils and autoreactive IgE in the pathogenesis of systemic lupus erythematosus. Curr Allergy Asthma Rep 11:378–387
47. Sokol CL, Chu NQ, Yu S, Nish SA, Laufer TM, Medzhitov R (2009) Basophils function as antigen-presenting cells for an allergen-induced T helper type 2 response. Nat Immunol 10:713–720
48. Caughey GH (2011) Mast cell proteases as protective and inflammatory mediators. Adv Exp Med Biol 716:212–234
49. Akahoshi M, Song CH, Piliponsky AM, Metz M, Guzzetta A, Abrink M, Schlenner SM, Feyerabend TB, Rodewald HR, Pejler G, Tsai M, Galli SJ (2011) Mast cell chymase reduces the toxicity of Gila monster venom, scorpion venom, and vasoactive

intestinal polypeptide in mice. J Clin Invest 121:4180–4191
50. Maurer M, Wedemeyer J, Metz M, Piliponsky AM, Weller K, Chatterjea D, Clouthier DE, Yanagisawa MM, Tsai M, Galli SJ (2004) Mast cells promote homeostasis by limiting endothelin-1-induced toxicity. Nature 432:512–516
51. Metz M, Piliponsky AM, Chen CC, Lammel V, Abrink M, Pejler G, Tsai M, Galli SJ (2006) Mast cells can enhance resistance to snake and honeybee venoms. Science 313:526–530
52. Malaviya R, Ikeda T, Ross E, Abraham SN (1996) Mast cell modulation of neutrophil influx and bacterial clearance at sites of infection through TNF-alpha. Nature 381:77–80
53. Piliponsky AM, Chen CC, Grimbaldeston MA, Burns-Guydish SM, Hardy J, Kalesnikoff J, Contag CH, Tsai M, Galli SJ (2010) Mast cell-derived TNF can exacerbate mortality during severe bacterial infections in C57BL/6-KitW-sh/W-sh mice. Am J Pathol 176:926–938
54. Piliponsky AM, Chen CC, Nishimura T, Metz M, Rios EJ, Dobner PR, Wada E, Wada K, Zacharias S, Mohanasundaram UM, Faix JD, Abrink M, Pejler G, Pearl RG, Tsai M, Galli SJ (2008) Neurotensin increases mortality and mast cells reduce neurotensin levels in a mouse model of sepsis. Nat Med 14:392–398
55. Cochrane DE, Carraway RE, Boucher W, Feldberg RS (1991) Rapid degradation of neurotensin by stimulated rat mast cells. Peptides 12:1187–1194
56. Piliponsky AM, Chen CC, Rios EJ, Treuting PM, Lahiri A, Abrink M, Pejler G, Tsai M, Galli SJ (2012) The chymase mouse mast cell protease 4 degrades TNF, limits inflammation, and promotes survival in a model of sepsis. Am J Pathol 181:875–886
57. Thakurdas SM, Melicoff E, Sansores-Garcia L, Moreira DC, Petrova Y, Stevens RL, Adachi R (2007) The mast cell-restricted tryptase mMCP-6 has a critical immunoprotective role in bacterial infections. J Biol Chem 282:20809–20815
58. Hendrix S, Kramer P, Pehl D, Warnke K, Boato F, Nelissen S, Lemmens E, Pejler G, Metz M, Siebenhaar F, Maurer M (2013) Mast cells protect from post-traumatic brain inflammation by the mast cell-specific chymase mouse mast cell protease-4. FASEB J 27:920–929
59. Waern I, Lundequist A, Pejler G, Wernersson S (2012) Mast cell chymase modulates IL-33 levels and controls allergic sensitization in dust-mite induced airway inflammation. Mucosal Immunol 6(5):911–920
60. Waern I, Karlsson I, Thorpe M, Schlenner SM, Feyerabend TB, Rodewald HR, Åbrink M, Hellman L, Pejler G, Wernersson S (2012) Mast cells limit extracellular levels of IL-13 via a serglycin proteoglycan-serine protease axis. Biol Chem 393:1555–1567
61. Hershko AY, Charles N, Olivera A, Alvarez-Errico D, Rivera J (2012) Cutting edge: persistence of increased mast cell numbers in tissues links dermatitis to enhanced airway disease in a mouse model of atopy. J Immunol 188:531–535
62. Hahn EL, Bacharier LB (2005) The atopic march: the pattern of allergic disease development in childhood. Immunol Allergy Clin North Am 25:231–246
63. McLachlan JB, Shelburne CP, Hart JP, Pizzo SV, Goyal R, Brooking-Dixon R, Staats HF, Abraham SN (2008) Mast cell activators: a new class of highly effective vaccine adjuvants. Nat Med 14:536–541
64. Paananen K, Kovanen PT (1994) Proteolysis and fusion of low density lipoprotein particles independently strengthen their binding to exocytosed mast cell granules. J Biol Chem 269:2023–2031
65. Pulendran B, Ono SJ (2008) A shot in the arm for mast cells. Nat Med 14:489–490
66. Galli SJ, Borregaard N, Wynn TA (2011) Phenotypic and functional plasticity of cells of innate immunity: macrophages, mast cells and neutrophils. Nat Immunol 12:1035–1044
67. Chmelar J, Chung KJ, Chavakis T (2013) The role of innate immune cells in obese adipose tissue inflammation and development of insulin resistance. Thromb Haemost 109:399–406
68. Xu JM, Shi GP (2012) Emerging role of mast cells and macrophages in cardiovascular and metabolic diseases. Endocr Rev 33:71–108
69. Anand P, Singh B, Jaggi AS, Singh N (2012) Mast cells: an expanding pathophysiological role from allergy to other disorders. Naunyn Schmiedeberg's Arch Pharmacol 385:657–670
70. Altintas MM, Nayer B, Walford EC, Johnson KB, Gaidosh G, Reiser J, De La Cruz-Munoz-N, Ortega LM, Nayer A (2012) Leptin deficiency-induced obesity affects the density of mast cells in abdominal fat depots and lymph nodes in mice. Lipids Health Dis 11:21
71. Liu J, Divoux A, Sun J, Zhang J, Clement K, Glickman JN, Sukhova GK, Wolters PJ, Du J,

Gorgun CZ, Doria A, Libby P, Blumberg RS, Kahn BB, Hotamisligil GS, Shi GP (2009) Genetic deficiency and pharmacological stabilization of mast cells reduce diet-induced obesity and diabetes in mice. Nat Med 15:940–945

72. Tanaka A, Nomura Y, Matsuda A, Ohmori K, Matsuda H (2011) Mast cells function as an alternative modulator of adipogenesis through 15-deoxy-delta-12, 14-prostaglandin J2. Am J Physiol Cell Physiol 301:C1360–C1367

73. Divoux A, Moutel S, Poitou C, Lacasa D, Veyrie N, Aissat A, Arock M, Guerre-Millo M, Clément K (2012) Mast cells in human adipose tissue: link with morbid obesity, inflammatory status, and diabetes. J Clin Endocrinol Metab 97:E1677–E1685

74. Wang X, Chen H, Zhang M, Liu Z (2012) Roles of mast cells and monocyte Chemoattractant Protein-1 in the renal injury of obesity-related glomerulopathy. Am J Med Sci 346(4):295–301

75. Fenger RV, Linneberg A, Vidal C, Vizcaino L, Husemoen LL, Aadahl M, Gonzalez-Quintela A (2012) Determinants of serum tryptase in a general population: the relationship of serum tryptase to obesity and asthma. Int Arch Allergy Immunol 157:151–158

76. Ward BR, Arslanian SA, Andreatta E, Schwartz LB (2012) Obesity is not linked to increased whole-body mast cell burden in children. J Allergy Clin Immunol 129:1164–1166

77. Poglio S, De Toni-Costes F, Arnaud E, Laharrague P, Espinosa E, Casteilla L, Cousin B (2010) Adipose tissue as a dedicated reservoir of functional mast cell progenitors. Stem Cells 28:2065–2072

78. Oka T, Kalesnikoff J, Starkl P, Tsai M, Galli SJ (2012) Evidence questioning cromolyn's effectiveness and selectivity as a 'mast cell stabilizer' in mice. Lab Investig 92:1472–1482

79. Shi MA, Shi GP (2012) Different roles of mast cells in obesity and diabetes: lessons from experimental animals and humans. Front Immunol 3:7

80. Iyer A, Lim J, Poudyal H, Reid RC, Suen JY, Webster J, Prins JB, Whitehead JP, Fairlie DP, Brown L (2012) An inhibitor of phospholipase A2 group IIA modulates adipocyte signaling and protects against diet-induced metabolic syndrome in rats. Diabetes 61:2320–2329

81. Kim S, Moustaid-Moussa N (2000) Secretory, endocrine and autocrine/paracrine function of the adipocyte. J Nutr 130:3110S–3115S

82. Weber C, Zernecke A, Libby P (2008) The multifaceted contributions of leukocyte subsets to atherosclerosis: lessons from mouse models. Nat Rev Immunol 8:802–815

83. Jeziorska M, McCollum C, Woolley DE (1997) Mast cell distribution, activation, and phenotype in atherosclerotic lesions of human carotid arteries. J Pathol 182:115–122

84. Kaartinen M, Penttila A, Kovanen PT (1994) Mast cells of two types differing in neutral protease composition in the human aortic intima. Demonstration of tryptase- and tryptase/chymase-containing mast cells in normal intimas, fatty streaks, and the shoulder region of atheromas. Arterioscler Thromb 14:966–972

85. Kovanen PT, Kaartinen M, Paavonen T (1995) Infiltrates of activated mast cells at the site of coronary atheromatous erosion or rupture in myocardial infarction. Circulation 92:1084–1088

86. Ramalho LS, Oliveira LF, Cavellani CL, Ferraz ML, de Oliveira FA, Miranda Correa RR, de Paula Antunes Teixeira V, De Lima Pereira SA (2013) Role of mast cell chymase and tryptase in the progression of atherosclerosis: study in 44 autopsied cases. Ann Diagn Pathol 17:28–31

87. Bot I, Bot M, van Heiningen SH, van Santbrink PJ, Lankhuizen IM, Hartman P, Gruener S, Hilpert H, van Berkel TJ, Fingerle J, Biessen EA (2011) Mast cell chymase inhibition reduces atherosclerotic plaque progression and improves plaque stability in ApoE−/− mice. Cardiovasc Res 89:244–252

88. Heikkila HM, Trosien J, Metso J, Jauhiainen M, Pentikainen MO, Kovanen PT, Lindstedt KA (2010) Mast cells promote atherosclerosis by inducing both an atherogenic lipid profile and vascular inflammation. J Cell Biochem 109:615–623

89. Sun J, Sukhova GK, Wolters PJ, Yang M, Kitamoto S, Libby P, MacFarlane LA, Mallen-St Clair J, Shi GP (2007) Mast cells promote atherosclerosis by releasing proinflammatory cytokines. Nat Med 13:719–724

90. Smith DD, Tan X, Raveendran VV, Tawfik O, Stechschulte DJ, Dileepan KN (2012) Mast cell deficiency attenuates progression of atherosclerosis and hepatic steatosis in apolipoprotein E-null mice. Am J Physiol Heart Circ Physiol 302:H2612–H2621

91. Bot I, Biessen EA (2011) Mast cells in atherosclerosis. Thromb Haemost 106:820–826

92. Bot I, de Jager SC, Zernecke A, Lindstedt KA, van Berkel TJ, Weber C, Biessen EA (2007)

Perivascular mast cells promote atherogenesis and induce plaque destabilization in apolipoprotein E-deficient mice. Circulation 115:2516–2525

93. Zhang J, Alcaide P, Liu L, Sun J, He A, Luscinskas FW, Shi GP (2011) Regulation of endothelial cell adhesion molecule expression by mast cells, macrophages, and neutrophils. PLoS One 6:e14525

94. Sun J, Sukhova GK, Yang M, Wolters PJ, MacFarlane LA, Libby P, Sun C, Zhang Y, Liu J, Ennis TL, Knispel R, Xiong W, Thompson RW, Baxter BT, Shi GP (2007) Mast cells modulate the pathogenesis of elastase-induced abdominal aortic aneurysms in mice. J Clin Invest 117:3359–3368

95. Lappalainen J, Lindstedt KA, Oksjoki R, Kovanen PT (2011) OxLDL-IgG immune complexes induce expression and secretion of proatherogenic cytokines by cultured human mast cells. Atherosclerosis 214:357–363

96. Wang J, Cheng X, Xiang MX, Alanne-Kinnunen M, Wang JA, Chen H, He A, Sun X, Lin Y, Tang TT, Tu X, Sjöberg S, Sukhova GK, Liao YH, Conrad DH, Yu L, Kawakami T, Kovanen PT, Libby P, Shi GP (2011) IgE stimulates human and mouse arterial cell apoptosis and cytokine expression and promotes atherogenesis in Apoe−/− mice. J Clin Invest 121:3564–3577

97. Coussens LM, Zitvogel L, Palucka AK (2013) Neutralizing tumor-promoting chronic inflammation: a magic bullet? Science 339:286–291

98. Oleinika K, Nibbs RJ, Graham GJ, Fraser AR (2013) Suppression, subversion and escape: the role of regulatory T cells in cancer progression. Clin Exp Immunol 171:36–45

99. Chang DZ (2012) Mast cells in pancreatic ductal adenocarcinoma. Onco Targets Ther 1:754–755

100. Cheema VS, Ramesh V, Balamurali PD (2012) The relevance of mast cells in oral squamous cell carcinoma. J Clin Diagn Res 6:1803–1807

101. de Souza DA Jr, Toso VD, de Cássia Campos MR, Lara VS, Oliver C, Jamur MC (2012) Expression of mast cell proteases correlates with mast cell maturation and angiogenesis during tumor progression. PLoS One 7: e40790

102. Gounaris E, Erdman SE, Restaino C, Gurish MF, Friend DS, Gounari F, Lee DM, Zhang G, Glickman JN, Shin K, Rao VP, Poutahidis T, Weissleder R, McNagny KM, Khazaie K (2007) Mast cells are an essential hematopoietic component for polyp development. Proc Natl Acad Sci U S A 104:19977–19982

103. Maltby S, Khazaie K, McNagny KM (2009) Mast cells in tumor growth: angiogenesis, tissue remodelling and immune-modulation. Biochim Biophys Acta 1796:19–26

104. Mangia A, Malfettone A, Rossi R, Paradiso A, Ranieri G, Simone G, Resta L (2011) Tissue remodelling in breast cancer: human mast cell tryptase as an initiator of myofibroblast differentiation. Histopathology 58:1096–1106

105. Nelissen S, Lemmens E, Geurts N, Kramer P, Maurer M, Hendriks J, Hendrix S (2013) The role of mast cells in neuroinflammation. Acta Neuropathol 125:637–650

106. Pittoni P, Tripodo C, Piconese S, Mauri G, Parenza M, Rigoni A, Sangaletti S, Colombo MP (2011) Mast cell targeting hampers prostate adenocarcinoma development but promotes the occurrence of highly malignant neuroendocrine cancers. Cancer Res 71:5987–5997

107. Rabenhorst A, Schlaak M, Heukamp LC, Forster A, Theurich S, von Bergwelt-Baildon-M, Büttner R, Kurschat P, Mauch C, Roers A, Hartmann K (2012) Mast cells play a protumorigenic role in primary cutaneous lymphoma. Blood 120:2042–2054

108. Ribatti D, Crivellato E (2011) Mast cells, angiogenesis and cancer. Adv Exp Med Biol 716:270–288

109. Souza LR, Fonseca-Silva T, Santos CC, Oliveira MV, Correa-Oliveira R, Guimaraes AL, De Paula AM (2010) Association of mast cell, eosinophil leucocyte and microvessel densities in actinic cheilitis and lip squamous cell carcinoma. Histopathology 57:796–805

110. Staser K, Yang FC, Clapp DW (2012) Pathogenesis of plexiform neurofibroma: tumor-stromal/hematopoietic interactions in tumor progression. Annu Rev Pathol 7:469–495

111. Cai SW, Yang SZ, Gao J, Pan K, Chen JY, Wang YL, Wei LX, Dong JH (2011) Prognostic significance of mast cell count following curative resection for pancreatic ductal adenocarcinoma. Surgery 149:576–584

112. Huang B, Lei Z, Zhang GM, Li D, Song C, Li B, Liu Y, Yuan Y, Unkeless J, Xiong H, Feng ZH (2008) SCF-mediated mast cell infiltration and activation exacerbate the inflammation and immunosuppression in tumor microenvironment. Blood 112:1269–1279

113. Soucek L, Lawlor ER, Soto D, Shchors K, Swigart LB, Evan GI (2007) Mast cells are required for angiogenesis and macroscopic

expansion of Myc-induced pancreatic islet tumors. Nat Med 13:1211–1218
114. Xia Q, Wu XJ, Zhou Q, Jing Z, Hou JH, Pan ZZ, Zhang XS (2011) No relationship between the distribution of mast cells and the survival of stage IIIB colon cancer patients. J Transl Med 9:88
115. Starkey JR, Crowle PK, Taubenberger S (1988) Mast-cell-deficient W/Wv mice exhibit a decreased rate of tumor angiogenesis. Int J Cancer 42:48–52
116. Chang DZ, Ma Y, Ji B, Wang H, Deng D, Liu Y, Logsdon CD, Hwu P (2011) Mast cells in tumor microenvironment promotes the in vivo growth of pancreatic ductal adenocarcinoma. Clin Cancer Res 17:7015–7023
117. Wasiuk A, Dalton DK, Schpero WL, Stan RV, Conejo-Garcia JR, Noelle RJ (2012) Mast cells impair the development of protective anti-tumor immunity. Cancer Immunol Immunother 61:2273–2282
118. Oldford SA, Haidl ID, Howatt MA, Leiva CA, Johnston B, Marshall JS (2010) A critical role for mast cells and mast cell-derived IL-6 in TLR2-mediated inhibition of tumor growth. J Immunol 185:7067–7076
119. Murata T, Aritake K, Matsumoto S, Kamauchi S, Nakagawa T, Hori M, Momotani E, Urade Y, Ozaki H (2011) Prostagladin D2 is a mast cell-derived antiangiogenic factor in lung carcinoma. Proc Natl Acad Sci U S A 108:19802–19807
120. Sinnamon MJ, Carter KJ, Sims LP, Lafleur B, Fingleton B, Matrisian LM (2008) A protective role of mast cells in intestinal tumorigenesis. Carcinogenesis 29:880–886
121. Pittoni P, Piconese S, Tripodo C, Colombo MP (2011) Tumor-intrinsic and -extrinsic roles of c-kit: mast cells as the primary off-target of tyrosine kinase inhibitors. Oncogene 30:757–769
122. Brown MA, Hatfield JK (2012) Mast cells are important modifiers of autoimmune disease: with so much evidence, why is there still controversy? Front Immunol 3:147
123. Costanza M, Colombo MP, Pedotti R (2012) Mast cells in the pathogenesis of multiple sclerosis and experimental autoimmune encephalomyelitis. Int J Mol Sci 13:15107–15125
124. Mican JM, Metcalfe DD (1990) Arthritis and mast cell activation. J Allergy Clin Immunol 86:677–683
125. Nakano S, Mishiro T, Takahara S, Yokoi H, Hamada D, Yukata K, Takata Y, Goto T, Egawa H, Yasuoka S, Furouchi H, Hirasaka K, Nikawa T, Yasui N (2007) Distinct expression of mast cell tryptase and protease activated receptor-2 in synovia of rheumatoid arthritis and osteoarthritis. Clin Rheumatol 26:1284–1292
126. Noordenbos T, Yeremenko N, Gofita I, van de Sande M, Tak PP, Cañete JD, Baeten D (2012) Interleukin-17-positive mast cells contribute to synovial inflammation in spondylarthritis. Arthritis Rheum 64:99–109
127. Tetlow LC, Woolley DE (1995) Distribution, activation and tryptase/chymase phenotype of mast cells in the rheumatoid lesion. Ann Rheum Dis 54:549–555
128. Pimentel TA, Sampaio AL, D'Acquisto F, Perretti M, Oliani SM (2011) An essential role for mast cells as modulators of neutrophils influx in collagen-induced arthritis in the mouse. Lab Investig 91:33–42
129. Shin K, Gurish MF, Friend DS, Pemberton AD, Thornton EM, Miller HR, Lee DM (2006) Lymphocyte-independent connective tissue mast cells populate murine synovium. Arthritis Rheum 54:2863–2871
130. Shiota N, Shimoura K, Okunishi H (2006) Pathophysiological role of mast cells in collagen-induced arthritis: study with a cysteinyl leukotriene receptor antagonist, montelukast. Eur J Pharmacol 548:158–166
131. Kneilling M, Hultner L, Pichler BJ, Mailhammer R, Morawietz L, Solomon S, Eichner M, Sabatino J, Biedermann T, Krenn V, Weber WA, Illges H, Haubner R, Röcken M (2007) Targeted mast cell silencing protects against joint destruction and angiogenesis in experimental arthritis in mice. Arthritis Rheum 56:1806–1816
132. Juurikivi A, Sandler C, Lindstedt KA, Kovanen PT, Juutilainen T, Leskinen MJ, Mäki T, Eklund KK (2005) Inhibition of c-kit tyrosine kinase by imatinib mesylate induces apoptosis in mast cells in rheumatoid synovia: a potential approach to the treatment of arthritis. Ann Rheum Dis 64:1126–1131
133. Eklund KK, Joensuu H (2003) Treatment of rheumatoid arthritis with imatinib mesylate: clinical improvement in three refractory cases. Ann Med 35:362–367
134. Mancardi DA, Jonsson F, Iannascoli B, Khun H, Van Rooijen N, Huerre M, Daëron M, Bruhns P (2011) Cutting edge: the murine high-affinity IgG receptor FcgammaRIV is sufficient for autoantibody-induced arthritis. J Immunol 186:1899–1903
135. Nigrovic PA, Binstadt BA, Monach PA, Johnsen A, Gurish M, Iwakura Y, Benoist C, Mathis D, Lee DM (2007) Mast cells contribute to initiation of autoantibody-mediated

arthritis via IL-1. Proc Natl Acad Sci U S A 104:2325–2330

136. Nigrovic PA, Malbec O, Lu B, Markiewski MM, Kepley C, Gerard N, Daëron M, Lee DM (2010) C5a receptor enables participation of mast cells in immune complex arthritis independently of Fcgamma receptor modulation. Arthritis Rheum 62:3322–3333
137. Tsuboi N, Ernandez T, Li X, Nishi H, Cullere X, Mekala D, Hazen M, Köhl J, Lee DM, Mayadas TN (2011) Regulation of human neutrophil Fcgamma receptor IIa by C5a receptor promotes inflammatory arthritis in mice. Arthritis Rheum 63:467–478
138. Zhou JS, Xing W, Friend DS, Austen KF, Katz HR (2007) Mast cell deficiency in kit(W-sh) mice does not impair antibody-mediated arthritis. J Exp Med 204:2797–2802
139. Pitman N, Asquith DL, Murphy G, Liew FY, McInnes IB (2011) Collagen-induced arthritis is not impaired in mast cell-deficient mice. Ann Rheum Dis 70:1170–1171
140. Feyerabend TB, Weiser A, Tietz A, Stassen M, Harris N, Kopf M, Radermacher P, Möller P, Benoist C, Mathis D, Fehling HJ, Rodewald HR (2011) Cre-mediated cell ablation contests mast cell contribution in models of antibody- and T cell-mediated autoimmunity. Immunity 35:832–844
141. Rajasekaran N, Solomon S, Watanabe T, Ohtsu H, Gajda M, Bräuer R, Illges H (2009) Histidine decarboxylase but not histamine receptor 1 or 2 deficiency protects from K/BxN serum-induced arthritis. Int Immunol 21:1263–1268
142. McNeil HP, Shin K, Campbell IK, Wicks IP, Adachi R, Lee DM, Stevens RL (2008) The mouse mast cell-restricted tetramer-forming tryptases mouse mast cell protease 6 and mouse mast cell protease 7 are critical mediators in inflammatory arthritis. Arthritis Rheum 58:2338–2346
143. Shin K, Nigrovic PA, Crish J, Boilard E, McNeil HP, Larabee KS, Adachi R, Gurish MF, Gobezie R, Stevens RL, Lee DM (2009) Mast cells contribute to autoimmune inflammatory arthritis via their tryptase/heparin complexes. J Immunol 182:647–656
144. Magnusson SE, Pejler G, Kleinau S, Abrink M (2009) Mast cell chymase contributes to the antibody response and the severity of autoimmune arthritis. FASEB J 23:875–882
145. Ji H, Ohmura K, Mahmood U, Lee DM, Hofhuis FM, Boackle SA, Takahashi K, Holers VM, Walport M, Gerard C, Ezekowitz A, Carroll MC, Brenner M, Weissleder R, Verbeek JS, Duchatelle V, Degott C, Benoist C, Mathis D (2002) Arthritis critically dependent on innate immune system players. Immunity 16:157–168
146. Xu D, Jiang HR, Kewin P, Li Y, Mu R, Fraser AR, Pitman N, Kurowska-Stolarska M, McKenzie AN, McInnes IB, Liew FY (2008) IL-33 exacerbates antigen-induced arthritis by activating mast cells. Proc Natl Acad Sci U S A 105:10913–10918
147. Millauer N, Zuercher AW, Miescher SM, Gerber HA, Seitz M, Stadler BM (1999) High IgE in rheumatoid arthritis (RA) patients is complexed with anti-IgE autoantibodies. Clin Exp Immunol 115:183–188
148. Kikuchi Y, Kaplan AP (2001) Mechanisms of autoimmune activation of basophils in chronic urticaria. J Allergy Clin Immunol 107:1056–1062
149. Miller JF (1975) Cellular basis of the immune response. Acta Endocrinol Suppl (Copenh) 194:55–76
150. Nakae S, Suto H, Iikura M, Kakurai M, Sedgwick JD, Tsai M, Galli SJ (2006) Mast cells enhance T cell activation: importance of mast cell costimulatory molecules and secreted TNF. J Immunol 176:2238–2248
151. Mekori YA (2004) The mastocyte: the "other" inflammatory cell in immunopathogenesis. J Allergy Clin Immunol 114:52–57
152. Bachelet I, Levi-Schaffer F, Mekori YA (2006) Mast cells: not only in allergy. Immunol Allergy Clin N Am 26:407–425
153. Kalesnikoff J, Galli SJ (2008) New developments in mast cell biology. Nat Immunol 9:1215–1223
154. Inamura N, Mekori YA, Bhattacharyya SP, Bianchine PJ, Metcalfe DD (1998) Induction and enhancement of Fc(epsilon)RI-dependent mast cell degranulation following coculture with activated T cells: dependency on ICAM-1- and leukocyte function-associated antigen (LFA)-1-mediated heterotypic aggregation. J Immunol 160:4026–4033
155. Frandji P, Tkaczyk C, Oskeritzian C, David B, Desaymard C, Mecheri S (1996) Exogenous and endogenous antigens are differentially presented by mast cells to CD4+ T lymphocytes. Eur J Immunol 26:2517–2528
156. Kambayashi T, Allenspach EJ, Chang JT, Zou T, Shoag JE, Reiner SL, Caton AJ, Koretzky GA (2009) Inducible MHC class II expression by mast cells supports effector and regulatory T cell activation. J Immunol 182:4686–4695

157. Gaudenzio N, Espagnolle N, Mars LT, Liblau R, Valitutti S, Espinosa E (2009) Cell-cell cooperation at the T helper cell/ mast cell immunological synapse. Blood 114:4979–4988
158. Stelekati E, Bahri R, D'Orlando O, Orinska Z, Mittrucker HW, Langenhaun R, Glatzel M, Bollinger A, Paus R, Bulfone-Paus S (2009) Mast cell-mediated antigen presentation regulates CD8+ T cell effector functions. Immunity 31:665–676
159. Frossi B, D'Inca F, Crivellato E, Sibilano R, Gri G, Mongillo M, Danelli L, Maggi L, Pucillo CE (2011) Single-cell dynamics of mast cell-CD4+ CD25+ regulatory T cell interactions. Eur J Immunol 41:1872–1882
160. Frossi B, Gri G, Tripodo C, Pucillo C (2010) Exploring a regulatory role for mast cells: 'MCregs'? Trends Immunol 31:97–102
161. Gri G, Piconese S, Frossi B, Manfroi V, Merluzzi S, Tripodo C, Viola A, Odom S, Rivera J, Colombo MP, Pucillo CE (2008) CD4+CD25+ regulatory T cells suppress mast cell degranulation and allergic responses through OX40-OX40L interaction. Immunity 29:771–781
162. Piconese S, Gri G, Tripodo C, Musio S, Gorzanelli A, Frossi B, Pedotti R, Pucillo CE, Colombo MP (2009) Mast cells counteract regulatory T-cell suppression through interleukin-6 and OX40/OX40L axis toward Th17-cell differentiation. Blood 114:2639–2648
163. Sibilano R, Frossi B, Suzuki R, D'Inca F, Gri G, Piconese S, Colombo MP, Rivera J, Pucillo CE (2012) Modulation of FcepsilonRI-dependent mast cell response by OX40L via Fyn, PI3K, and RhoA. J Allergy Clin Immunol 130(751–760):e752
164. Sibilano R, Gri G, Frossi B, Tripodo C, Suzuki R, Rivera J, MacDonald AS, Pucillo CE (2011) Technical advance: soluble OX40 molecule mimics regulatory T cell modulatory activity on FcepsilonRI-dependent mast cell degranulation. J Leukoc Biol 90:831–838
165. Tripodo C, Gri G, Piccaluga PP, Frossi B, Guarnotta C, Piconese S, Franco G, Vetri V, Pucillo CE, Florena AM, Colombo MP, Pileri SA (2010) Mast cells and Th17 cells contribute to the lymphoma-associated pro-inflammatory microenvironment of angioimmunoblastic T-cell lymphoma. Am J Pathol 177:792–802
166. Merluzzi S, Frossi B, Gri G, Parusso S, Tripodo C, Pucillo C (2010) Mast cells enhance proliferation of B lymphocytes and drive their differentiation toward IgA-secreting plasma cells. Blood 115:2810–2817
167. Sokol CL, Barton GM, Farr AG, Medzhitov R (2008) A mechanism for the initiation of allergen-induced T helper type 2 responses. Nat Immunol 9:310–318
168. Tang H, Cao W, Kasturi SP, Ravindran R, Nakaya HI, Kundu K, Murthy N, Kepler TB, Malissen B, Pulendran B (2010) The T helper type 2 response to cysteine proteases requires dendritic cell-basophil cooperation via ROS-mediated signaling. Nat Immunol 11:608–617
169. Hammad H, Plantinga M, Deswarte K, Pouliot P, Willart MA, Kool M, Muskens F, Lambrecht BN (2010) Inflammatory dendritic cells--not basophils--are necessary and sufficient for induction of Th2 immunity to inhaled house dust mite allergen. J Exp Med 207:2097–2111
170. Rodriguez Gomez M, Talke Y, Goebel N, Hermann F, Reich B, Mack M (2010) Basophils support the survival of plasma cells in mice. J Immunol 185:7180–7185
171. Charles N, Watford WT, Ramos HL, Hellman L, Oettgen HC, Gomez G, Ryan JJ, O'Shea JJ, Rivera J (2009) Lyn kinase controls basophil GATA-3 transcription factor expression and induction of Th2 cell differentiation. Immunity 30:533–543
172. Denzel A, Maus UA, Rodriguez Gomez M, Moll C, Niedermeier M, Winter C, Maus R, Hollingshead S, Briles DE, Kunz-Schughart LA, Talke Y, Mack M (2008) Basophils enhance immunological memory responses. Nat Immunol 9:733–742
173. Wakahara K, Baba N, Van VQ, Begin P, Rubio M, Ferraro P, Panzini B, Wassef R, Lahaie R, Caussignac Y, Tamaz R, Richard C, Soucy G, Delespesse G, Sarfati M (2012) Human basophils interact with memory T cells to augment Th17 responses. Blood 120:4761–4771
174. Wakahara K, Van VQ, Baba N, Begin P, Rubio M, Delespesse G, Sarfati M (2013) Basophils are recruited to inflamed lungs and exacerbate memory Th2 responses in mice and humans. Allergy 68:180–189
175. Goldszmid RS, Trinchieri G (2012) The price of immunity. Nat Immunol 13:932–938
176. Rivera J, Tessarollo L (2008) Genetic background and the dilemma of translating mouse studies to humans. Immunity 28:1–4

Part II

Obtaining the Cells: Purification, Culture, and Counting

Chapter 2

Purification of Basophils from Peripheral Human Blood

Franco H. Falcone and Bernhard F. Gibbs

Abstract

The purification of basophils from peripheral blood has represented a formidable challenge for researchers since they were discovered by Paul Ehrlich in 1879. From the first published attempts in the late 1960s, it took half a century to develop robust protocols able to give sufficient numbers of pure, functionally unimpaired basophils. The existing protocols for basophil purification exploit those properties of basophils which distinguish them from other cell types such as their localization in blood, density, and the presence or absence of surface markers. Purification techniques have been used in various combinations and variations to achieve a common goal in mind: to obtain a pure population of human basophils in sufficient numbers for downstream studies. The arduous way leading up to the modern protocols is summarized in this historical retrospective. A fast protocol for purification of basophils to near homogeneity is also described.

Key words Basophils, Basophilic granulocytes, Cell separation, Elutriation, Immunomagnetic selection, Percoll, Ficoll

1 Introduction

1.1 A Brief History of Basophil Purification Protocols

Basophilic granulocytes are among the rarest leukocytes in human peripheral blood, making up less than 1% of total white blood cell counts in healthy individuals. Their rarity, in combination with other factors, such as the lack of suitable selection markers and appropriate tools for purification, has hampered research with primary basophils for several decades [1]. While protocols for leukocyte cell separation were developed in the 1950s and 1960s, the first protocols attempting to obtain "pure" populations of basophils were published in 1967 and subsequent years. Many different protocols have been published since then, with varying degrees of success in terms of the recovery and purity obtained by these procedures.

An ideal purification protocol should fulfil most if not all of the following criteria: it should be fast (finished within a few hours of blood taking), have a high recovery and final purity (close to 100%), not lead to activation or priming of the cells (i.e., not alter their phenotype), should not require expensive specialized equipment,

Bernhard F. Gibbs and Franco H. Falcone (eds.), *Basophils and Mast Cells: Methods and Protocols*, Methods in Molecular Biology, vol. 2163, https://doi.org/10.1007/978-1-0716-0696-4_2,

and be financially affordable. Historically, most published protocols have failed one or more such criteria. Early protocols were characterized either by very low recoveries, low purities, or both.

Here, we shall briefly review how the protocols evolved from their modest origin as protocols for enrichment of basophils in the second half of the 1960s to sophisticated protocols resulting in near 100% pure basophil preparations in the twenty-first century.

1.2 The 1960s and 1970s: First Attempts at Enriching Basophils

To the best of our knowledge, the first documented attempts to enrich human basophils were published in 1967 by Sampson and Archer [2] and 1 year later by Parwaresch [3]. The Parwaresch protocol describes the use of a specially shaped siliconized glass tube curved at the bottom and open at the top into which 2 mL of heparinized blood are filled and spun in a centrifuge. After centrifugation, the tip of the closed curved end at the bottom of the glass tube is cut open with a scalpel blade and the open tube slowly immersed vertically into a mercury bath. The mercury enters the glass tube from the small opening in the bottom and pushes the blood upward toward the open end, from where fractions are collected onto microscope slides. The purity and recovery of the method are not reported but are likely to have been very low. The Sampson protocol used dextran sedimentation of the buffy coat layer obtained from centrifuged blood followed by a centrifugation in an albumin solution with a specific gravity of 1.085, resulting in an enriched basophil population with a recovery of 25% and purities ranging from 5% to 20%. Higher purities of up to 50% were obtained with blood from patients with chronic myeloid leukemia. Pruzansky and Patterson [4] described a two-step protocol in which leukocytes were first obtained by Ficoll–sodium metrizoate density gradient centrifugation, and basophils selectively eluted from a glass bead column, resulting in 20% pure basophils.

1.3 The 1980s: The 50% Purity Barrier Is Overcome, but Recoveries Remain Low

In comparison with the relatively poor outcomes associated with the first published purification protocols, the next technique published by RP Day in 1972 [5] seemed well ahead of its time both in terms of its achieved recovery (74%) and average purity (93%). The protocol consisted of two sequential density gradient centrifugations in Hypaque/Ficoll (1.080 g/cm^3) and Hypaque (1.065 g/cm^3), respectively, from which a pellet was retrieved and separated in a third equilibrium density gradient centrifugation step (1.065 and 1.085 g/cm^3). Subsequent attempts by other laboratories however failed to reproduce these results [6–8] and did not achieve purities or recoveries beyond 40%. For a while, no protocols seemed to be able to achieve purities higher than around 50% [9–11].

A notable exception from this trend was the purification protocol published by Kauffmann et al., who were able to obtain 70% average purity (recovery ranging from 8% to 20%) in one single discontinuous Percoll gradient centrifugation step [12]. A similar single-step Percoll gradient technique employed by Leonard however resulted in a mean basophil purity of only 19% (range, 5–53%) [13].

While these protocols did not yet achieve the high purities and yields of later years, they enabled researchers to perform for the first time important functional studies with human basophils. In particular, the protocol developed by MacGlashan and Lichtenstein in 1980 [7] represents a milestone in the history of basophil purification, enabling several subsequent studies which led to new insights into functional properties of this cell type. The protocol started with a density centrifugation step with Hypaque–Ficoll, in which the separation medium had been adjusted to a density of up to 1.086 g/mL leading to higher retention of basophils compared with the standard density of 1.077 g/mL. The cells were then passively sensitized with penicillin-specific IgE obtained from patients allergic to penicillin, bound to Sepharose coupled to penicillin, and eluted with monovalent hapten (*N*-ε-Benzylpenicilloyl-*N*-ε-formyl-L-lysine, BPO-FLYS). The use of EDTA in the buffer, in combination with low temperatures, and the use of the monovalent hapten for elution ensured that the obtained basophils were not pre-activated, hence were suitable for functional studies.

Purities only increased when protocols using adhesion to glass beads as final step or multiple repeated density centrifugations were abandoned and replaced by other, more efficient selection methods. Weil and coauthors described a two-step positive selection protocol using flow cytometry and FITC-labeled anti-IgE antibodies, resulting in 97–99% purity but only 15% recovery [14]. A downside of this method was that due to the nature of the selection, the basophils could not be used for functional studies, as pointed out by Bjerke and coauthors [15].

Toll and coauthors published a three-step protocol consisting of Dextran sedimentation to remove red blood cells, followed by Ficoll–Hypaque centrifugation [8]. The final step relied on an anti-IgE Sepharose column, which enabled affinity purification of basophils. Basophils bound to the anti-IgE Sepharose were eluted by high buffer flow rates or mechanical agitation. Although this protocol, due to its IgE-based positive selection principle and increased mechanical stress, which is known to increase the risk of degranulation [15], could be expected to affect the cells' functionality, the authors reported no functional impairment in histamine release tests, purities ranging between 80 and 95%, and a recovery of around 30%.

A second flow cytometric procedure was published in 1986 by Pendy et al. [16] Ficoll-purified leukocytes containing basophils were subjected to lactic acid stripping, a treatment which was shown by Pruzansky to dissociate IgE from its high-affinity receptor [17]. Stripped basophils were then sensitized with FITC-labeled mouse anti-DNP IgE and sorted by flow cytometry. As expected, this positive selection protocol resulted in high purities (98–99%) but low recoveries (29%, range 18–35%). Viability was partially affected (≥88%). Functional data were not reported. Warner and colleagues further refined MacGlashan's 1980 protocol consisting of Percoll gradient centrifugation and positive selection on penicillin-Sepharose after passive sensitization with penicillin-specific IgE [18]. Final purity obtained was 64% (52–92%) and recovery 48% (28–64%). This improved protocol substantially improved the purity while moderately improving (5–10%) the yield, but most importantly, it preserved functional integrity of the obtained cell population.

Another flow cytometry protocol reported high purities (95–98%), but these were obtained from chronic myeloid leukemia patients and umbilical cord blood and thus are not representative of what can be achieved from peripheral blood of healthy donors [19].

One of the main problems when purifying basophils is the accompanying high number of contaminating leukocytes. This issue was initially overcome by exploiting the expression of the high-affinity IgE receptor on basophils, in combination with positive selection protocols. Positive selection however bears the risk of impairing basophil functionality. An alternative technology which can be used to remove the bulk of contaminants is the use of countercurrent centrifugal elutriation (*see* **Note 1**). Most protocols using elutriation reported basophil purities around 4–10% after the elutriation step, although more recent protocols reported higher purities due to modified elutriation protocols, e.g., 17.7% [20] or 12.6% [21].

The first application of the principle of elutriation to the purification of basophils was published by De Boer in 1986 [22]. Percoll (density of 1077 g/cm^3) gradients were used as first step, followed by elutriation, and a final Percoll centrifugation (1067 g/cm^3). Final purity is reported as 69% (45–87%) and the recovery 51% (31–80%) as judged by histamine content or 60% by Alcian blue counts. The authors point to the fact that basophil-specific gravity appears to increase when Ficoll–Isopaque is used, possibly due to the phagocytic uptake of Isopaque [23]. The specific gravity of basophils was reported to range from 1.065 to 1.075 g/cm^3, which is lower than the range for histamine containing cells in blood (1.065 to 1.085 g/cm^3) reported by Graham [24].

Interestingly, in the same year Miroli and coauthors published a one-step Ficoll–Paque centrifugation protocol with a final purity of 9% (range 6.4–12.5%) and a recovery of 82% (range 65–95%) [25]. This protocol clearly shows the limitations of one-step protocols but, as pointed out by the authors, represents a good first step due to its low cost and robustness. The protocols published by Miroli and De Boer disagree regarding the distribution of basophils in the various layers after centrifugation of blood on Ficoll–Paque.

The protocol published in 1989 by Lett-Brown et al. was the first attempt to use negative selection [26]. Leukocytes were obtained by Percoll density centrifugation, and contaminating cells were partially removed using monoclonal antibodies to B-cell and T-cell antigens and removal of these cells with an anti-mouse immunoglobulin. This technique resulted in 75% purity and 54% recovery. Purified basophils appeared functionally unimpaired judging by the low spontaneous release and normal histamine content of 1.1 pg/cell.

By the end of the 1980s, all the modules necessary for successful basophil purification were in place and had been tested in various combinations. The first real breakthroughs en route to obtaining pure basophils however only became possible after development of the new technology of immunomagnetic selection, as described in the next section.

1.4 The 1990s: Immunomagnetic Selection Makes First Breakthroughs Possible

As results obtainable by protocols published in the 1980s were still unsatisfactory in terms of achieved purity, recovery, or both, new purification techniques were tested in the 1990s. Schroeder and Hanrahan describe the use of neuraminidase-treated red blood cells coated with TNP hapten [27]. Basophils obtained by a Percoll gradient centrifugation were first incubated with anti-TNP purified mouse monoclonal IgE. The passively sensitized basophils were then incubated with the TNP-bearing red blood cells at 4 °C. The rosettes with attached basophils were isolated by another Percoll centrifugation and red blood cells removed by hypotonic lysis. This protocol resulted in a mean purity of 83.5% (range 66.8–92.6) and a recovery of 58%, which compared very favorably with previous protocols. However, basophils obtained by this protocol were functionally impaired, as their maximum histamine release was only half of the basophils before rosetting. This may reflect both the positive selection procedure and the effects of hypotonic lysis on basophils.

Mul and coworkers improved the protocol by De Boer and Roos [22] by adding a final immunomagnetic purification step [28]. After Percoll centrifugation, cells underwent elutriation. The elutriation fraction containing the basophils was again purified using a discontinuous Percoll gradient (1.068 and 1.075 g/cm^3), resulting in 70% pure basophils. The cells obtained from the Percoll gradient were used for the final immunomagnetic negative selection step. Magnetic bead-cell rosettes were induced by coating beads with goat anti-mouse IgG and incubating the cells with mouse monoclonal antibodies to antigens expressed on contaminating cells (CD2, CD14, CD16 and CD19). Contaminating cells were then removed with a magnet. This protocol, albeit long and laborious from today's point of view, successfully increased the purity from 70% to 94.2% (range, 87–98%) with an excellent recovery of 80.2% (range 72–85%). The basophils were fully responsive to anti-IgE and fMLP stimulation.

Bjerke's protocol [15] also used immunomagnetic selection as the final step after Percoll gradient centrifugation (45.6% pure), resulting in 97.8% pure basophils with a recovery of 66.2%, which were fully functional. This protocol, because of its brevity and the high purity and recovery achieved, can be considered the first major breakthrough in basophil purification. Around the same time, Mita et al. used a similar two-step protocol consisting of Percoll gradients followed by immunomagnetic selection which resulted in an average purity of at least 80% [29].

Gibbs et al. obtained 67.3% pure basophils using a three-step protocol consisting of Ficoll–Paque density centrifugation, elutriation, and negative immunomagnetic selection with Dynabeads [20]. Although basophil functionality was intact, the recovery was low (21.8%). Very similar results were obtained with the same protocol by Falcone et al. [30]. A subsequent protocol published by the same group, using the first available commercial kit for basophil immunomagnetic selection, achieved a considerable improvement in terms of purity (97.6%) and recovery (49.7%), high viability (99.6%), and unimpaired functionality [21]. Taken together, the main advantage of protocols including an elutriation step is that they allow the processing of larger volumes of blood, which would be cost-prohibitive for most laboratories if using Percoll and immunomagnetic separation alone. The downside is that not many laboratories are equipped with a countercurrent centrifugal elutriator. Relatively high purities, albeit lower than in the aforementioned protocols, can also be achieved without elutriation, as shown by the protocol of McEuen [31], consisting of Percoll density centrifugation followed by immunomagnetic negative selection with a commercially available kit. This resulted in 82.4% pure basophils (median, 95%, range 35–100%) and a mean recovery of 51.2%. The highest purities were achieved in the samples in which the first step had achieved a 10% or higher enrichment. Purities of 10% or higher are routinely achieved by elutriation, again pointing to the benefits of this technique.

Flow cytometry was also used by several protocols using this technology for the final purification step. Tanimoto and colleagues obtained a relatively high basophil purity of 84.7% (range 77.3–90%) by negative selection, but recoveries were low (16%, range 11–19.9%) [32]. A two-step protocol by Kepley et al. [33] consisting of Percoll density gradient centrifugation (32% purity; range 20–45%) followed by flow cytometry solely based on size and granularity, i.e., light-scattering properties, resulted in 97% purity (range 95–99%). While this protocol had the advantage of not necessitating the use of expensive antibodies and resulted in unimpaired basophil functionality, it showed the characteristic problem of flow cytometry-based approaches: the low recovery (23%; range 5–28%). The four-step protocol published by Willheim [34] consisted of Ficoll–Paque centrifugation, elutriation, immunomagnetic cell sorting to remove contaminating monocytes, and fluorescent cell sorting after CDw17 and CD14 labeling (basophils were $CD17^+/CD14^-$). This protocol achieved the highest purities (99.4%, range: 98.7–99.9%) compared with any other protocol published before then. However, as could probably be expected from such a complex, laborious four-step protocol involving flow cytometry, recoveries were very low (mean 12.8%, range 8–25%).

1.5 The 2000s: New Emphasis on Rapid or Simplified Protocols

As the aims of obtaining almost homogeneous basophil populations, high recovery, and pristine functionality had been achieved in the 1990s (although usually not together in the same protocol), protocols published in the 2000s focused on practical aspects such as duration of the protocol, simplicity of procedure, and preservation of basophil functionality (as reflected by the use of the terms "rapid" or "simplified" in the titles of the corresponding publications). The protocol by Tsang et al. [35] consists of Ficoll–Histopaque density centrifugation followed by immunomagnetic purification with a commercial kit and results in 96.5% (95–98%) pure basophils with a mean yield of 40.8% (21–56%). A more advanced version of this commercial kit, in combination with Heta-Sep precipitation, resulted in the rapid two-step purification protocol by Gibbs et al. [36] who obtained a mean purity of 99.3% (range 97–100%) with a mean recovery of 75.6% (range 39–100%). This is arguably the quickest protocol available to date, with viable, functional pure basophils obtained in less than 90 min.

An interesting innovative approach combining multistep Percoll density gradients with a miniaturized separation apparatus was described by Shiono and coauthors [37]. In less than 2 h, functionally active basophils can be enriched using a continuous flow-through cell separation method based on a five-step Percoll gradient with densities ranging from 1.071 to 1.090 g/mL. The authors were able to obtain enriched basophils (fraction 4 mean purity was 60.2%, range 50.8–71.5%, $n = 6$) from small amounts of blood (9 mL) in under 2 h. While the low purity and number of basophils may preclude this technique from being useful for some functional studies, which require homogeneous basophils in larger amounts, it may be useful for other, e.g., diagnostic applications. The main advantage is that it requires no antibodies and can be used for generation of other subcellular fractions other than basophils.

Figure 1 summarizes the development of the purification protocols over the last 45 years. It is clear that with the exception of the early 1972 protocol [5], purities increased steadily over the years. Overall, there is no clear relationship between the number of steps in the protocol (indicated by the different colours) or the yield (diameter of bubbles) with the final purity (y-axis). However, it is also clear that protocols including flow cytometry, despite resulting in high final purities, have the lowest recoveries (means range 13–23%) and are typically associated with reduced viability (~90%). Whether the partial viability impairment is due to the obligatory red blood cell lysis step is not clear.

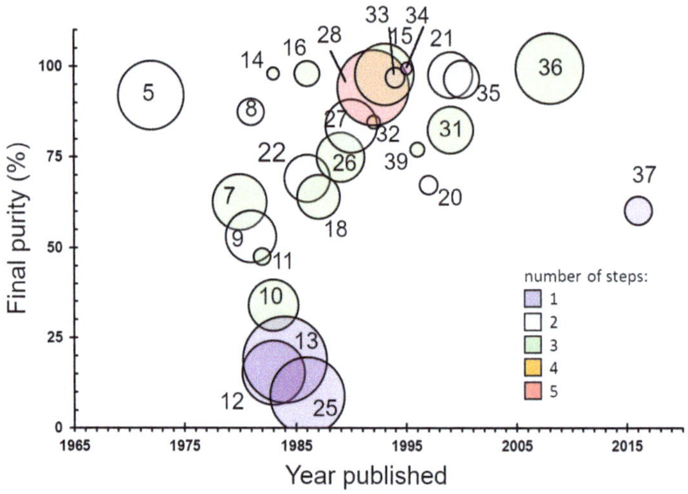

Fig. 1 Bubble plot linking the year of publication (*x*-axis), the final purity (*y*-axis), and the recovery (bubble diameter) of basophil purification of most protocols described in this chapter [5, 7–16, 18, 20–22, 25–38]. The color, as shown in the key, indicates the number of steps in the protocol. Only protocols describing purification from peripheral blood of healthy (in some cases, atopic or atopic dermatitis) donors were included. Protocols purifying basophils from CML patients, or protocols where recovery and purity had not been clearly reported, were not considered. Toll et al. [8] reported their obtained basophil purity to be 80–95% but did not specify a mean or median purity and was herein assumed to be 87.5%

1.6 Concluding Remarks

Our short history of basophil purification has shown that a plethora of techniques has been developed for purification of human basophils. While some early techniques such as affinity chromatography for positive selection and adhesion to glass beads have now been abandoned, several techniques can be used in different combinations to obtain pure basophils. So how to decide which protocol is the best?

The first question to answer is whether the purification is necessary. Some protocols allow assessment of basophil activation in whole blood or leukocyte preparations, e.g., when measuring the release of histamine, a preformed mediator that is to a large extent restricted to basophils in peripheral blood, or direct measurement of basophil activation via upregulation of surface activation markers such as CD63 or CD203c (*see* chapters 15–17 about basophil activation tests). However certain applications, like signal transduction studies using Western blotting, or gene transcription profiling using RT-PCR, will require highly purified cells, as well as high recovery.

Despite all these advances, the low number of basophils in peripheral blood of healthy donors remains the limiting factor.

2 Materials

2.1 Reagents

All solutions should be sterile and used chilled unless stated otherwise.

1. EasySep™ Human Basophil Enrichment Kit (STEMCELL Technologies).
2. RoboSep Isolation buffer: DPBS without Ca^{2+}/Mg^{2+} with 2% FBS (heat inactivated; v/v) and 1 mM EDTA (from 500 mM EDTA stock solution, pH 7.3) (STEMCELL Technologies).
3. HetaSep™ cell isolation medium (STEMCELL Technologies).
4. Accustain™ May-Grünwald stain 0.25% w/v in methanol (Sigma-Aldrich).
5. 3% (v/v) Acetic acid with methylene blue (STEMCELL Technologies).
6. 0.4% (w/v) Trypan blue (STEMCELL Technologies).

2.2 Supplies

1. 10 mL BD Vacutainer sodium heparin tubes (NH 170 I.U).
2. 5 mL BD Falcon polystyrene round bottomed tubes.
3. Superfrost standard microscope slides.
4. Shandon Filter Cards (Thermo Fisher Scientific).
5. Shandon Cytoclip Stainless-Steel Slide Clip (Thermo Fisher Scientific).
6. Cytofunnels disposable, single (Thermo Fisher Scientific).

2.3 Equipment

1. EasySep™ Magnet for isolating up to 2.5×10^8 cells (immunomagnetic column-free magnet) (STEMCELL Technologies).
2. CytoSpin 2, 3, or 4 cytocentrifuge (Thermo Scientific Shandon).
3. Centrifuge.

3 Methods

3.1 Isolation of Basophils from Peripheral Blood

1. Collect peripheral blood by venipuncture in three or four blood collection tubes containing Na-heparin.
2. Pool the blood into a sterile 50 mL tube, add one part of HetaSep to five parts of blood, then invert carefully, without shaking.
3. Split the HetaSep–blood mixture between three sterile 15 mL tubes, and centrifuge at $110 \times g$ for 6 min with the brake off (*see* **Note 2**).

4. Let stand an additional 10 min at room temperature (*see* **Note 3**).
5. *Carefully* collect and pool the phases into a sterile 50 mL tube, containing the white blood cells without disturbing the red blood cell layer (*see* **Note 4**).
6. Top up the pooled upper phases to 50 mL with chilled RoboSep isolation buffer, and wash twice, using at least one wash at 120 × *g*, 10 min, followed by a second wash with chilled RoboSep buffer at 450 × *g*, 7 min (*see* **Note 5**).
7. After removing the supernatant, resuspend cell pellet in a total volume of 0.5 mL of chilled RoboSep buffer.
8. Mix 20 μL of cell suspension with 180 μL of 3% acetic acid/methylene blue, and count cells in a hemocytometer (*see* **Note 6**).
9. Carefully resuspend the cell pellet at 5×10^7 cells/mL or, if less than 25 million cells, a total volume of 0.5 mL of chilled RoboSep separation buffer, and transfer to a sterile round-bottomed 5 mL standard FACS tube (*see* **Note** 7).
10. Add 50 μL/mL of Human Basophil Enrichment Cocktail (i.e., for 500 μL, add 25 μL of cocktail), mix well, and let stand at RT or 4 °C for 10 min.
11. Resuspend EasySep® Nanoparticles by vigorously pipetting more than five times. Do not Vortex. Add 100 μL/mL (i.e., for 500 μL, add 50 μL of nanoparticles) of EasySep® Nanoparticles, and mix well without vortexing.
12. Incubate at RT or 4 °C for 10 min.
13. Top-up suspension to 2.5 mL with separation buffer; gently resuspend the cells before transferring the tube containing the cell suspension to an EasySep purple magnet, and incubate for 5 min at RT.
14. Decant unbound cells into a fresh FACS tube. Do not feel tempted to shake or blot off any drops that may remain hanging from the mouth of the tube (*see* **Note 8**).
15. Remove the empty tube from the EasySep magnet, and place the new tube (containing the supernatant from **step 14**) into the magnet. Set aside for 5 min at RT.
16. Repeat **steps 14** and **15** for a total of 3× 5-min separations in the magnet. To improve recovery, at the expense of purity, only two isolations may be done.
17. Wash cells with isolation buffer or the desired assay medium.
18. Take aliquots for purity (cytospin staining using May–Grünwald's) and standard Trypan blue viability determination (*see* **Note 9**).

3.2 Preparation of Cytospins

1. Assemble cytofunnels with Filter Card and glass slide, which you have labeled with a pencil on the frosted area (*see* **Note 10**). Add 100 μL of cell suspension in cell culture medium containing FCS or RoboSep buffer (*see* **Note 11**).
2. Load cell suspension into cytofunnel and spin 1 min 500 rpm.
3. Disassemble cytospin units; allow slides to air-dry at RT for 10 min or until dry.
4. Cover whole slide with sufficient volume of May–Grünwald solution, and stain for 3 min.
5. Pour off staining solution, rinse with deionized water, and allow to dry.
6. Examine cytospins under light microscope at 40× magnification or to see basophil morphology at 100× with oil immersion. Count a minimum of 500 cells for evaluation of purity. Basophils should appear fully granulated as shown in Fig. 2.

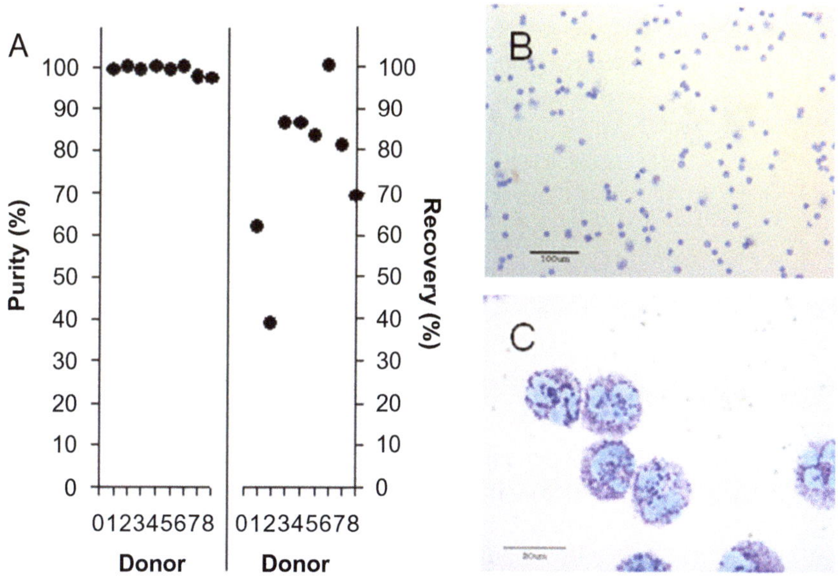

Fig. 2 (**a**) Percentage purity and recovery of basophils purified from fresh peripheral blood with the new two-step procedure ($n = 8$). Recovery was determined by Alcian blue staining of cells both before and after magnetic cell sorting. Cytospin preparations stained with May–Grünwald stain showing 100% pure basophils at 100× magnification (**b**) and at 1000× magnification (**c**), showing that basophils are fully granulated. (Reproduced from [36] with permission from John Wiley & Sons (RightsLink)

4 Notes

1. Elutriators are technical devices, which allow the separation of particles with varying densities. Countercurrent centrifugal elutriators are special centrifuge-like devices manufactured by Beckman Instruments, which allow a separation of a cell suspension into various fractions with different densities. The process of separation is considered gentle and does not require expensive reagents. The special design of the centrifuge results in the separation buffer flowing in the direction opposite to the centrifugal force. By stepwisely increasing the flow rate (or decreasing the centrifugal force), cells are sequentially washed into collection tubes. It is not possible to obtain pure basophils by this protocol alone as their densities overlap in part with lymphocytes and neutrophils. However, it works well in generating a partially enriched population (>10% purity in our hands) which can be used for, e.g., immunomagnetic selection.
2. Recovery from HetaSep is generally improved when using the minimal-sized tube for total blood/HetaSep volumes; therefore, splitting a 30 mL sample between 3× 15 mL tubes would be ideal.
3. The success of the separation can be influenced by several factors which are difficult to control, ranging from the ambient temperature to the status of the blood donor (e.g., fasting or after a fatty meal) to the age of the blood sample. It is important to see a neat separation between the RBC layer and the WBC layer above.
4. We do this first using a sterile disposable cell culture 5- or 10-mL pipette followed by a pastette, making sure that the air is expelled before introducing into the liquid. You can also use pastettes only.
5. The slow centrifugation step is needed to remove the bulk of platelets.
6. This technique allows a so-called nucleated cell count. All cells dissolve, but the cell nuclei remain intact, which are stained light blue by the methylene blue dye. This allows a quick count without interference from red blood cells, which have no nuclei and are often difficult to distinguish from white blood cells in a hemocytometer in inexperienced hands.
7. 500 μL Is the minimum volume for the purple EasySep magnet, but we recommend starting with at least 50 million cells (i.e., 1 mL at the recommended concentration of 50 million per mL) for optimal performance.

8. As this is a negative selection protocol, the unwanted cells will remain attached to the magnet via the magnetic beads. The decanted buffer should only contain basophils.
9. If necessary, cell recovery can be determined using Alcian blue staining (which permits for rapid basophil-specific staining) in a hemocytometer, as described by Gilbert and Ornstein [39].
10. Do not use a marker pen. This will wash off during staining and ruin your cytospin.
11. Cytospins work best with at least 100,000 cells. However, this would mean sacrificing a large number of purified basophils just for purity determination. We therefore routinely load fewer cells onto the cytospins.

References

1. Falcone FH, Haas H, Gibbs BF (2000) The human basophil: a new appreciation of its role in immune responses. Blood 96:4028–4038
2. Sampson D, Archer GT (1967) Release of histamine from human basophils. Blood 29:722–736
3. Parwaresch MR (1968) A new method for the enrichment of leukocytes, particularly of basophils. Blut 17:260–265
4. Pruzansky JJ, Patterson R (1970) Decrease in basophils after incubation with specific antigens of leukocytes from allergic donors. Int Arch Allergy Appl Immunol 38:522–526
5. Day RP (1972) Basophil leucocyte separation from human peripheral blood: a technique for their isolation in high purity and high yield. Clin Allergy 2:205–212
6. Ishizaka T, De Bernardo R, Tomioka H et al (1972) Identification of basophil granulocytes as a site of allergic histamine release. J Immunol 108:1000–1008
7. MacGlashan DW, Lichtenstein LM (1980) The purification of human basophils. J Immunol 124:2519–2521
8. Toll JB, Wikberg JE, Andersson RG (1981) Purification of human basophils by affinity chromatography on anti-IgE-sepharose 6MB. Allergy 36:411–417
9. Pruzansky JJ, Patterson R (1981) Enrichment of human basophils. J Immunol Methods 44:183–190
10. Landry FJ, Findlay SR (1983) Purification of human basophils by negative selection. J Immunol Methods 63:329–336
11. Raghuprasad PK (1982) A rapid simple method of basophil purification by density centrifugation on Percoll. J Immunol 129:2128–2133
12. Kauffman HF, Levering PR, De Vries K (1983) A single centrifugation step method for the simultaneous separation of different leukocytes with special reference to basophilic leukocytes. J Immunol Methods 57:1–7. https://doi.org/10.1016/0022-1759(83)90058-3
13. Leonard EJ, Roberts RL, Skeel A (1984) Purification of human blood basophils by single step isopycnic banding on Percoll. J Leukoc Biol 35:169–177
14. Weil GJ, Leiserson WM, Chused TM (1983) Isolation of human basophils by flow microfluorometry. J Immunol Methods 58:359–363
15. Bjerke T, Nielsen S, Helgestad J et al (1993) Purification of human blood basophils by negative selection using immunomagnetic beads. J Immunol Methods 157:49–56
16. Pendy LM, Arra SJ, Anselmino LM, Thomas LL (1986) A flow cytometric procedure for the isolation of antigenically responsive human basophils. J Immunol Methods 91:59–63
17. Pruzansky JJ, Grammer LC, Patterson R, Roberts M (1983) Dissociation of IgE from receptors on human basophils. I. Enhanced passive sensitization for histamine release. J Immunol 131:1949–1953
18. Warner JA, Peters SP, Lichtenstein LM et al (1989) Differential release of mediators from human basophils: differences in arachidonic acid metabolism following activation by unrelated stimuli. J Leukoc Biol 45:558–571
19. Bodger MP, Newton LA (1987) The purification of human basophils: their immunophenotype and cytochemistry. Br J Haematol 67:281–284

20. Gibbs BF, Noll T, Falcone FH et al (1997) A three-step procedure for the purification of human basophils from buffy coat blood. Inflamm Res 46:137–142. https://doi.org/10.1007/s000110050537
21. Haisch K, Gibbs BF, Körber H et al (1999) Purification of morphologically and functionally intact human basophils to near homogeneity. J Immunol Methods 226:129–137
22. de Boer M, Roos D (1986) Metabolic comparison between basophils and other leukocytes from human blood. J Immunol 136:3447–3454
23. Splinter TA, Beudeker M, Van Beek A (1978) Changes in cell density induced by isopaque. Exp Cell Res 111:245–251
24. Graham HT, Lowry OH, Wheelwright F et al (1955) Distribution of histamine among leukocytes and platelets. Blood 10:467–481
25. Miroli AA, James BM, Spitz M (1986) Single step enrichment of human peripheral blood basophils by Ficoll-Paque centrifugation. J Immunol Methods 88:91–96
26. Lett-Brown MA, Robinson L, Juneja HS, Grant JA (1989) Purification of human basophils. Their response to anti-IgE. J Immunol Methods 117:163–167
27. Schroeder JT, Hanrahan LR (1990) Purification of human basophils using mouse monoclonal IgE. J Immunol Methods 133:269–277
28. Mul FP, Knol EF, Roos D (1992) An improved method for the purification of basophilic granulocytes from human blood. J Immunol Methods 149:207–214
29. Mita H, Akiyama K, Hayakawa T et al (1993) Purification of human blood basophils and leukotriene C4 generation following calcium ionophore stimulation. Prostaglandins Leukot Essent Fatty Acids 49:783–788
30. Falcone FH, Dahinden CA, Gibbs BF et al (1996) Human basophils release interleukin-4 after stimulation with *Schistosoma mansoni* egg antigen. Eur J Immunol 26:1147–1155. https://doi.org/10.1002/eji.1830260528
31. McEuen AR, Buckley MG, Compton SJ, Walls AF (1999) Development and characterization of a monoclonal antibody specific for human basophils and the identification of a unique secretory product of basophil activation. Lab Investig 79:27–38
32. Tanimoto Y, Takahashi K, Takata M et al (1992) Purification of human blood basophils using negative selection by flow cytometry. Clin Exp Allergy 22:1015–1019
33. Kepley C, Craig S, Schwartz L (1994) Purification of human basophils by density and size alone. J Immunol Methods 175:1–9
34. Willheim M, Agis H, Sperr WR et al (1995) Purification of human basophils and mast cells by multistep separation technique and mAb to CDw17 and CD117/c-kit. J Immunol Methods 182:115–129
35. Tsang S, Hayashi M, Zheng X et al (2000) Simplified purification of human basophils. J Immunol Methods 233:13–20
36. Gibbs BF, Papenfuss K, Falcone FH (2008) A rapid two-step procedure for the purification of human peripheral blood basophils to near homogeneity. Clin Exp Allergy 38:480–485. https://doi.org/10.1111/j.1365-2222.2007.02919.x
37. Shiono H, Matsui T, Okada T, Ito Y (2016) Single-step enrichment of basophils from human peripheral blood by a novel method using a Percoll density gradient. J Sep Sci 39:3062–3071. https://doi.org/10.1002/jssc.201600329
38. Yamaguchi M, Hirai K, Ohta K et al (1996) Nonreleasing basophils convert to releasing basophils by culturing with IL-3. J Allergy Clin Immunol 97:1279–1287. https://doi.org/10.1016/S0091-6749(96)70196-3
39. Gilbert HS, Ornstein L (1975) Basophil counting with a new staining method using alcian blue. Blood 46:279–286

Chapter 3

Mast Cell Purification Protocols

Jasmine Farrington, Elizabeth P. Seward, and Peter T. Peachell

Abstract

Studying a tissue-specific mast cell can be of particular benefit given the heterogeneity that is known to exist among mast cells isolated or developed from different sources. Methods for isolating mast cells from a variety of tissues have been in existence for a number of years although, over time, these methodologies have been refined. We have had considerable experience studying mast cells isolated from human lung tissue. It is for this reason that, in this chapter, we provide detailed methods for the isolation and purification of human lung mast cells. However, it should be noted that the methods that are described in this chapter are generally applicable to the isolation of mast cells from different tissues, and this will also be discussed.

Key words Mast cells, CD117, Dynabeads®

1 Introduction

Methods for the isolation of mast cells from animal and human tissues have been in existence for quite some time [1]. The methods that have been described show commonalities in that, in order to isolate mast cells, the tissue is physically and enzymatically disrupted. The details of the disruption can differ from method to method, but nonetheless the principles are the same. This initial process leads to a mixed cell preparation of which mast cells comprise a proportion. Mast cells can be studied in a mixed cell preparation especially if straightforward functional studies are anticipated. For example, simple mediator release experiments are possible especially if stimuli (e.g., anti-IgE) are used and outputs (e.g., histamine) are measured that are likely to be mast cell-specific. However, further purification may be necessary if more discrete biochemical studies are required.

In this chapter, we describe in detail methods for the isolation and purification of human lung mast cells (HLMC). We also provide some general details to show how these methods are applicable to the isolation of mast cells from other tissues.

Bernhard F. Gibbs and Franco H. Falcone (eds.), *Basophils and Mast Cells: Methods and Protocols*, Methods in Molecular Biology, vol. 2163, https://doi.org/10.1007/978-1-0716-0696-4_3,

2 Materials

2.1 Buffers and Culture Media

1. Digestion buffer: Dulbecco's Modified Eagle's Medium (DMEM) (with added 4-(2-hydroxyethyl)-1-piperazineethanesulfonic acid (HEPES) and GlutaMAX; GIBCO) containing 2% antibiotic-antimycotic solution, 1% nonessential amino acids, and 10% fetal calf serum (FCS).
2. Cell culture-treated collagenase type I-A.
3. Hyaluronidase.
4. HBSS protein solution: 85% Hanks' Balanced Salt Solution (HBSS), 2% FCS, 10% horse serum, and 1% bovine serum albumin.
5. HLMC culture medium: DMEM (with added HEPES and GlutaMAX) containing 1% antibiotic-antimycotic solution, 1% nonessential amino acids, 10% FCS, and supplemented with 100 ng/mL human stem cell factor, 50 ng/mL IL-6, and 10 ng/mL IL-10.

2.2 Mast Cell Staining

1. Kimura stain (*see* **Note 1**): 11 mL of 0.05% toluidine blue solution, 0.8 mL of 0.03% light green "SF yellowish" dye, 0.5 mL of 4 mg/mL saponin in 50% ethanol, and 5 mL of phosphate buffered saline at pH 6.4. The 0.05% toluidine blue solution is prepared as follows: 0.05 g toluidine blue is dissolved in 50 mL 1.8% NaCl solution, 22 mL ethanol, and 28 mL dH_2O.

2.3 Magnetic Cell Sorting

1. Magnetic particle concentrator (MPC-1; Dynal).
2. Magnetic Dynabeads®.
3. Mouse antihuman CD117 antibody.
4. HBSS.
5. FCS.
6. Phycoerythrin (PE)-tagged anti-CD117 IgG_1 antibody.

3 Methods

We obtain non-lesional tissue from surgical lung resections. This requires appropriate ethical approval. The tissue is then enzymatically digested using a modification of methods that have been described elsewhere ([2, 3]; *see* **Note 2**). Mast cells are isolated and further purified using the Dynal® magnetic bead system using methods similar to those described [4]. All procedures are carried out aseptically.

3.1 Conjugation of Dynabeads®

1. To coat sheep anti-mouse IgG conjugated Dynabeads® with mouse anti-human CD117 antibody before use, rinse 100 μL of bead mixture three times in HBSS (containing 2% FCS) using an MPC-1 magnet (Dynal).
2. After the final wash, resuspend beads in 0.4 mL of HBSS protein solution, and add 8 μL of mouse anti-human CD117 antibody.
3. Incubate the mixture at 4 °C for 2 h with continuous rotation.
4. Wash beads three times using the MPC-1 magnet as described above; then resuspend in 100 μL of HBSS protein solution. If coated beads are not to be used immediately, add 0.1% sodium azide as a preservative, and store for up to 2 weeks at 4 °C.
5. Before use, remove sodium azide by washing the appropriate amount of beads three times in 8 mL of HBSS (containing 2% FCS).
6. Recover beads by using the magnet. After the final wash, resuspend the beads in 1 mL of cold HBSS protein solution.

3.2 Tissue Dispersal and Digestion

1. Place lung tissue in a small amount of digestion buffer.
2. Chop tissue into small fragments using surgical scissors, and rinse over a 100-μm nylon gauze with DMEM digestion buffer (containing 2% FCS).
3. Chop tissue fragments vigorously for a further 3–6 min to generate a consistent pulp. Throughout the chopping process, keep the tissue moist by adding DMEM digestion buffer (containing 2% FCS).
4. Rinse the pulp twice over 100 μm gauze, and then transfer to a sterile pot containing 4 mL of digestion buffer per gram of chopped tissue.
5. Store chopped tissue overnight at 4 °C, with a loosely attached lid to allow oxygenation (*see* **Note 3**).
6. The following day, allow the tissue suspension to equilibrate to room temperature (19–22 °C) for 40 min before enzymatic digestion.
7. Once equilibrated, add the following to the tissue: cell culture-treated collagenase type I-A (3 mg/g of tissue) and hyaluronidase (3.76 mg/g of tissue; *see* **Note 4**). To aid disruption, place a sterile magnetic bead in the pot, and stir the mixture gently for 75 min at 37 °C in a humidified CO_2 incubator.
8. After this incubation, obtain further physical disruption by plunging the mixture 30 times with a 50 mL syringe.
9. After physical and enzymatic disruption, filter the dispersed tissue three times by washing over a 100-μm gauze.

10. Wash the dispersed cells three times with fresh DMEM (containing 2% FCS) by centrifugation (480 × *g*, 8 min) with volumes of 200–400, 100, and 50 mL of buffer for successive centrifugations.
11. After these washings, mast cell numbers are established. The mixed cell population is resuspended in typically 5–10 mL of buffer. Mast cells are identified using Kimura stain. Incubate an aliquot of the cells with the stain for 5 min (ratio of 1:10; e.g., 10 μL cells, 90 μL stain) and count the cells using a hemocytometer (*see* **Note 5**).

3.3 Mast Cell Purification

1. Resuspend the mixed cell population in 4 mL of HBSS protein solution, and incubate with continuous rotation for 30 min at 4 °C (*see* **Note 6**).
2. Make up the total volume of the cell suspension to 50–100 mL with cold (4 °C) HBSS (containing 2% FCS).
3. Strain through 100-μm sterile cell filters (6–10 depending on volume).
4. Centrifuge cells (480 × *g*, 8 min).
5. In the meantime, wash an appropriate volume of coated Dynal beads in 8 mL of HBSS (containing 2% FCS) using the MPC-1 magnet; after the wash, resuspend the beads in 1 mL of HBSS protein solution. The bead-to-mast cell ratio used is 5:1, which equates to 125 μL of bead suspension for 10×10^6 mast cells.
6. After centrifugation, resuspend the cells in 2 mL of HBSS protein solution, and then mix with the beads, and incubate with continuous rotation at 4 °C for 2 h.
7. After incubation, make up the bead-cell suspension to a total volume of 10 mL with cold HBSS (containing 2% FCS), and transfer to a 15 mL tube.
8. Separate the mixture using an MPC-1 magnet by inserting the tube in the magnet for 3 min, discarding the supernatant, removing the tube from the magnet, and resuspending the magnetically adherent cells in 10 mL of HBSS (containing 2% FCS).
9. Repeat this cycle for a total of three washes. The principle of the immunomagnetic bead separation is shown in Fig. 1 (*see* **Note 7**).
10. After the last wash, resuspend the isolated mast cells conjugated to the beads in 1 mL of HLMC culture medium with cytokines, and count the cells using the Kimura stain. Additional culture medium can be added as appropriate.

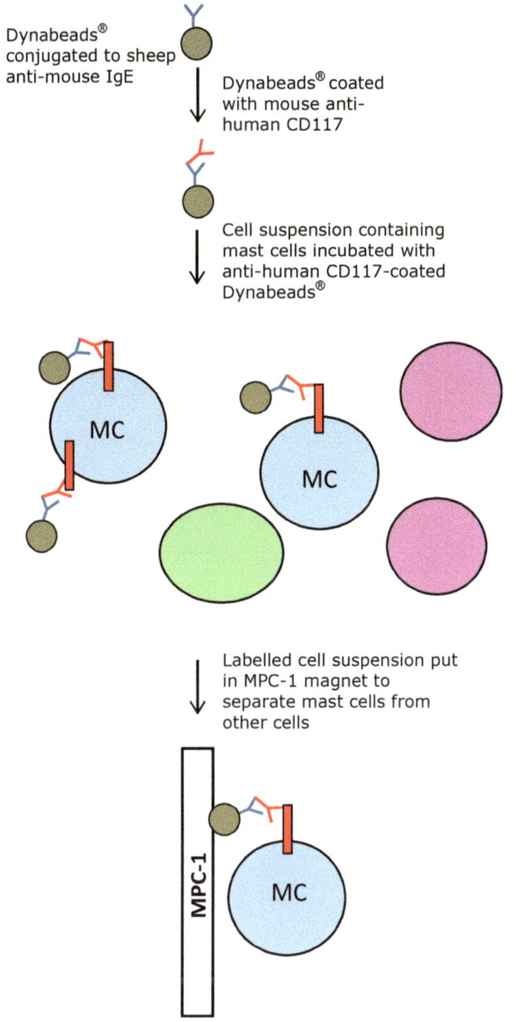

Fig. 1 Schematic diagram showing mast cell (MC; blue) purification, utilizing Dynabeads® technology. Dynabeads® are conjugated to sheep anti-mouse IgG and then precoated with mouse anti-human CD117. The mixed cell population produced from the digestion step is then incubated with the coated beads. Mast cells (which expressing CD117 receptors) attach to the Dynabeads® and adhere to the MPC-1, whereas other cells (shown in green and purple) do not. Contaminant cells are then washed away and mast cells recovered following removal from the magnetic field (adapted, with permission, from a diagram created by Claire Tree-Booker)

11. Carry out full medium exchange within 3 days of the isolation and purification process, and, thereafter, exchange the culture medium weekly.
12. Cells can be cultured for longer time periods in T12.5 culture flasks, at a density of 100,000–300,000 cells per mL at 37 °C, in a humidified CO_2 incubator. A full or half exchange of media

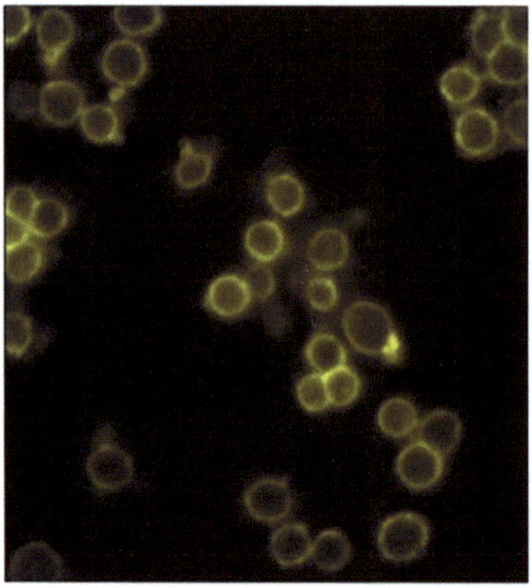

Fig. 2 Human lung mast cells stained with PE-tagged antibody to the mast cell surface antigen, CD117. The excitation wavelength was 488 nm, and fluorescent images were taken at 578 nm. The picture shows live cells at 75 days post-isolation. The fluorescently tagged antibody labeled 98% of the cells

and transfer to a new tissue culture flask is carried out every 7 days. Purity of the cultured mast cells is determined using the Kimura stain and further verified by using phycoerythrin (PE)-tagged anti-CD117 IgG_1 antibody (Miltenyi Biotec). Figure 2 shows an example of staining of mast cells with PE-tagged anti-CD117 following long-term culture.

4 Notes

1. In this chapter, a Kimura stain has been described for the detection of mast cells. However, alternative stains can be used to detect mast cells including Alcian blue [5].
2. A number of alternative methods have been used for the digestion of human lung to obtain mast cells as well as those described in this chapter. These vary in the types of buffer used, enzyme cocktails employed, and mechanisms of physical disruption. In principle, similar protocols have been described for the isolation of mast cells from alternative tissues including the human skin [6–8] and gut [9, 10] and from a variety of animal tissues [1].
3. This step aids the elimination of bacteria and other microorganisms decreasing the likelihood of infection in cell preparations and cultures.

4. Titrating the activity and suitability of various batches of collagenase and FCS can be quite a valuable investment in terms of ensuring good yields of functional mast cells.
5. These methods usually generate ~1 × 10^6 mast cells per g of lung tissue, and the purity is between 85 and 95%.
6. This stage acts as a blocking step to help prevent nonspecific protein binding.
7. Alternative methods have been described for the purification of mast cells including the use of discontinuous Percoll gradients, countercurrent elutriation (requires customization of a dedicated centrifuge and a specialized rotor head and chamber), and alternative immunomagnetic bead separation protocols. These methods are described elsewhere [11–15]. Recently, utilizing methods described in this chapter as a basis, an optimized protocol for the isolation and analysis of human lung mast cells has been reported [16].

Acknowledgments

The authors are grateful to Mr. J. Rao, Mr. J. Edwards, Mr. D. Hopkinson, and Mr. A. Ahmed (Cardiothoracic Surgery) and Dr. S.K. Suvarna, Dr. P. Kitsanta, and Dr. J. Bury (Histopathology) at the Northern General Hospital, Sheffield, for their invaluable help in providing lung tissue specimens.

References

1. Ali H, Pearce FL (1985) Isolation and properties of cardiac and other mast cells from the rat and Guinea-pig. Agents Actions 16:138–140
2. Cruse G, Kaur D, Yang W, Duffy SM, Brightling CE, Bradding P (2005) Activation of human lung mast cells by monomeric immunoglobulin E. Eur Respir J 25:858–863
3. Sanmugalingam D, Wardlaw AJ, Bradding P (2000) Adhesion of human lung mast cells to bronchial epithelium: evidence for a novel carbohydrate-mediated mechanism. J Leukoc Biol 68:38–46
4. Okayama Y, Hunt TC, Kassel O, Ashman LK, Church MK (1994) Assessment of the anti-c-kit monoclonal antibody YB5.B8 in affinity magnetic enrichment of human lung mast cells. J Immunol Methods 169:153–161
5. Gilbert HS, Ornstein L (1975) Basophil counting with a new staining method using alcian blue. Blood 46:279–286
6. Benyon RC, Lowman MA, Church MK (1987) Human skin mast cells: their dispersion, purification and secretory chracterization. J Immunol 138:861–867
7. Lawrence ID, Warner JA, Cohan VL, Hubbard WC, Kagey-Sobotka A, Lichtenstein LM (1987) Purification and characterization of human skin mast cells. Evidence for human mast cell heterogeneity. J Immunol 139:3062–3069
8. Bastan R, Peirce MJ, Peachell PT (2001) Regulation of immunoglobulin E-mediated secretion by protein phosphatases in human basophils and mast cells of skin and lung. Eur J Pharmacol 430:135–141
9. Lowman MA, Rees PH, Benyon RC, Church MK (1988) Human mast cell heterogeneity: histamine release from mast cells dispersed from skin, lung, adenoids, tonsils, and colon in response to IgE-dependent and nonimmunologic stimuli. J Allergy Clin Immunol 81:590–597
10. Fox CC, Dvorak AM, Peters SP, Kagey-Sobotka A, Lichtenstein LM (1985) Isolation

and characterization of human intestinal mucosal mast cells. J Immunol 135:483–491

11. Weston MC, Anderson N, Peachell PT (1997) Effects of phosphodiesterase inhibitors on human lung mast cell and basophil function. Br J Pharmacol 121:287–295
12. Lewis A, Wan J, Baothman B, Monk PN, Suvarna SK, Peachell PT (2013) Heterogeneity in the responses of human lung mast cells to stem cell factor. Clin Exp Allergy 43:50–59
13. Church MK, Hiroi J (1987) Inhibition of IgE-dependent histamine release from human dispersed lung mast cells by anti-allergic drugs and salbutamol. Br J Pharmacol 90:421–429
14. Schulman ES, MacGlashan DW, Peters SP, Schleimer RP, Newball HH, Lichtenstein LM (1982) Human lung mast cells: purification and characterisation. J Immunol 129:2662–2667
15. Ishizaka T, Conrad DH, Schulman ES, Sterk AR, Ishizaka K (1983) Biochemical analysis of initial triggering events of IgE-mediated histamine release from human lung mast cells. J Immunol 130:2357–2362
16. Ravindran A, Rönnberg E, Dahlin JS, Mazzurana L, Säfholm J, Orre AC, Al-Ameri M, Peachell P, Adner M, Dahlén SE, Mjösberg J, Nilsson G (2018) An optimized protocol for the isolation and functional analysis of human lung mast cells. Front Immunol 9:2193. https://doi.org/10.3389/fimmu.2018.02193

Chapter 4

Generation of a Human Allergic Mast Cell Phenotype from $CD133^+$ Stem Cells

Hans Jürgen Hoffmann

Abstract

Cultured human mast cells are a useful tool for research into innate immune responses as well as allergic mechanisms. Mast cells cultured from peripheral blood can provide information on immune mechanisms of known, selected individuals. With the method presented here, eight million mast cells can be cultured from ca. one million stem cells purified from one unit (450 mL) of human peripheral blood. Culture with IgE and IL4 optimizes an allergic phenotype of the mast cells.

Key words Mast cell, Human stem cells, Buffy coat, Cultured mast cells, SCF, CD63, FcεRI

1 Introduction

Mast cells are tissue resident cells first described by Ehrlich in 1878 that derive from bone marrow-derived hematopoietic stem cells. These $CD34^+$ $C117^+$ $CD133^+$ cells released into circulation can be purified and cultured to resemble tissue mast cells. Mature mast cells contain large amounts of preformed heparin, histamine, and a number of cytokines and continue to express the stem cell factor receptor CD117. The method we describe to culture mast cells can be used to obtain sizeable numbers of mast cells from identified or anonymous donors [1–4].

Mast cells are present in most tissues, characteristically proximal to blood vessels, nerves, and epithelia. Early work [5] identified two subtypes of mast cells based on the expression of chymase. Recent work has described distinguishable phenotypes of mast cells associated with anatomical compartments of the human lung [6] that were identified by protein markers (FcεRI, IL9-R, HDAC, and others). Mast cells can be purified from tissue or cultured from purified stem cells. In both instances, they are dependent on SCF [7].

Bernhard F. Gibbs and Franco H. Falcone (eds.), *Basophils and Mast Cells: Methods and Protocols*, Methods in Molecular Biology, vol. 2163, https://doi.org/10.1007/978-1-0716-0696-4_4,

2 Materials

2.1 Isolation and Culture of CD133⁺ Human Stem Cells

1. Human peripheral blood (100–450 mL), or fresh buffy coat, can be citrated or in EDTA (*see* **Note 1**).
2. Ficoll Paque (GE Healthcare).
3. Sterile 50 mL tubes.
4. Sterile PBS.
5. Low-speed centrifuge.
6. AC133$^+$ Cell Isolation Kit (Miltenyi).
7. MACS Buffer (Miltenyi).
8. LS Separation Columns (Miltenyi).
9. StemSpan culture medium (STEMCELL Technologies, Vancouver, Canada).
10. Penicillin/streptomycin.
11. Fetal bovine serum (FBS).
12. Human recombinant stem cell factor (rhSCF) (R&D Systems, Abingdon, UK).
13. Human recombinant IL-3 (rhIL-3; R&D Systems, Abingdon, UK).
14. Human recombinant IL-6 (rhIL-6; R&D Systems, Abingdon, UK).
15. Human recombinant IL-4 (rhIL-4; PeproTech EC Ltd., London, UK).
16. Human monoclonal myeloma IgE, produced by various companies.
17. CM: Stem Span culture medium with 50 ng/mL rhIL-6, 100 ng/mL SCF and 100 μg/mL penicillin/streptomycin.

2.2 Reagents for Determining Mast Cell Purity and Reactivity

2.2.1 May–Grünwald–Giemsa Staining

1. May–Grünwald solution (Merck, Darmstadt, Germany).
2. Giemsa (Merck) diluted 1/10 in PBS, pH 6.4.
3. Microslides and coverslips.

2.2.2 Detection of Density of FcεRI, Mast Cell Reactivity, and Kit Surface Receptor Expression by Flow Cytometry

1. Polystyrene round-bottom tubes (5 mL).
2. PBS +0.1% BSA (4 °C).
3. Anti-CD63 FITC (BioLegend, San Diego, CA, clone H5C6).
4. Anti-CD203c PE (BioLegend, San Diego, CA, clone NP4D6).
5. QiFiKit (Dako, Glostrup, Denmark).
6. Anti-human IgE (BD Pharmingen, San Jose, CA, clone G7–18).

3 Methods

All work is to be done in a sterile lab bench, except mast cell activation tests and staining (Subheadings 3.2 and 3.3).

3.1 Purification of CD133⁺ Stem Cells from Peripheral Blood

1. Bring Ficoll and PBS to room temperature (22 °C) and aliquot 13 mL Ficoll into a sterile 50 mL tube for each 24 mL blood to be processed (*see* **Note 2**).
2. Dilute blood with 0.25 volumes of PBS in a sterile flask, and carefully layer 30 mL blood/PBS on Ficoll (*see* **Note 3**). Centrifuge 30 min at room temperature and 450 × *g* without acceleration or brake to preserve the integrity of the gradient interface.
3. Using a sterile pipette or pastettete, remove the plasma layer above the interface. Collect PBMC at the interface, and wash with 20 mL cold PBS for each gradient tube. Centrifuge 5 min at 4 °C and 450 × *g*. Pool all cells into one 10 mL tube, and wash them again in 10 mL MACS buffer, and count the number of leukocytes in the sample.
4. To count the cells, dilute 20 μL cells with 80 μL Trypan blue; count in a Bürker–Türk or Neubauer-modified chamber.
5. Resuspend all cells in 300 μL MACS Buffer (maximally 10^8 cells per 300 μL) in a 10 mL tube, and chill on ice.
6. Add 100 μL Fc blocking reagent and 100 μL CD133 microbeads per 300 μL cell suspension.
7. Incubate on ice for 2 min and at 4 °C for 28 min. Dilute cells with 10 mL MACS Buffer, and centrifuge for 5 min at 4 °C and 450 × *g*. Resuspend in 500 μL MACS buffer per 10^8 cells.
8. Wipe the MidiMACS with 70% ethanol and place it in the sterile bench.
9. Fill each vial with 300 μL cells up to 10 mL with MACS buffer. Centrifuge for 5 min at 450 × *g* and 4 °C. Pour off the supernatant; then resuspend the pellet in 5 mL MACS buffer.
10. Apply 3 mL MACS solution to an LS Column in a MACS magnet. When the MACS solution has reached the top of the column, add the cell suspension. Collect the eluate in a clean tube. Wash the column with 3 × 5 mL MACS buffer, collecting the eluate each time. Remove the column from the magnet, and pass 5 mL MACS buffer through to elute retained cells from the column. Count the number of cells in the eluate in a Bürker–Türk chamber or other hemocytometer. Centrifuge all tubes for 5 min at 450 × *g* and 4 °C. There should be no pellet in the second last tube and all CD133⁺ cells in the last tube.

3.2 Expansion of Stem Cells and Maturation of Mast Cells

1. Resuspend $CD133^+$ stem cells in CM with rhIL-3 at 5×10^5 cells/mL, and culture at 37 °C in a humidified CO_2 incubator.
2. Inspect the cultures twice a week for growth and sterility (*see* **Note 4**). If the medium appears spent, as indicated by an orange or even yellow color, add some more CM (10% of volume in well) to continue supporting growth until the next inspection or medium change. Young cultures should be seen and supplemented daily.
3. Change medium on a weekly basis. Transfer all cells and medium to a sterile centrifuge tube, and centrifuge for 5 min at 450 × *g* and 4 °C. Remove the supernatant. Resuspend up the cell pellet in 1 mL CM, and count in a Bürker–Türk chamber. Resuspend the cells at 5×10^5 cells/mL in the appropriate medium, and transfer to a fresh well or flask of appropriate size.
4. IL-3 should be added for the first 3 weeks to expand cell lines. In weeks 4–6, the cells can be cultured in CM only. In weeks 7–8, FBS (10%), 10 ng/mL rhIL-4, and 80 kU/L IgE are added to mature mast cells (*see* **Note 5**).
5. At the end of the culture period, evaluate mast cell quality, purity, and reactivity as described in the next section.

3.3 Mast Cell Staining with May–Grüwald–Giemsa by Microscopy

1. Prepare a cytospin with 10^3 to 10^4 mast cells. Centrifuge at 1000 rpm for 5 min. Air-dry until cells appear dry.
2. Stain with May–Grünwald solution for 4 min, and wash in PBS for 4 min.
3. Rinse in water, stain with Giemsa solution for 8 min, and rinse again in water.
4. Air-dry cytospins, and examine mast cells in a light microscope to confirm appearance of granules in the cytoplasm.

3.4 Mast Cell Reactivity and Phenotyping by Flow Cytometry

1. Prepare 5 log dilutions of anti-IgE or (APC-labelled) anti-FcεRI antibody in PBS starting with 1 ng/10 μL, as well as a blank tube (*see* **Note 6**).
2. Add 10 μL mast cell preparation containing at least 5000 cells, as well as 80 μL CM medium.
3. Incubate at 37 °C for 30 min to activated mast cells.
4. Add antibodies to CD63 (FITC) and CD203c (PE), and incubate at 4 °C for 20 min (*see* **Note 4**).
5. Wash with 2 mL PBS; centrifuge at 450 × *g* and 4 °C.
6. Acquire data on a flow cytometer; plot a histogram of CD63 for all cells for all tubes, and set a threshold at 2% of the blank. Plot the fraction of positive cells against anti-IgE or anti-FcεRI concentration (*see* **Note 7**).

7. Add titrated amounts of antibody to FcεRI and CD117 to aliquots of mast cell culture. Use QiFikit according to manufacturer's instructions to determine the density of receptors on cells.

4 Notes

1. We have no experience with purifying CD133$^+$ cells for mast cell culture from heparinized blood.
2. Typical yields of stem cells per mL blood are 3800 for fresh blood, 2500 for buffy coats, and blood 24 h old (unpublished observations).
3. This can take some practice. It is important to avoid disturbing the surface of the Ficoll layer. Find a comfortable and relaxed position, e.g., resting the elbow on the laminar flow bench edge. Use a pipetting aid which allows you to set the speed of dispensing to low, and use a constant flow rate, holding the tube at an angle of approx. 45–60°. The interface between Ficoll and diluted blood may not be disturbed.
4. Optimization of antibody concentrations should be performed by titration prior to the start of experiment.
5. If there was a sign of contamination, the flask was removed, or the well was emptied and cleansed with 70% ethanol in water.
6. IgE can be obtained from myeloma or as recombinant product. We have used recombinant human IgE [8].
7. If APC-labeled anti-FcεRI is used to stimulate cells, the fraction of cells expressing FcεRI as well as the relationship between FcεRI expression and activation can be examined.

References

1. Andersen HB, Holm M, Hetland TE et al (2008) Comparison of short term in vitro cultured human mast cells from different progenitors—peripheral blood-derived progenitors generate highly mature and functional mast cells. J Immunol Methods 336:166–174
2. Holm M, Andersen HB, Hetland TE et al (2008) Seven week culture of functional human mast cells from buffy coat preparations. J Immunol Methods 336:213–221
3. Hoffmann HJ, Frandsen PM, Christensen LH et al (2012) Cultured human mast cells are heterogeneous for expression of the high-affinity IgE receptor FcεRI. Int Arch Allergy Immunol 157:246–250
4. Krohn IK, Lund G, Frandsen PM et al (2013) Mast cell FcεRI density and function dissociate from dependence on soluble IgE concentration at very low and very high IgE concentrations. J Asthma 50:117–121
5. Irani AM, Schwartz LB (1994) Human mast cell heterogeneity. Allergy Proc 15:303–308
6. Andersson CK, Mori M, Bjermer L et al (2009) Novel site-specific mast cell subpopulations in the human lung. Thorax 64:297–305

7. Kirshenbaum AS, Goff JP, Kessler SW et al (1992) Effect of IL-3 and stem cell factor on the appearance of human basophils and mast cells from CD34[+] pluripotent progenitor cells. J Immunol 148:772–777

8. Christensen LH, Holm J, Lund G et al (2008) Several distinct properties of the IgE repertoire determine effector cell degranulation in response to allergen challenge. J All Clin Immunol 122:298–304

Chapter 5

Generation and Culture of Peripheral CD34$^+$ Stem Cell-Derived Mast Cells (PSCMCs)

Yanyan Luo, Stefan Frischbutter, Jörg Scheffel, and Frank Siebenhaar

Abstract

Mast cells (MCs) are long-living tissue-resident cells that play an important role in inflammatory and allergic reactions. In vitro models of mast cell functions have allowed better understanding of the function of mast cells and mast cell-mediated disorders. In this unit, we describe a protocol used for the generation and culture of peripheral CD34$^+$ stem cell-derived mast cells (PSCMCs). It provides a useful tool for the investigation of the biology of human MCs in vitro.

Key words Stem cell-derived mast cells, CD34, CD117

1 Introduction

Mast cells (MC) are widely regarded as important effector cells in several immune responses [1]. They are long-living tissue-resident cells that play an important role in inflammatory and allergic reactions. In addition, MCs contribute to the pathogenesis of a multitude of diseases that include MC-mediated disorders such as urticaria, type I allergies, and mastocytosis as well as autoimmune and other inflammatory conditions [2].

MCs are derived from committed hematopoietic lineage cells of the bone marrow and other peripheral stem cell source as derivatives of hematopoietic progenitor cells that migrate into peripheral tissues where they complete their maturation. In humans, mast cells that are absent in the blood in their fully granulated mature state, however, can be derived from CD34$^+$ and CD34$^+$ CD117$^+$ progenitor cells derived from the peripheral blood by in vitro culture [3]. Therefore, the peripheral blood is a readily available source for obtaining CD34$^+$ cells for the generation and culture of MCs. The peripheral CD34$^+$ stem cell-derived mast cells (PSCMCs) exhibit a typical MC phenotype, robust histamine release in response to

Bernhard F. Gibbs and Franco H. Falcone (eds.), *Basophils and Mast Cells: Methods and Protocols*, Methods in Molecular Biology, vol. 2163, https://doi.org/10.1007/978-1-0716-0696-4_5,

stimulation with anti-FcεRI or IgE/anti-IgE, and increased proliferation and differentiation in response to IL-1β or IFN-γ [4].

2 Material

2.1 Isolation and Expansion of CD34+ Cells

1. Whole blood in heparin blood tubes (total volume ≥ 50 mL).
2. EasySep™ Human CD34 Positive Selection Kit II (STEMCELL, Catalog #17856).
3. RoboSep Buffer: 500 mL PBS + 10 mL heat-inactivated FCS and 1 mL 0.5 M EDTA (*see* **Note 1**).
4. 50 mL Falcon tubes.
5. Ficoll–Paque Premium sterile solution.
6. SepMate-50 tubes for density gradient centrifugation.
7. 5 mL Polystyrene round-bottom tubes.
8. Big Easy EasySep Magnet.
9. 24-Well Non-TC Plate.
10. StemSpan CD34+ Expansions Supplement (10×) (STEMCELL, Catalog #02691).
11. StemSpan SFEM (STEMCELL, Catalog. #09650).

2.2 Purification, Differentiation, and Culture of MCs

1. StemSpan SFEM (STEMCELL Technologies, Catalog. #09650).
2. 24-Well non-TC plate.
3. 6-Well non-TC plate.
4. CD117 MicroBead Kit (Miltenyi Biotec).
5. MACS Buffer (500 mL PBS w/o Ca^{2+}/Mg^{2+} supplemented with 25 mL heat-inactivated FCS and 5 mL 100x Pen/Strep).
6. IL-3, IL-6, SCF, and LDL.
7. CD34+ Expansion Medium: StemSpan™ CD34+ Expansion Supplement (10×) at 1:10 dilution with SFEM.

3 Methods

3.1 Isolation of CD34+ Cells with CD34 Positive Selection Kit II (Day 0)

1. Collect fresh heparin anticoagulated peripheral blood in 50 mL Falcon tubes. Add 50 μL RosetteSep/mL of sample, and mix them carefully by pipetting.
2. Incubate 20 min at room temperature (RT).
3. Dilute the samples with the same volume RoboSep Buffer.

4. Lay 15 mL Ficoll Solution in SepMate-50 tubes (without air bubbles), and overlay the diluted blood samples very carefully (*see* **Note 2**).
5. Centrifuge for 10 min, 1200 × *g* at RT with max., brake on.
6. Transfer supernatants carefully to new 50 mL Falcon tube. Fill up to 50 mL with RoboSep Buffer. Resuspend slowly by pipetting up and down.
7. Centrifuge for 10 min, 120 × *g* at RT without brake. Discard the supernatants by pipetting, and pool all pellets in a 50 mL Falcon tubes, and fill up with RoboSep Buffer to 50 mL.
8. Centrifuge for 10 min 300 × *g*, at RT with low brake (4–6).
9. Discard the supernatants by pipetting. Pool the samples, and fill up to 750 μL with RoboSep Buffer. Transfer the sample to one 14 mL polystyrene round-bottom tube.
10. Add 50 μL/mL sample (volume from **step 8**) CD34 Selection cocktail (vortex before usage), and incubate for 10 min at room temperature. Slowly resuspend the sample; do not vortex.
11. Vortex the RapidSphere Beads, and add 50 μL/mL sample.
12. Incubate for 3 min at room temperature.
13. Fill up the sample to 3 mL with RoboSep Buffer, and incubate the tube in the Magnet for 3 min at RT.
14. Pick up the magnet, and in one continuous motion, invert the magnet together with the tube, pouring off the supernatant. Remove the tube from the magnet; this tube contains the isolated cells. Resuspend the cells very gently.
15. Add 2.5 mL RoboSep Buffer and repeat **Steps 11** and **12** four times.
16. Resuspend and culture the cells in 1 mL CD34$^+$ Expansion Medium in 24-well non-TC plate.

3.2 Expansion and Differentiation of Mast Cells (Day 0–35)

1. From day 0 to day 7, culture the CD34$^+$ cells in CD34 expansion medium for 1 week at 37 °C and 5% CO_2.
2. On day 8, switch to SFEM containing 20 ng/mL IL-3 and 50 ng/mL SCF; culture the cells in 6-well non-TC plate, and feed them twice a week.
3. On day 35, FACS should show that about 30–40% of the cells are double positive for CD117 and FcεRI.

3.3 Purification of Mast Cells with the CD117 MicroBead Kit (Day 35)

1. Dilute the cell culture 1:1 with MACS Buffer.
2. Centrifuge the cells for 3 min at 300 × *g* at RT.
3. Discard the supernatant by pipetting, and resolve the cells in 500 μL MACS Buffer.

4. Add 25 μL anti-CD117 Beads (vortex before usage) and 25 μL Fc Block, and resuspend carefully. Incubate for 20 min on ice.
5. Fill up with MACS Buffer to 2.5 mL, and centrifuge as before (3 min at 300 × *g* at RT).
6. Discard the supernatant by pipetting, and resolve the cells in 500 μL MACS Buffer. Place the cells one ice while preparing the separation column.
7. Place one MS MACS Column in the suitable MACS Separator, and equilibrate with 500 μL MACS Buffer (see Protocol for "CD117 MicroBead Kit" from Miltenyi Biotec for more details).
8. Add the cells to the equilibrated MACS column, and wait until the whole sample passed through.
9. Wash the column with 3× 500 μL MACS Buffer.
10. Remove the column from the separator, and place it in a new 15 mL Falcon tube. Pipette 5 mL MACS Buffer into the column, and flush out the retained CD117+ cells by firmly pushing the plunger into the column.
11. Centrifuge for 3 min at 300 × *g* at RT, and resuspend in SFEM containing 50 ng/mL IL-6 and 125 ng/mL SCF.

3.4 Culture of Mast Cells (from Day 35)

1. Culture the MCs in SFEM containing 50 ng/mL IL-6 and 125 ng/mL SCF; feed cells twice a week (*see* **Note 3**).

4 Notes

1. It is not necessary to prepare the buffer fresh every time, but it should be kept sterile and in the fridge.
2. SepMate tubes have an insert, which facilitates pipetting of blood without mixing with the Ficoll layer. Using these tubes (similar tubes are available also from other companies) speeds up the process of overlaying. Overlaying Ficoll with diluted blood without the use of these devices is also possible (and cheaper) but requires a steady hand and more pipetting practice.
3. Generally, those MCs could be maintained for at least 6 months and retain their MC phenotype and function.

References

1. Galli SJ, Maurer M, Lantz CS (1999) Mast cells as sentinels of innate immunity. Curr Opin Immunol 11(1):53–59
2. Siebenhaar F, Redegeld FA, Bischoff SC, Gibbs BF, Maurer M (2018) Mast cells as drivers of disease and therapeutic targets. Trends Immunol 39(2):151–162
3. Dahlin JS, Malinovschi A, Öhrvik H, Sandelin M, Janson C, Alving K, Hallgren J (2016) Lin CD34hi CD117int/hi FcεRI+ cells in human blood constitute a rare population of mast cell progenitors. Blood 127(4):383–391
4. Schmetzer O, Valentin P, Smorodchenko A, Domenis R, Gri G, Siebenhaar F, Metz M, Maurer M (2014) A novel method to generate and culture human mast cells: peripheral CD34+ stem cell-derived mast cells (PSCMCs). J Immunol Methods 413:62–68

Chapter 6

Modulating the Human Basophil Phenotype During Its Development and Maturation: Basophils Derived from In Vitro Cultures of CD34$^+$ Progenitor Cells

Donald MacGlashan Jr.

Abstract

Historically, the human basophil that is studied experimentally comes from peripheral blood. But there is evidence that only a short portion of the basophil life cycle related to IgE-mediated function occurs in the blood. The same evidence suggests that IgE-mediated functionality is present for 5–7 days in the bone marrow (or other tissues) during which the cell modulates its phenotype according to local conditions. It is suggested that to properly understand the nature of basophil behavior, a better understanding of its biology during maturation would be helpful. For example, one highly suggestive line of evidence for the relevance of understanding the maturation period is related to the change in basophil phenotype that occurs during treatment of patients with omalizumab. During this treatment, the intrinsic reactivity or sensitivity of the basophils is significantly increased despite, or perhaps because of, the dramatic reduction in FcεRI expression that accompanies this treatment. One of the critical signaling enzymes to increase expression selectively in basophils during treatment is SYK, which is one of the earliest signaling tyrosine kinases involved in translating the aggregation of FcεRI into secretion from the cell. Treatment with omalizumab increases SYK expression, and this observation focuses some attention of how SYK expression is regulated. It is possible that the key regulation occurs during maturation of the basophil. Regardless of the mechanisms operative in this particular treatment, understanding the process of maturation and the extrinsic factors that influence it may lead to better understanding of disease processes. Therefore, this chapter will discuss and present techniques to work with maturing human basophils.

Key words Basophils, CD34, IgE, Culture, FcεRI, Histamine, Stem cells

1 Introduction

By the early 1960s, it was established that there was a circulating leukocyte that could be induced to release histamine after stimulation with allergens to which the individual was clinically sensitive (notably, positive skin tests and allergic symptoms) [1]. There were many interesting characteristics described for this circulating cell that related its behavior to allergic disease, but it was not until the early 1970s that this cell, expressing a high-affinity IgE receptor

Bernhard F. Gibbs and Franco H. Falcone (eds.), *Basophils and Mast Cells: Methods and Protocols*, Methods in Molecular Biology, vol. 2163, https://doi.org/10.1007/978-1-0716-0696-4_6, © Springer Science+Business Media, LLC, part of Springer Nature 2020

and packaging histamine in granules, was identified as the basophil [2]. By practical necessity, this cell was studied from peripheral blood, and several decades of study have refined our understanding of circulating basophil biology. However, to this day, there is very little known about the human basophil before it emerges into the circulation. This may represent a critical absence of understanding.

The life span or, perhaps more correctly, the residence time of a circulating human basophil is only imperfectly known. The few studies in the literature suggest times ranging between 12 h and 2–3 days [3, 4]. Likewise, the time required to produce a mature basophil in the bone marrow is poorly known. For reasons to be discussed in further detail below, one estimate is that from the moment the early basophil begins expressing FcεRI, 5–7 days are required to reach the maturity that allows the cell to emerge into the circulation (if coming from the bone marrow rather than tissues). If this is true, then from the perspective of those interested in IgE-mediated behaviors, the FcεRI-expressing cell that eventually becomes a mature circulating basophil exists for longer in the bone marrow than the short window of time it circulates. Furthermore, the maturing cell probably behaves differently by virtue of signaling and regulatory pathways that are not fully established during development. If the modulatory influences that determine the expression of atopy are to be better understood, it stands to reason that a better understanding of the characteristics of the maturing basophil, rather than mature basophil, is needed.

This concern is brought into greater focus when we wish to understand how the non-cross-linking anti-IgE antibody, omalizumab, works. If this antibody is distributed throughout the bone marrow space [5], then it is capable of modulating the maturing basophil, and this has immediate consequences for the therapeutic benefit of this treatment. As it turns out, the use of this therapeutic may have revealed some important aspects of basophil maturation, and its use has focused further attention on the life of the basophil before it enters the circulation. Other therapeutics, both chemical and biological, may also operate during the maturation of basophils (and, of course, other cells).

1.1 Omalizumab and the Maturing Basophil

Omalizumab is a therapeutic antibody with specificity for IgE. The antibody was designed to bind to an epitope on IgE that is hidden when IgE is bound to both FcεRI and FcεRII but also blocks IgE binding to FcεRI if the antibody binds to free IgE [6]. The inability of the antibody to bind to FcεRI-bound IgE (*see* **Note 1**) disallows this anti-IgE antibody from inducing anaphylaxis by mast cell or basophil activation of any kind. To date, there has never been an indication that this antibody can activate any cell type to which IgE is bound. The sporadic hypersensitivity reactions in the clinic that led to a black box label for this therapeutic appear to result from a class effect similar to many therapeutic antibodies that are administered at high doses, although it is not possible to rule out activation for secretion. In the earliest studies of treatment with omalizumab, concurrent studies on regulation of FcεRI expression suggested that reducing free IgE levels with omalizumab would result in downregulation of FcεRI [7–10]. Therefore, basophils were studied for changes in both surface IgE and FcεRI densities. It was found that both cell surface IgE and FcεRI decreased with a "half-life" of 2–3 days [9]. This observation was unexpected, because other studies had shown that for naturally expressed FcεRI on basophils, the dissociation half-life of IgE bound to FcεRI was 10 days [11, 12]. These observations led to the synthesis of a model whereby the turnover rate of a cell became a determinant of the rate that FcεRI expression changed during treatment with omalizumab [5, 13]. The key to this model is that if the cell turnover rate is faster than the dissociation rate of IgE from FcεRI, then the apparent rate of loss of expression is a reflection of the cell turnover rate and maturation time of the cell rather than simple receptor-IgE dissociation kinetics. An expanded version of this model was explored using mice [5], and some of the implicit assumptions in the model shown to be valid. Using the premises established in the model, the results from the early omalizumab studies indicated that the time between the appearance of the first FcεRI on a maturing basophil and its emergence into the circulation was, as noted above, 5–7 days [9, 13]. The fast decrease in FcεRI expression resulted from a <24-h turnover rate of the circulating basophil. The explanation for the rapid kinetics is that omalizumab immediately reduces free IgE levels and in the bone marrow an early basophil that has not yet accumulated FcεRI (normally due to the presence of free IgE) *never* upregulates receptor, and since it emerges after several days of maturation, observers of the circulating basophil see a cell that has low receptor expression levels. This behavior is distinct from that of skin mast cells [14] whose turnover rate is on the order of months.

Although work in the Denburg Laboratory has long considered the hematopoietic development of basophils and eosinophils in patients with atopy [15–19], the omalizumab studies focused additional attention on the maturing basophil and how it behaves with respect to the absence of IgE and upregulation of FcεRI due to the absence of free IgE. It is easy to imagine that the moment a CD34 progenitor commits to becoming a basophil and turns on expression of FcεRI that a number of related changes need to occur. FcεRI expression can reach quite remarkable levels; up to one million FcεRI per basophil has been observed although densities are typically in the 100,000 to 500,000 range. The regulatory apparatus that prevents spontaneous secretion due to random chance aggregation of FcεRI must compensate for these high densities [20]. Numerical modeling of the number of spontaneous aggregates that form at different densities of FcεRI show that when the densities exceed 200,000 per cell, there are enough random long-lived spontaneous aggregates to potentially drive IgE-mediated activation [21]. Possibly, without suppressive mechanisms in place, this could be a problem for the maturing cell. Therefore, it seems likely that the simple act of not upregulating FcεRI expression in the absence of free IgE might alter the development of basophil during maturation. Recent studies have confirmed this suspicion. When an allergic/asthmatic patient is placed on omalizumab, not only do FcεRI levels decrease in response to reduced free IgE, but the intrinsic sensitivity of the cell increases markedly [22–24]. For example, a typical basophil needs approximately 2000 allergen-specific IgE molecules on its cell surface to initiate histamine release that is 50% of its optimal response (label this number the natural ED50 = *e*ffective *d*ensity for 50% of the maximal response). During treatment with omalizumab, the ED50 decreases to 170 allergen-specific IgE molecules per cell, i.e., 12 times more sensitive to stimulation. This change in intrinsic sensitivity appears to influence the efficacy of treatment with omalizumab [23], and it is therefore an important process to understand.

1.2 Expression of SYK in Basophils and Other Leukocytes

SYK is a critical signaling element, and it effectively provides the enzymatic activity needed for signaling to an otherwise enzymatically dead receptor (no enzymatic activities have been described for any subunit of FcεRI) by binding to the phosphorylated gamma subunit. Given this behavior, one would expect that SYK expression might track with FcεRI expression levels. It does not. Upregulation of FcεRI in mature peripheral blood basophils, ex vivo, does not change SYK expression. In fact, in human basophils, the ratio of FcεRI and SYK expression is on the order of 5:1 to 25:1 [25]. SYK expression appears to be rate-limiting. Several indirect lines of evidence have been emerging that suggest that the ratio of FcεRI/SYK is important for activation [26, 27]. In addition to this unexpectedly low level of SYK expression, its expression is highly variable and predicts the IgE-mediated responsiveness of a given basophil preparation [25]. The responsiveness of basophils appears to be carefully tuned by SYK expression levels. Consistent with this close relationship between SYK expression and IgE-mediated release is the observation that treatment with omalizumab increases expression of SYK levels in a manner concordant with the increase in IgE-mediated release that occurs [22, 23]. Notably, only SYK expression in basophils is altered. In addition, the increase in SYK expression is inversely proportional to the starting SYK level. This observation focuses attention on how SYK is regulated, although the full extent of the changes that occur during treatment with omalizumab are not yet known.

These low levels of SYK expression are unique to basophils among the polymorphonuclear leukocytes, and with the exception of αβ T cells (which express the family member ZAP-70 instead of

SYK), the low level is unique among all other leukocytes [28]. The "sister" leukocyte, the eosinophil, expresses 10–12 times the level of SYK of basophils. Monocytes and dendritic cells express 30 times more SYK than basophils. Furthermore, there is very little natural variation among subjects in other leukocytes. It is reasonable to speculate that there are unique regulatory characteristics in basophils for the modulation of SYK levels.

1.3 Models of Basophil Maturation

These various observations suggested that one of the processes that distinguishes a maturing basophil from other leukocytes is its regulation of SYK expression. The bone marrow is not a readily accessible compartment in humans (at least not in a routine manner from preselected donors). To get a handle on the process of basophil maturation, a model of basophil maturation from CD34 progenitors has been explored [29–31]. There are some subtle procedural differences in the various published reports, and given that these procedures are each imperfect, the procedural differences might be important for generating the basophil. However, there is a common theme among the procedures; the cultures contain IL-3. Thus, the model is very simple and yet appears to capture some features of basophil maturation and generates cells with many phenotypic characteristics of peripheral blood basophils. Figure 1 shows some of these features: expression of FcεRI (*see* Fig. 1a) and the receptor for FMLP (*see* Fig. 1b) that permit stimulated histamine release (*see* Fig. 1c), expression of IL3R (CD123, *see* Fig. 1d), and the relative absence of c-kit (CD117) compared to culture-derived mast cells (*see* Fig. 1e).

Incubation of CD34-positive progenitors with IL-3 for 3 weeks results in a mixture of basophil-like cells. Most of these cells express FcεRI, although the morphological features are highly heterogeneous, as is the granulation; between 70% and 80% are stained by Alcian blue [32] but with significant heterogeneity. By a variety of criteria, these cells are more basophil-like than mast cell-like (indeed, there is little evidence of the mast cell phenotype) and more basophil-like than eosinophil-like. The list of differentiating features is extensive; mRNA profiling shows gene expression profiles that are heavily dominated by expression levels that are observed in mature peripheral blood basophils rather than mature peripheral blood eosinophils [33]. Fig. 2 shows one of the transcript comparisons that points to these cells being more basophil-like than eosinophil-like. As noted, these cells do not show markers of mast cells, but neither are there mRNA transcripts consistent with mast cells [33].

But the cells are not perfect representations of mature peripheral blood basophils. For example, SYK expression in CD34-derived basophils is 10–12 times greater than peripheral blood mature basophils (*see* Fig. 3a). This SYK observation led to studies that explored whether culture conditions would modify the

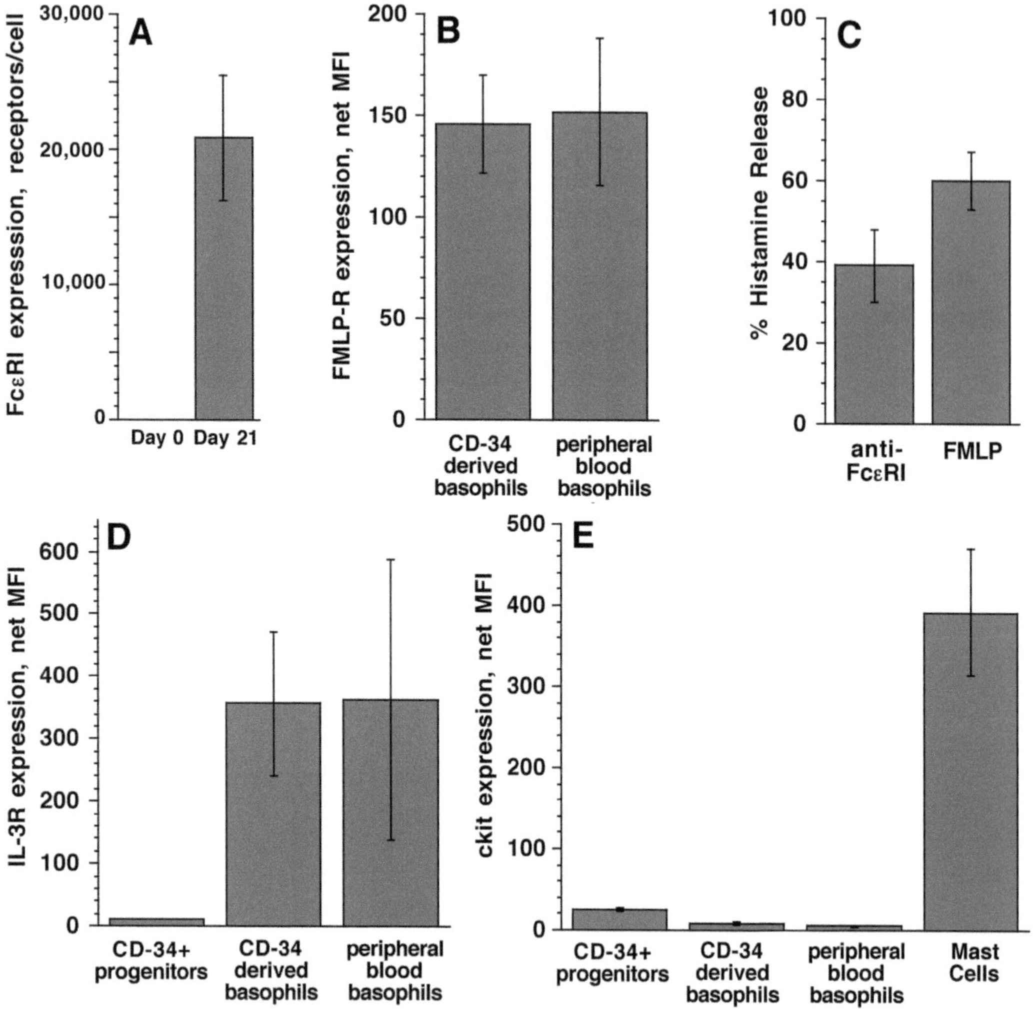

Fig. 1 Previously published [28] partial characterization of CD34-derived basophils (CD34B) relative to mature peripheral blood basophils. (Panel **a**) expression of FcεRI (as measured by flow cytometry); (panel **b**) expression of FMLP-receptor (as measured by flow cytometry); (panel **c**) histamine release from CD34B induced by stimulation with optimal concentrations of either anti-FcεRI Ab or 1 μM FMLP; (panel **d**) expression of IL-3 receptor (CD123); (panel **e**), expression of CD117 (SCF receptor). The mast cells were cultured-derived as well

phenotype of the CD34-derived basophils and, in particular, modify expression of SYK. This difference led to a search for a culture method that might change this discordant phenotypic characteristic. A discovered method that reduced SYK expression in the CD34B involved the addition of low concentrations of Flt3L (FMS-like tyrosine kinase 3 ligand) and SCF (stem cell factor) (low relative to typically used concentrations in the 100 ng/mL range) (*see* Fig. 3b).

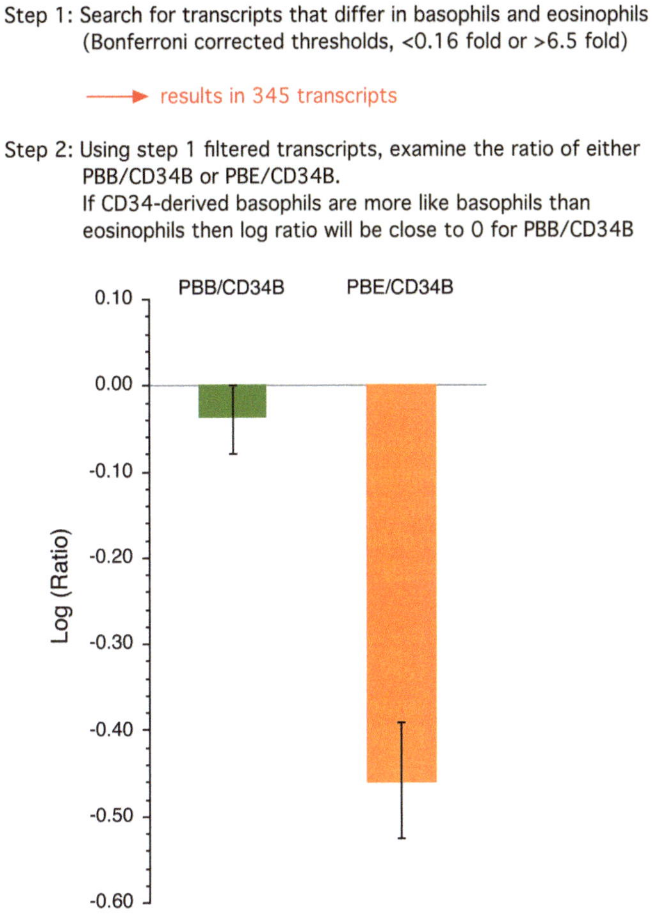

Fig. 2 Comparison of peripheral blood basophils, peripheral blood eosinophils, and CD34-derived basophils (method A). Working with the transcriptomes of each cell types, the figure explains the algorithm used for determining if the CD34B cells were more like peripheral blood basophils or eosinophils

1.4 Culture Differences

As noted, the cells generated in these cultures are heterogeneous in size, granularity, and nuclear morphology. A rough set of categories (large vs. small, strong Alcian blue positivity vs. weak, etc.) can be enumerated to monitor qualitative differences between either the culture methods or between cultures of a single method. By and large, both methods of culture generate a wide variety of morphologically different cells, 70–80% of which stain with Alcian blue to various degrees of intensity. Cells that are morphologically indistinguishable from mature basophils in peripheral blood represent a small fraction of the total cells when *method A* (*see* Subheading 3.1 below) is used. With *method B* (*see* Subheading 3.2 below), there are tenfold more of these cells, and although there remain only 70–80% of the cells staining with Alcian blue, there are two- to threefold more of these Alcian blue cells when using method B (*see*

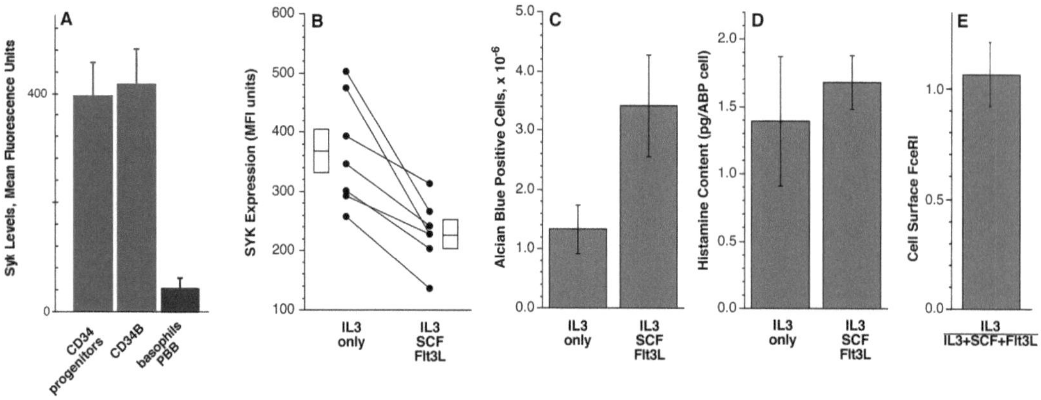

Fig. 3 Comparisons of the two culture methods. (Panel **a**) SYK expression in CD34$^+$ progenitors, CD34-derived basophils (method A), and peripheral blood basophils. (Panel **b**) expression of SYK in culture-derived basophils used *method A* (*see* Subheading 3.1) (IL-3 alone) or *method B* (*see* Subheading 3.2) (IL-3 plus SCF and Flt3L). (Panel **c**) increased production of Alcian-blue positive cells using method B; (panel **d**) similar histamine content between the two culture methods; and (panel **e**) similar expression of FcεRI between the two methods

Fig. 3c). With these basophils markers, method A and method B are very similar in outcomes although method B generates these cells somewhat faster (14 days instead of 17–21 days). For example, FcεRI expression and histamine content didn't change (*see* Fig. 3d, e). A more complete comparison at the transcript and DNA level can be found in a recent published study [33].

It is in this context that two methods of culture are described in this chapter. The two culture methods below are distinguished by differences in SYK expression. In addition to the two culture methodologies, there are two sources of CD34 cells that can be used as starting material for culture. One method is simple because the CD34$^+$ cells are purchased from a commercial supplier. The second method enriches these cells from peripheral blood obtained by leukapheresis. These cultures are incomplete although for some questions they represent the current best technology for studying the basophil maturation process.

2 Materials

2.1 Supplies

1. PIPES, NaCl, KCl, BSA, EDTA, D-glucose, erythrosin B (Sigma-Aldrich, St. Louis, MO, USA).
2. StemPro-34 serum-free medium (Thermo Fisher, Waltham, MA, USA).
3. Nutrient supplement (shipped with StemPro-34, thaw once, and split into 250 μL aliquots for media preparation (do not freeze–thaw more than once)).
4. GlutaMAX (Thermo Fisher, Waltham, MA, USA).

5. Human serum albumin (ICN, Aurora, OH, USA (now MP Biomedical)).
6. Percoll (Pharmacia, Piscataway, NJ, USA).
7. rhIL-3 (BioSource, Camarillo, CA, USA).
8. Flt-3 Ligand (R&D Systems, Oakville, ON, Canada).
9. Stem Cell Factor, SCF (R&D Systems, Oakville, ON, Canada).
10. Human CD34 MicroBead Kit (Miltenyi Biotec, Auburn, CA, USA).
11. $CD34^+$ progenitor stems (commercially prepared, multiple sources, e.g., AllCells (Quincy, MA, USA), or Fred Hutchinson Cancer Research Center (Seattle, WA, USA)).
12. FBS (fetal bovine serum) (multiple sources, heat-inactivated).
13. Penicillin/streptomycin, 100× stock (=10,000 units/mL penicillin and 10 mg/mL streptomycin).

2.2 Buffers

All buffers are prepared sterile and all procedures performed in a sterile biological hood.

1. 10× PIPES, a 10×solution of PIPES (250 mM PIPES, 1100 mM NaCl, 50 mM KCl, pH at 7.4).
2. 1× PIPES, a tenfold dilution of 10× PIPES.
3. PAG buffer dilute 100 mL 10× PIPES with 900 mL of high-quality water, 0.1% glucose (w/v), and 0.003% HSA (w/v) (human serum albumin).
4. PAG-EDTA (ethylenediamine *N*,*N*,*N'*,*N'*-tetra-acetic acid) consisting of PAG supplemented with 1 mM EDTA.
5. EB, PAG containing 0.25% BSA instead of HSA and 1 mM EDTA.
6. PBS (commercial molecular biology grade, any source).
7. PBS-FCS, PBS with 1% FCS, (99 mL of PBS + 1 mL FBS, made fresh each time and 0.2 μm filtered using, e.g., Nalgene Rapid-Flow units).
8. StemPro Media, 10 mL StemPro media, 250 μL supplement (40×), 100 μL L-glutaMAX (200 mM), and 100 μL penicillin streptomycin (100×); Subheading 3.1—5 μL rhIL-3 (10 μg/mL in PBS, diluted from 100 μg/mL stored stock) or Subheading 3.2—5 μL rhIL-3 (10 μg/mL) plus 8 μL Flt3L (10 μg/mL in PBS, diluted from 100 μg/mL stored stock) plus 4 μL SCF (10 μg/mL in PBS, diluted from 100 μg/mL stored stock). Filter the combined components with a syringe-based 0.2 μm filter.
9. 1× Percoll; dilute stock Percoll with 10× PIPES solution, 900 mL Percoll plus 100 mL 10× PIPES.

10. 62% Percoll, 62 mL of 1× Percoll plus 38 mL 1× PIPES.
11. 53% Percoll, 53 mL of 1× Percoll plus 47 mL of 1× PIPES.

2.3 Equipment

1. Separation magnet Midi-MACS mounted on a MACS Multi-Stand (Miltenyi Biotec, Auburn, CA, USA).
2. MACS LS column (Miltenyi Biotec, Auburn, CA, USA).
3. Filtration kits, 0.20 μm, Nalgene Rapid-Flow (75 mm filter unit).
4. Cell Culture flask, 25 and 75 mm^2, canted neck.
5. Hemocytometer for cell counting.

3 Methods

3.1 Using Commercially Prepared CD34$^+$ Progenitors

1. Thaw 1 mL aliquot of cells rapidly in a 37 °C water bath (should require only 30–60 s to thaw).
2. Transfer cells to a 50 mL polypropylene tube (and rinse vial with 0.1 mL PBS-FCS).
3. Sequentially dilute the thawed cells as follows:
   (a) Drip in slowly 1 mL PBS-FCS (1 mL should require approximately 10 s to add) with gentle agitation; sit for 3 min (room temperature).
   (b) Drip in slowly 2 mL PBS-FCS (2 mL should require approximately 20 s to add); sit for 3 min.
   (c) Drip in slowly 4 mL PBS-FCS (4 mL should require approximately 40 s to add); sit for 3 min.
   (d) Drip in slowly 8 mL PBS-FCS (8 mL should require approximately 80 s to add); sit for 3 min.
   (e) Drip in slowly 16 mL PBS-FCS (16 mL should require approximately 160 s to add); sit for 3 min.
4. Centrifuge at 470 × *g* for 10 min at room temperature.
5. Aspirate supernatant leaving approximately 0.25 mL residual buffer, gently "rake" the cell pellet to bring it back into suspension, and gently add 30 mL PBS-FCS.
6. Centrifuge at 470 × *g* for 10 min at room temperature.
7. Aspirate supernatant leaving approximately 0.25 mL residual buffer, gently "rake" the cell pellet to bring it back into suspension, and add 10 mL of *complete StemPro media* (*see* **Note 2**). Transfer suspended cells to a 75 mm^2 culture flask.
8. If desired, a sample may be taken for a cell count.
9. Culture in a 5% CO_2 incubator at 37 °C for 2 weeks (method B) or 2.5–3 weeks (method A).

10. In many cases, a low G centrifugation (40G for 1 min in a 70 μL volume) is used to remove a significant fraction of dead cells from the harvested cultures prior to use in subsequent assays. Previous studies of these cells established that dead cells do not contribute mRNA; therefore, cell lysis procedures for mRNA can be based on the live cell counts. However, protein remains present in dead cells, so lysis for protein measurements is based on the total cell count.

3.2 Using CD34⁺ Cells Obtained from Leukapheresis Packs

This method requires a two-step Percoll gradient (sterile) and the Miltenyi CD34 isolation kit. All buffers are sterile, and all work is performed in a biological hood (including the transfer of the leukapheresis cells from their "software" apparatus).

3.2.1 Percoll of Leukapheresis Cells (Processed Blood Enriched with Leukocytes)

1. Dilute leukapheresis cells 1:1 with sterile 0.9% isotonic saline and 1 mM EDTA. The volume of cells varies considerably between packs, typically 50–100 mL of leukapheresis cells (*see* **Note 3**).
2. Establish a two-step gradient of Percoll in a 50 mL polypropylene tube that can be capped, 12 mL lower layer of 62% Percoll, and 12 mL upper layer of 53% Percoll. These gradients can be established by many different techniques, but the goal is to maintain a "tight" interface between the two layers. The Percoll has been equilibrated to room temperature prior to the next steps.
3. Overlay 25 mL of the diluted cells onto the gradient.
4. Centrifuge at room temperature at 450 × *g* for 20–25 min.
5. Recover the upper interface ($CD34^+$ cells are found in the topmost interface), i.e., the layer between the diluted cell solution and the 53% Percoll layer (*see* **Note 4**). To recover the upper layer, first aspirate about 80% of the solution at the top of the tube. Using a plastic pipette, transfer the layer to another BSA-treated 50 mL tube (*see* **Note 5**).
6. Dilute the recovered cells with PAG-EDTA.
7. Centrifuge the cells for 8 min at 350 × *g* at room temperature.
8. Aspirate the supernatant, and resuspend the cells with PAG buffer (no EDTA).
9. Centrifuge the cells for 8 min at 350 × *g* at room temperature.
10. Aspirate the supernatant and resuspend the cells with 0.5 mL PAG.

3.2.2 Positive Selection of CD34 Cells

1. Count the cells using a hemocytometer.
2. Recover 1 billion total cells, place in a 15 mL tube, centrifuge for 8 min, and resuspend the pellet in 0.5 mL EB.

3. Add 50 μL reagent A (CD34 kit) and 100 μL of reagent B (anti-CD34 Ab) to the suspension.
4. Incubate on ice for 20 min.
5. Apply the entire solution to a LS column in its magnet; allow the full solution to pass into the column bed before rinsing the column with approximately 10 mL of EB.
6. Remove the LS column from the magnet, and suspend over a 15 mL tube.
7. Allow 10 mL of EB to pass through the column to recover the CD34 cells.
8. Centrifuge the tube for 8 min at 350 × *g* at 4 °C.

3.2.3 Culture

1. Aspirate the supernatant, and resuspend the cells in 0.5 mL of StemPro to take a sample for counting, and a sample, if desired, to determine the purity of the CD34 cells by flow cytometry (*see* **Note 6**). Miltenyi provides a specific antibody for this purpose.
2. The cells are resuspended in either Subheadings 3.1 or 3.2 StemPro media and cultured in a 75 mm^2 culture flask (as above).
3. Culture in a 5% CO_2 incubator at 37 °C for 2 weeks (Subheading 3.1) or 2.5–3 weeks (Subheading 3.2).

In many cases, a low G centrifugation (40 × *g* for 1 min in a 70 μL volume) is used to remove a significant fraction of dead cells from the harvested cultures prior to use in subsequent assays. Previous studies of these cells established that dead cells do not contribute mRNA; therefore, cell lysis procedures for mRNA can be based on the live cell counts. However, protein remains present in dead cells, so lysis for protein measurements is based on the total cell count.

4 Notes

1. Studies by Eggel et al. [34] demonstrated that omalizumab could interact with IgE bound to FcεRI. This interaction is not traditional and does not lead to aggregation but at very high concentrations (>1 mg/mL) can induce dissociation of IgE from FcεRI. This has been called facilitated dissociation. We have shown that at therapeutic concentrations, there is little, if any, facilitated dissociation of IgE [35], but there is a possibility that the more rapid decrease in basophil FcεRI during treatment results from such a process. The key experiment was whether other omalizumab-like therapeutics would (1) not show facilitated dissociation and yet (2) still show the rapid

loss of FcεRI expression in peripheral blood basophils. Such an antibody now exists, the Novartis antibody ligelizumab. This antibody does not show facilitated dissociation (unpublished meeting report from the laboratory of Jardetzky et al.) but does still show the rapid loss of FcεRI expression on basophils [36]. Therefore, it appears that the turnover model remains a correct way to visualize what is occurring during treatment with omalizumab.

2. Note that the media can be made by Subheading 3.1 (IL-3 alone) or Subheading 3.2 (IL-3, Flt3L, SCF).
3. These packs typically contain approximately 2–5 billion leukocytes.
4. The interface layer between 53% and 62% Percoll contains basophils. Recovery of cells is optimized if all the tubes used for handling the cells are treated with EB solution overnight followed by complete aspiration to a dry tube prior to use. This is not always necessary as it requires a meaningful fraction of the recovered cells. The cell count provides a starting point for the later determination of recoveries, but the count doesn't change the final resuspension in StemPro media unless it is evident that the count is too high. Counts may range between 0.2 and 3.0 million cells (average = 1.7 million). If more than five million cells are recovered from 1 billion starting cells, then it is likely that the CD34 cells are more impure than desired.
5. Treating polypropylene tubes with a solution of EB for 24 h before their use has been found to significantly eliminate the loss of cells that adhere to the tube walls. This is especially important for the low number of cells that result from these procedures. All tube sizes are treated, from 50 mL tubes to microtubes. Before use, aspirate all the EB from the tube, spin the tube once and aspirate the residual EB.
6. The purity of CD34 progenitor cells from these preparations varies widely, from 25% to 75%. The variability doesn't allow a strong expectation for the number of cells later recovered from culture. Determining the input purity can aid in setting up consistent cultures based on CD34 count rather than total cell number.

References

1. Lichtenstein LM, Osler AG (1964) Studies on the mechanisms of hypersensitivity phenomena. Ix. Histamine release from human leukocytes by ragweed pollen antigen. J Exp Med 120:507–530
2. Ishizaka T, De BR, Tomioka H, Lichtenstein LM, Ishizaka K (1972) Identification of basophil granulocytes as a site of allergic histamine release. J Immunol 108:1000–1008
3. Parwaresch MR (1976) The human blood basophil. Springer-Verlag, New York
4. Arndt D, Parwaresch MR (1977) Kinetics of human basophilic granulocytes. Folia Haematol Int Mag Klin Morphol Blutforsch 104:24–36
5. Zaidi AK, MacGlashan DW (2010) Regulation of Fc epsilon RI expression during murine

basophil maturation: the interplay between IgE, cell division, and Fc epsilon RI synthetic rate. J Immunol 184:1463–1474
6. Sun LK, Liou NC, Sun NC, Gossett LA, Sun C, Davis FM, MacGlashan DW Jr, Chang TW (1991) Transfectomas expressing both secreted and membrane-bound forms of chimeric IgE with anti-viral specificity. J Immunol 146:199–205
7. Stallman PJ, Aalberse RC (1977) Quantitation of basophil-bound IgE in atopic and nonatopic subjects. Int Arch Allergy Appl Immunol 54:114–120
8. Hsu C, MacGlashan J, D.W. (1996) IgE antibody up-regulates high affinity IgE binding on murine bone marrow derived mast cells. Immunol Lett 52:129–134
9. MacGlashan DW Jr, Bochner BS, Adelman DC, Jardieu PM, Togias A, McKenzie-White J, Sterbinsky SA, Hamilton RG, Lichtenstein LM (1997) Down-regulation of FceRI expression on human basophils during in vivo treatment of atopic patients with anti-IgE antibody. J Immunol 158:1438–1445
10. Lantz CS, Yamaguchi M, Oettgen HC, Katona IM, Miyajama I, Kinet JP, Galli SJ (1997) IgE regulates mouse basophil FceRI expression in vivo. J Immunol 158:2517–2521
11. MacGlashan DW Jr, Xia HZ, Schwartz LB, Gong JP (2001) IgE-regulated expression of FceRI in human basophils: control by regulated loss rather than regulated synthesis. J Leuk Biol 70:207–218
12. Gomez G, Jogie-Brahim S, Shima M, Schwartz LB (2007) Omalizumab reverses the phenotypic and functional effects of IgE-enhanced fc epsilonRI on human skin mast cells. J Immunol 179:1353–1361
13. MacGlashan D (2004) Loss of receptors and IgE in vivo during treatment with anti-IgE antibody. J Allergy Clin Immunol 114:1472–1474
14. Beck LA, Marcotte GV, MacGlashan D, Togias A, Saini S (2004) Omalizumab-induced reductions in mast cell Fce psilon RI expression and function. J Allergy Clin Immunol 114:527–530
15. Denburg JA, Richardson M, Telizyn S, Bienenstock J (1983) Basophil/mast cell precursors in human peripheral blood. Blood 61:775–780
16. Denburg JA, Messner H, Lim B, Jamal N, Telizyn S, Bienenstock J (1985) Clonal origin of human basophil/mast cells from circulating multipotent hemopoietic progenitors. Exp Hematol 13:185–188
17. Denburg JA, Telizyn S, Belda A, Dolovich J, Bienenstock J (1985) Increased numbers of circulating basophil progenitors in atopic patients. J Allergy Clin Immunol 76:466–472
18. Otsuka H, Dolovich J, Befus AD, Telizyn S, Bienenstock J, Denburg JA (1986) Basophilic cell progenitors, nasal metachromatic cells, and peripheral blood basophils in ragweed-allergic patients. J Allergy Clin Immunol 78:365–371
19. Wood LJ, Sehmi R, Gauvreau GM, Watson RM, Foley R, Denburg JA, O'Byrne PM (1999) An inhaled corticosteroid, budesonide, reduces baseline but not allergen-induced increases in bone marrow inflammatory cell progenitors in asthmatic subjects. Am J Respir Crit Care Med 159:1457–1463
20. Schweitzer-Stenner R, Pecht I (2005) Death of a dogma or enforcing the artificial: monomeric IgE binding may initiate mast cell response by inducing its receptor aggregation. J Immunol 174:4461–4464
21. MacGlashan D Jr (2017) FceRI density and spontaneous secretion from human basophils. PLoS One 12:e0179734
22. Zaidi AK, Saini SS, Macglashan DW Jr (2010) Regulation of Syk kinase and FcRbeta expression in human basophils during treatment with omalizumab. J Allergy Clin Immunol 125 (902–908):e7
23. MacGlashan DW Jr, Savage JH, Wood RA, Saini SS (2012) Suppression of the basophil response to allergen during treatment with omalizumab is dependent on 2 competing factors. J Allergy Clin Immunol 130 (1130–1135):e5
24. Macglashan DW Jr, Saini SS (2013) Omalizumab increases the intrinsic sensitivity of human basophils to IgE-mediated stimulation. J Allergy Clin Immunol 132(906-11):e1–e4
25. MacGlashan DW Jr (2007) Relationship between Syk and SHIP expression and secretion from human basophils in the general population. J Allergy Clin Immunol 119:626–633
26. Nag A, Faeder JR, Goldstein B (2010) Shaping the response: the role of FcepsilonRI and Syk expression levels in mast cell signaling. IET Syst Biol 4:334–347
27. Havard S, Scola AM, Kay LJ, Ishmael SS, Macglashan DW Jr, Peachell PT (2011) Characterization of syk expression in human lung mast cells: relationship with function. Clin Exp Allergy 41:378–388
28. Ishmael SS, MacGlashan DW Jr (2010) Syk expression in peripheral blood leukocytes, CD34+ progenitors, and CD34-derived basophils. J Leukoc Biol 87:291–300
29. Arock M, Schneider E, Boissan M, Tricottet V, Dy M (2002) Differentiation of human

basophils: an overview of recent advances and pending questions. J Leukoc Biol 71:557–564

30. Kepley CL, Pfeiffer JR, Schwartz LB, Wilson BS, Oliver JM (1998) The identification and characterization of umbilical cord blood-derived human basophils. J Leukoc Biol 64:474–483
31. Langdon JM, Schroeder JT, Vonakis BM, Bieneman AP, Chichester K, Macdonald SM (2008) Histamine-releasing factor/translationally controlled tumor protein (HRF/TCTP)-induced histamine release is enhanced with SHIP-1 knockdown in cultured human mast cell and basophil models. J Leukoc Biol 84:1151–1158
32. Gilbert HS, Ornstein L (1975) Basophil counting with a new staining method using alcian blue. Blood 46:279–286
33. Peng X, Zhao M, Gao L, Sen R, MacGlashan J, D.W. (2020) Identifying regulatory pathways of SYK expression in human basophils. J Allergy Clin Immunol 145(3):947–957
34. Eggel A, Baravalle G, Hobi G, Kim B, Buschor P, Forrer P, Shin JS, Vogel M, Stadler BM, Dahinden CA, Jardetzky TS (2014) Accelerated dissociation of IgE-FcepsilonRI complexes by disruptive inhibitors actively desensitizes allergic effector cells. J Allergy Clin Immunol 133(1709-19):e8
35. MacGlashan D Jr (2015) Facilitated dissociation of IgE versus cell replacement. J Allergy Clin Immunol 135:294–295
36. Arm JP, Bottoli I, Skerjanec A, Floch D, Groenewegen A, Maahs S, Owen CE, Jones I, Lowe PJ (2014) Pharmacokinetics, pharmacodynamics and safety of QGE031 (ligelizumab), a novel high-affinity anti-IgE antibody, in atopic subjects. Clin Exp Allergy 44:1371–1385

Chapter 7

Generation of Mast Cells from Murine Stem Cell Progenitors

Emily J. Swindle

Abstract

Mouse bone marrow-derived mast cells (mBMMCs) are an invaluable tool for the study of mast cell function as they represent a primary source of mature mast cells. They can be sourced from wild-type, knockout, and transgenic mice and are used to repopulate mast cell-deficient mice. This method describes the isolation of mast cell hematopoietic progenitors from the bone marrow of mouse femurs and their subsequent culture in an IL-3-rich culture medium. After 4 weeks in culture, mBMMCs are obtained in high number and are of high purity. Assessment of their granularity by toluidine staining and IgE receptor expression by flow cytometry is also described. These cells are a useful tool in the determination of in vitro and in vivo mast cell function in innate and adaptive immunity.

Key words Mast cell, BMMC, IgE, FcεRI, Degranulation.

1 Introduction

Mouse mast cells develop from committed mast cell progenitors (MCp) which originate in the bone marrow and are identified as lineage-negative (Lin-), $CD34^+$, $CD117^+$ (c-Kit), integrin $\beta7^+$, $T1/ST2^+$, and $Fc\varepsilon RI^-$ cells [1]. These MCp then leave the bone marrow and migrate through the blood as Lin^- $CD117^{hi}$ $T1/ST2^+$ integrin $\beta7^{hi}$ $CD16/32^{hi}$ cells [2] before entering peripheral tissue where their growth, development, and survival require the growth factor stem cell factor (SCF). There are two major phenotypes of mast cells depending on their location in specific tissue (connective or mucosal tissue) and the presence of granule-associated proteases (mouse mast cell protease (MCP) or human tryptase or chymase) [3]. Mouse connective tissue mast cells contain high levels of MCP4, MCP5, and MCP6, whereas mouse mucosal mast cells have MCP1 and MCP2 [4]. Depending on the cytokines used for their generation, mBMMCs can be generated which are either of the mucosal (IL-3 only) or connective tissue type (IL-3 and SCF) [5]. They express the high-affinity IgE receptor (FcεRI) and can be

Bernhard F. Gibbs and Franco H. Falcone (eds.), *Basophils and Mast Cells: Methods and Protocols*, Methods in Molecular Biology, vol. 2163, https://doi.org/10.1007/978-1-0716-0696-4_7,

passively sensitized with IgE and activated following antigen stimulation.

The following methods describe the isolation of stem cell progenitors from mouse bone marrow and the subsequent expansion and differentiation of these progenitors into mature mast cells in an IL-3 rich culture medium. After 4 weeks, mature mBMMCs are obtained in high numbers (approx. 100 million cells per mouse (2 femurs) and of high purity (>99%). Depending on the background strain of mouse used to generate mBMMCs (BALB/c or c57/b6), we have observed reduced numbers of mature mast cells with lower FcεRI expression and reduced IgE-dependent activation from c57/b6 mice compared to BALB/c mice, which may be related to the different levels of MCp found in these strains [2]. While this protocol generates mBMMC using IL-3-rich media, other protocols exist which include SCF in combination with IL-3 to generate mBMMCs [6]. Assessment of the granularity and purity of these mast cell cultures is also described by toluidine blue staining and analysis of FcεRI and CD117 (KIT) receptor expression by flow cytometry, respectively. These methods can be used to obtain BMMCs from wild-type, knockout, and transgenic mice for further investigations into mast cell activation, chemotaxis, and adhesion.

2 Materials

2.1 Isolation of Mouse Bone Marrow Progenitors

1. Mice (2–6 months old).
2. Scissors.
3. Forceps.
4. Ethanol solution (70% in ddH_2O).
5. BMMC culture medium (1000 mL): to 825 mL of RPMI 1640, add 100 mL of FCS (10%) and the following 4 mM L-glutamine, 100 U/mL penicillin/100 μg/mL streptomycin, 25 mM HEPES, 1 mM sodium pyruvate, 1× nonessential amino acids, 50 μM β-mercaptoethanol, and 30 ng/mL murine IL-3 (PeproTech). Mix thoroughly, filter through 0.2 μm filter, and store at 4 °C for up to 2 weeks. Prior to use pre-warm the culture medium at 37 °C.
6. Conical tube (20 mL).
7. Syringe needle (25G).
8. Syringe (30 mL).
9. 6-Well plate.
10. Tissue culture flasks (125 cm^2).

2.2 Determination of BMMC Purity

2.2.1 Toluidine Staining

1. MOTA's fixative: to a 100 mL glass bottle, add ddH_2O (50 mL) and lead acetate (basic, 4 g), and with continuous mixing, add glacial acetic acid (1–2 mL) until the lead acetate is fully solubilized. Finally, add the 100% ethanol (50 mL), and store at room temperature for maximum of 2 months.
2. Toluidine blue solution: to a 50 mL conical tube, add 0.25 g toluidine blue, 35 mL ddH_2O, and 15 mL 100% ethanol. Adjust the pH of the solution to less than 1 with 1–2 mL HCl, and store at room temperature.
3. Microslides and coverslips.
4. Permount Mounting Medium (Fisher Scientific).
5. Cytospin (Shandon).

2.2.2 Flow Cytometry

1. Polystyrene round-bottom tubes (5 mL).
2. PBS containing 0.1% BSA (4 °C).
3. Anti-mouse FcεRI-FITC (MAR-1 clone, eBioscience, #11–5898).
4. Hamster IgG-FITC.
5. Anti-mouse CD117-PE (2B8 clone, BD Biosciences Pharmingen, # 553355).
6. Rat IgG_1-PE.

3 Methods

3.1 Bone Marrow Cell Isolation and Mast Cell Culture

3.1.1 Isolation and Culture of Bone Marrow Progenitors

1. Using 70% ethanol solution, sterilize scissors and forceps (*see* **Note 1**). Sacrifice a mouse, and rinse in 70% ethanol. Cut away the skin from around the back leg area, and remove both femurs using the scissors and forceps. Place the femurs in BMMC culture medium (10 mL) (*see* **Note 2**).
2. Add 10 mL BMMC culture medium to a well of a 6-well plate, and place the femurs into the well. To another well, add 5 mL culture medium, fill a syringe with medium, and attach a 25G needle to it. Using forceps, hold the femur, and using the needle, flush the bone marrow out of the femur into the culture medium (*see* **Note 3**).
3. Transfer the flushed bone marrow cells from the 6-well plate to a 125 cm^2 culture flask. Rinse the well with 5 mL fresh medium, and add this to the same culture flask. Add a further 40 mL of culture medium to the flask, and culture the cells in an incubator for 3–4 days.

3.1.2 Expansion of Bone Marrow Cells and Maturation of BMMCs

1. After 3–4 days, change the medium of the bone marrow culture twice a week (e.g., Monday and Friday) by transferring the cell suspension to a sterile conical tube (50 mL) and centrifuge (300 × *g*, 5 min, room temperature). Remove the supernatant, and resuspend the cell pellet in 50 mL fresh culture medium, and transfer to a new culture flask (T125cm^2) (*see* **Note 4**). After 4 weeks, the culture consists of mature BMMCs. Before proceeding to use in experiments, the purity of the BMMCs is evaluated by toluidine blue staining of cytospin preparations (see section below) and FcεRI and CD117 expression by flow cytometry (see section below) (*see* **Note 5**).

3.2 Determination of BMMC Purity

3.2.1 Toluidine Blue Staining of Cytospin Preparations

1. Take 100,000 cells per sample and centrifuge (300 × *g*, 5 min, room temperature).
2. Resuspend BMMC at 1 × 10^6/mL, and transfer 100 μL of BMMC in duplicate to a cytospin chamber, and centrifuge at 500 rpm for 5 min (*see* **Note 6**).
3. Immerse cells (on microslide) in Mota's fixative (15 min, room temperature); then gently wash with ddH_2O.
4. Allow to air-dry (15 min), and then immerse each slide in toluidine blue solution (20 min, room temperature).
5. Gently wash slide with ddH_2O, air-dry, and overlay with a coverslip using Permount.
6. Visualize cells under microscope for staining of granules by toluidine blue.

3.2.2 Surface Expression of FcεRI and CD117 (Kit) by Flow Cytometry

1. Take 50,000 cells per sample (2 samples required; isotype and specific antibody), and centrifuge (300 × *g*, 5 min, room temperature).
2. Resuspend cells at 0.5 × 10^6 cells/mL in prechilled PBS containing 0.1% BSA (4 °C), and transfer 100 μL aliquots into polystyrene round-bottom tubes (5 mL).
3. Add 1 μL of either anti-FcεRI-FITC and anti-CD117-PE or their respective isotype controls (hamster IgG-FITC and rat IgG_1-PE) to each tube, and incubate for 30 min at 4 °C in the dark (*see* **Note** 7).
4. Wash off excess antibodies by adding PBS containing 0.1% BSA (1 mL) to each tube, and centrifuge (300 × *g*, 5 min, 4 °C). Resuspend cell pellet in 100 μL PBS containing 0.1% BSA, and analyze 10,000 cells on a flow cytometer (*see* **Note 8**).

4 Notes

1. All procedures, including bone removal and femur lavage, are performed in a laminar flow hood. All reagents are sterilized by filtration and all surgical equipment sterilized by immersion in 70% alcohol prior to use.
2. Remove as much attached muscle from the femurs as possible with scissors; this makes it easier to flush the bone.
3. The bone will become clear when the red marrow is removed.
4. It is important to replace the culture flasks every time the medium is changed to remove contaminating adherent cells. This is especially important in the first 2 weeks of culture when there are many adherent cells present. After 2 weeks of culture, adjust the cell density to 1×10^6/mL when the bone marrow cells start to proliferate.
5. The BMMCs should not be used after 7 weeks in culture as their reactivity begins to decrease.
6. A pale white spot should appear on slide indicating cells have dried.
7. Prior optimization of antibody concentrations should be performed by titration prior to the start of experiment.
8. Typically, BMMC samples are greater than 99% double positive for CD117 and FcεRI.

Acknowledgments

Financial support for work in the authors' laboratories was provided by the Faculty of Medicine, University of Southampton, and the Asthma, Allergy, and Inflammation Research charity.

References

1. Chen CC, Grimbaldeston MA, Tsai M, Weissman IL, Galli SJ (2005) Identification of mast cell progenitors in adult mice. Proc Natl Acad Sci U S A 102:11408–11413
2. Dahlin JS, Heyman B, Hallgren J (2013) Committed mast cell progenitors in mouse blood differ in maturity between Th1 and Th2 strains. Allergy 68:1333–1337
3. Pejler G, Rönnberg E, Waern I, Wernersson S (2010) Mast cell proteases: multifaceted regulators of inflammatory disease. Blood 115:4981–4990
4. Stevens RL, Friend DS, McNeil HP, Schiller V, Ghildyal N, Austen KF (1997) Strain-specific and tissue-specific expression of mouse mast cell secretory granule proteases. Proc Natl Acad Sci U S A 91:128–132
5. Caslin HL, Kiwanuka KN, Haque TT, Taruselli MT, MacKnight HP, Paranjape A, Ryan JJ (2018) Controlling mast cell activation and homeostasis: work influenced by bill Paul that continues today. Front Immunol 9:868
6. Shelburne CP, McCoy ME, Piekorz R, Sexl V, Roh KH, Jacobs-Helber SM, Gillespie SR, Bailey DP, Mirmonsef P, Mann MN, Kashyap M, Wright HV, Chong HJ, Bouton LA, Barnstein B, Ramirez CD, Bunting KD, Sawyer S, Lantz CS, Ryan JJ (2003) Stat5 expression is critical for mast cell development and survival. Blood 102(4):1290–1297

Chapter 8

Integration of the Human Dermal Mast Cell into the Organotypic Co-culture Skin Model

Jonghui Kim, Sven Guhl, Magda Babina, Torsten Zuberbier, and Metin Artuc

Abstract

The organotypic co-culture skin model has been providing an advanced approach to in vitro investigations of the skin. Mast cells, containing various mediators such as tryptase and chymase, are thought to contribute to many physiological and pathological events of the skin interactively with other cells. Here, we introduce an organotypic co-culture skin model which successfully integrates human dermal mast cells for further study of mast cell interactions with fibroblasts and keratinocytes.

Key words Organotypic co-culture skin model, Skin equivalent, Dermis equivalent, Human dermal mast cells, Fibroblasts, Keratinocytes

1 Introduction

The organotypic co-culture skin model, first introduced by Bell and coworkers in 1981 [1] and modified subsequently by others [2, 3], consists of a fibroblast-populated dermis equivalent (DE) and a superficial keratinocytic layer. It has provided an advanced approach to in vitro investigations of the skin. Despite the simplified composition of the organotypic co-culture system, it presents not only an in vivo-like morphologic pattern of epidermal differentiation, which is not seen in the standard 2D culture, but also a similar phenotypic expression of various epidermal differentiation markers which closely resemble the original tissue [4–7]. The organotypic skin model has become a powerful tool for investigations of processes like keratinocyte growth regulation [8], dynamics of basement membrane formation [9], and dermal–epidermal junction morphogenesis [10].

Mast cells (MCs) are bone marrow-derived cells, distributed widely throughout the body, particularly in the subepithelial tissue near blood vessels and nerves, and are thought to contribute to

Bernhard F. Gibbs and Franco H. Falcone (eds.), *Basophils and Mast Cells: Methods and Protocols*, Methods in Molecular Biology, vol. 2163, https://doi.org/10.1007/978-1-0716-0696-4_8,

physiological and pathological events of the skin interactively with other cells. Some studies indicated that MCs enhance the proliferation of fibroblasts and may impede that of keratinocytes via their specific mediators tryptase and chymase [11–13]. Conversely, fibroblasts and keratinocytes also have effects on mast cell growth via growth factors such as stem cell factor (SCF), nerve growth factor, and granulocyte-macrophage colony-stimulating factor [14–17].

In order to further understand the interplay between MCs and other resident tissue cells, the development of a well-established organotypic co-culture skin model, integrating human dermal MCs, would be desirable. Several studies used organotypic co-culture skin models, including other MCs such as bone marrow-derived interleukin-3-independent murine mast cells (C1MC/C571 cell line) [18] and leukemic mast cells (HMC-1 cell line) [19, 20], respectively. Later, human dermal MCs, which are more difficult to culture, were first integrated into the organotypic co-culture system by our group in 2002 [21] to study MC-keratinocyte interactions. Our previous work required further improvement, e.g., to eliminate the problem that keratinocytes themselves and also the organotypic co-cultures with mast cells were kept in medium containing serum (FBS) and cholera toxin, which was unfavorable.

Thus, we modified our previous protocol and could integrate human dermal mast cells into a serum-and cholera toxin-free organotypic skin culture.

2 Materials

Prepare all materials for cell isolation, and culture under sterile conditions, and store all solutions and media at 4 °C unless indicated otherwise.

2.1 Cell Isolation

1. Fetal calf serum (FCS): Heat inactivated for 30 min at 56 °C to inactivate heat labile complement proteins.
2. Phosphate-buffered saline without Ca^{2+}/Mg^{2+} (PBS−).
3. Phosphate-buffered saline with Ca^{2+}/Mg^{2+} (PBS+).
4. Dispase solution: Dilute dispase type 1 (BD Bioscience) in PBS+ to a final concentration of 3 U/mL.
5. Dispersion medium: 5 mL of penicillin/streptomycin (penicillin 10.000 IU/ streptomycin 10.000 μg/mL), 10 mL of FCS, 1.25 mg of amphotericin B (250 μg/mL), and 2.5 mL of $MgSO_4$ (1 M) in 500 mL of PBS+. Shortly before use, 10–15 mg of collagenase type IV (Worthington Biochem, Freehold, NJ, USA), 5–7.5 mg of hyaluronidase type 1S

(Sigma), and DNase 1 (end concentration in the dispersion medium 10 μg/mL) (Roche Applied Science, Basel, Switzerland) are diluted in 10 mL of the prepared solution.

6. 0.05% Trypsin/0.01% EDTA. Store aliquots at −20 °C.
7. Soybean trypsin inhibitor (SBTI): Dissolve 100 mg of SBTI in 100 mL PBS−. Store 10 mL-aliquots at −20 °C.
8. Shaking water bath (37 °C).
9. Centrifuge.
10. Automated cell counter.
11. Incubator (37 °C and 5% CO_2).

2.2 Purification of Dermal Mast Cells

1. Sieves (mesh size 40, 100, and 300 μm).
2. Pre-separation filter (30 μm; Miltenyi Biotec).
3. MACS buffer: 500 mL of PBS−, 50 mL of FCS, 10 mL of EDTA (100 mM).
4. Sandoglobulin®, an Intravenous Immunoglobulin (IVIG) preparation (30.000 μg/mL; CSL Behring, Germany).
5. Human AB serum (Invitrogen).
6. Anti-human CD117 MicroBead kit (Miltenyi Biotec).
7. AutoMACS Miltenyi Biotec or another suitable MACS separator.
8. Trypan blue.
9. Acidic toluidine blue (0.1% in 0.5 N HCl).

2.3 Cell Culture

1. Keratinocyte medium: KBM-Gold™ (Lonza) supplemented with KGM-Gold SingleQuots (Lonza).
2. Fibroblast medium: 500 mL of low-glucose Dulbecco's Modified Eagle Medium with L-glutamine (DMEM) (Gibco) supplemented with 5 mL of 100x penicillin/streptomycin and 25 mL of FCS.
3. Stem cell factor (SCF) (50 μg/mL; PeproTech).
4. Interleukin-4 (IL-4) (20 μg/mL; PeproTech).
5. Mast cell medium: 500 mL of Iscove's Modified Dulbecco's Medium (IMDM) supplemented with 5 mL of nonessential amino acids, 5 mL of penicillin/streptomycin, and 50 mL of FCS. Add SCF and IL-4 directly before use (end concentration 100 ng/mL and 20 ng/mL, respectively).
6. Serum-free freezing medium (Promo Cell, cat. no. C-29910).
7. Dimethylsulfoxide (DMSO), cell culture grade.
8. Cryo 1 °C Freezing Chamber (Nalgene).

2.4 Organotypic Co-culture

1. Collagen type I from rat tail (3.6 mg/mL; First Link Ltd., UK).
2. 10× Concentrated low-glucose Dulbecco's Modified Eagle Medium without L-glutamine, $NaHCO_3$, and folic acid.
3. 1 M NaOH.
4. Matrigel™ Matrix (BD Biosciences).
5. Ascorbic acid dissolved in distilled water at 50 mg/mL.
6. Metal grid (specially made): 20 × 20 × 5 mm (width × length × height).
7. Nylon mesh: Cut the nylon mesh into squares of approximately 20 × 20 mm. Wash and disinfect the small nylon mesh squares with 100% ethanol, and dry them in a laminar flow cabinet. Mix 1.1 mL of collagen type 1 and 50 mL of 0.03 N acetic acid, and incubate the small nylon mesh squares with this mixture on a small Petri dish (diameter 5 cm) for 30 min at room temperature. Incubate them with approximately 25 mL of 0.1% glutaraldehyde for a further 30 min at room temperature. Rinse four times with PBS−. Store in 70% ethanol until use (*see* **Note 1**). Dry them in a laminar flow cabinet before use.
8. Organotypic culture medium: KBM-Gold™ supplemented with ascorbic acid (end concentration 50 μg/mL), SCF (end concentration 2.5 ng/mL), $CaCl_2$ (end concentration 1.5 mM), 0.1% bovine serum albumin (BSA), and KGM-Gold SingleQuots except Bovine Pituitary Extract.

2.5 Preparation of Frozen Sections

1. Liquid nitrogen container (Messer Griesheim).
2. Liquid nitrogen.
3. Cryomold 4557 (Tissue-Tek®, cat. no. 62534-25).
4. Cryotome FSE cryostat (Thermo Fisher Scientific).
5. Glass slides (Super Frost).

2.6 Immunohistochemistry

1. PAP-Pen (SCI, Germany).
2. Tris-buffer (10×): Dissolve 9 g of Tris-base, 68.5 g of Tris–HCl, and 87.5 g of NaCl in distilled water to a total volume of 1 L. Dilute 1 volume of the Tris-buffer (10×) in nine volumes of deionised water before use.
3. Peroxidase blocking reagent (Dako).
4. Avidin/Biotin kit (Vector Laboratories).
5. Fc receptor block solution: Mix 25 mL of RPMI 1640, 5 mL of FCS, and 50 μL of Sandoglobulin®. Store 1-mL aliquots at −20 °C until use.
6. Antibody diluting solution: Dilute 2 mL of RPMI 1640 and 2 mL of FCS in 17.5 mL of deionised water. Store 1 mL aliquots at −20 °C until use.

7. Biotinylated secondary antibodies and streptavidin-HRP (Dako).
8. Dako AEC+ High Sensitivity Substrate Chromogen, Ready-to-Use (Dako).
9. Hematoxylin: Strain through paper filter before every use.
10. Mounting medium.
11. Anti-tryptase mouse monoclonal IgG1, c = 0.2 mg/mL (clone AA1).
12. Anti-cytokeratin 10 rabbit monoclonal IgG (clone EP1607IHCY).
13. Anti-transglutaminase1 mouse monoclonal IgG2a (clone B. C1).
14. Anti-ß4 integrin mouse monoclonal IgG1 (clone ASC-8).
15. 5-Bromo-2′-deoxyuridine (BrdU) (Sigma).
16. Anti-BrdU mouse monoclonal IgG1, c = 1 mg/mL (clone AB-3).
17. Anti-chymase antibody (Clone CC1).

3 Methods

Breast skin and foreskin, obtained from individuals undergoing breast reduction surgery and circumcision respectively, are the source of the tissue material. For the isolation of human dermal mast cells, keratinocytes, and fibroblasts, we modify methods originally described by other investigators [22–25]. A skin equivalent (SE) has three parts: (1) an acellular layer to avoid escaping of cells from the dermis equivalent to the culture medium, (2) the dermis equivalent containing fibroblasts and human dermal mast cells, and (3) the epidermal layer consisting of keratinocytes. Subsequently, the SE (skin equivalent) is air-lifted, and the level of the culture medium is adjusted leaving the epithelial cells exposed to the air. Half of the culture medium is renewed every 2 days for a further 10-day organotypic culture. Staining is obtained with a labeled streptavidin biotin method.

We utilize the serum-free keratinocyte growth medium (without the bovine pituitary extract) which was first applied for the organotypic co-culture by Stark and coworkers [26]. According to our observation, all three cell subsets adapted well under this culture medium condition, and we could obtain stable results.

All co-culture procedures should be carried out in a safety workbench and especially procedures with collagen and Matrigel™ Matrix on ice, unless otherwise specified.

3.1 Mast Cell Isolation, Purification, and Culture

1. Remove fat tissue from the skin tissue. Spread the skin tissue on a Petri dish (diameter 14 cm). Prepare the skin tissue by cutting it into an approximate rectangle of 2 × 0.5 cm, then cut these small pieces vertically at intervals of approximately 1–2 mm by leaving a slight connection on a 2-cm edge.
2. Incubate the prepared skin tissue overnight with the dispase solution in a 50-mL centrifuge tube at 4 °C.
3. Remove the epidermal layer thoroughly using tweezers (*see* **Note 2**).
4. Cut the dermis into small pieces with operating scissors as finely as possible, reducing to a soft doughy state (*see* **Notes 3** and **4**).
5. Incubate the dermis in a shaking water bath with the pre-warmed dispersion medium for approximately 60 min at 37 °C (*see* **Notes 5** and **6**).
6. Strain the solution with the leftover dermis and released cells through two sieves (mesh size 300 and 100 μm) (*see* **Note 7**). Centrifuge the filtrate at 250 × *g* for 10 min at 4 °C to remove the dispersion medium (*see* **Note 8**).
7. Resuspend the cell pellet in PBS- and centrifuge at 250 × *g* for 10 min at 4 °C.
8. Remove the supernatant, and resuspend the cell pellet in MACS buffer; strain the suspension through a sieve (mesh size 40 μm), and centrifuge at 250 × *g* for 10 min at 4 °C.
9. Remove the supernatant, resuspend the cell pellet in MACS buffer, and determine the cell number.
10. Strain the cell suspension through the pre-separation filter (30 μm) to remove cell clumps.
11. Centrifuge the cell suspension at 250 × *g* for 10 min at 4 °C, and remove the supernatant completely.
12. Resuspend the cell pellet in MACS buffer (350 μL per 10^8 cells), add human AB Serum (150 μL per 10^8 cells) and Sandoglobulin® (10 μL per 10^8 cells), mix well, and incubate for 15 min at 4 °C.
13. Add CD117 MicroBeads (25 μL for every 500 μL cell suspension), and incubate for further 11 min at 4 °C.
14. Wash the cells by adding MACS buffer, centrifuge at 250 × *g* for 10 min at 4 °C, and remove the supernatant completely.
15. Resuspend the cell pellet at 10^8 cells per 500 μL of MACS buffer (*see* **Note 9**).
16. Place the column (large cell separation column) into the magnetic field of the suitable MACS separator. Choose the MACS column according to the number of the total cells (*see* **Note 10**).

17. Apply the cell suspension onto the column. Collect flow-through containing unlabeled cells (the negative fraction) (*see* **Note 11**).
18. Remove the column from the separator and place it in a collection tube.
19. Pipet the appropriate amount of MACS buffer onto the column, and flush out the labeled cells by pushing the plunger into the column.
20. Add 10 mL of the Iscove's Modified Dulbecco's Medium to the suspension. Check the viability with trypan blue staining. Determine MC purity with acidic toluidine blue.
21. Centrifuge at 250 × *g* for 10 min at 4 °C, remove the supernatant, and resuspend the cell pellet in the mast cell medium at 1×10^6/mL. Add 2 μL of SCF and 2 μL of IL-4 for 1 mL of the cell suspension.
22. Place the culture into a humidified incubator with 5% CO_2 at 37 °C.
23. Add 2 μL of SCF and 2 μL of IL-4 for each mL of the cell suspension twice a week.
24. For long-term culture, renew half of the medium once a week (semi-depletion), and add 1 μL of SCF and 1 μL of IL-4 for each mL of the cell suspension. Alternatively, add the same amount of SCF and IL-4 twice a week without medium renewal.

3.2 Fibroblast Isolation and Culture

1. Cut the de-epidermalized dermis with a scalpel into squares of approximately 3 × 3 mm, and place in a culture dish (10 cm diameter) (*see* **Note 12**).
2. Leave the dish open under the safety workbench for up to 5 min so that the dermis pieces adhere to the surface of the dish.
3. Add fibroblast medium, and submerge the small dermis pieces in the medium, avoiding release from the dish.
4. Place the dishes in a humidified 37 °C incubator containing 5% CO_2.
5. Change the medium every 3 days.
6. Trypsinize fibroblasts at confluence (*see* **Note 13**). For trypsinization of the fibroblasts, remove the culture medium, and wash the cells with PBS—twice. Add 4 mL of 0.05% Trypsin/ 0.01% EDTA and incubate for 3–5 min at 37 °C. Tap the culture flask gently to release the cells. Neutralize the trypsin with the same volume of fibroblast medium. Determine the total number of fibroblasts and percent viability using trypan blue staining. Centrifuge the cell suspension at 250 × *g* for 10 min at 4 °C. Remove the supernatant.

7. Seed 1×10^6 cells into a 175-cm^2 cell culture flask, and add 20 mL of fibroblast medium. Renew the medium every 3 days. Split or cryopreserve the cells at confluence. For cryopreservation of the fibroblasts, resuspend the trypsinized fibroblasts in the pre-cooled FCS-DMSO-mixture (volume ratio of FCS:/MSO is 9:1) at 1×10^6/mL. Dispense 1 mL of the cell suspension into a cryogenic storage vial. Place the vial into a cryo 1 °C freezing chamber, and store it at −80 °C overnight. Transfer the vial into liquid nitrogen or a cryogenic freezer.

3.3 Keratinocyte Isolation and Culture

1. Rinse the 10–15 small pieces of freed epidermis for a short time in PBS− (*see* **Note 14**).
2. Incubate them with 5 mL of 0.05% Trypsin/0.01% EDTA for 15 min at room temperature.
3. Add the same volume of SBTI solution and mix gently for a short time.
4. Remove the epidermal layers by straining once through a sieve (mesh size 100 μm).
5. Determine the total number of keratinocytes and percent viability using trypan blue staining.
6. Centrifuge at 250 × *g* for 10 min at 4 °C.
7. Remove the supernatant. Add PBS− and wash the cells gently. Centrifuge once more at 250 × *g* for 10 min at 4 °C.
8. Remove the supernatant, and suspend 1×10^6 keratinocytes in 20 mL of the keratinocyte medium.
9. Place the cell suspension into a collagen-coated 175-cm^2 cell culture flask at 37 °C with 5% CO_2 (*see* **Note 15**).
10. Renew the medium every 3 days. Split or cryopreserve the cells at about 70–80% confluency.
11. Trypsinization of the keratinocytes: Remove the culture medium, and wash the cells with PBS− once. Add 4 mL of 0.05% Trypsin/0.01% EDTA, and incubate for 3–5 min at 37 °C. Tap the culture flask gently to release the cells. Neutralize the trypsin with the same volume of the SBTI solution. Determine the total number of keratinocytes and percent viability using trypan blue. Centrifuge the cell suspension at 250 × *g* for 10 min at 4 °C. Remove supernatant.
12. Cryopreservation of the keratinocytes: Suspend the trypsinized keratinocytes in the serum-free freezing medium at 10×10^6/mL. Dispense 1 mL of the cell suspension into a cryogenic storage vial. Place the vial into a cryo 1 °C freezing chamber, and store it at −80 °C overnight. Transfer the vial into liquid nitrogen or a cryogenic freezer.

3.4 Skin Equivalent (SE)

1. Prepare a suspension of fibroblasts and MCs together in pre-cooled FCS at 2.5×10^6/mL each.
2. Mix 10 volumes of collagen type I and 1 volume of 10× DMEM. Adjust the pH of the mixture to 7.0 by dropwise addition of 1 M NaOH prior to mixing with 1 volume of Matrigel™ Matrix. Leave the collagen-Matrigel-mixture on ice for 20 min (*see* **Note 16**).
3. Coat the surface of a 24-well tissue culture plate with 200 μL of the previously prepared collagen-Matrigel mixture evenly, and solidify at 37 °C for 10 min (*see* **Notes 17** and **18**).
4. Mix 100 μL of the fibroblast-MC suspension and 600 μL of the collagen-Matrigel-mixture in a 2-mL Eppendorf tube. Pipet the mixture onto the 24-well tissue culture plate above the acellular collagen-Matrigel layer (*see* **Note 17**).
5. Solidify the mixture at 37 °C for 30 min. Then, add 1 mL of the keratinocyte medium with ascorbic acid (end concentration 50 μg/mL) and SCF (end concentration 12.5 ng/mL), and incubate at 37 °C for further 30 min (*see* **Note 19**).
6. Prepare a suspension of keratinocytes in pre-warmed (37 °C) keratinocyte medium with ascorbic acid (end concentration 50 μg/mL) and SCF (end concentration 12.5 ng/mL) at 5.0×10^5 (*see* **Note 20**). Gently (but completely) remove the medium from the 24-well tissue culture plate, and carefully add 1 mL of the prepared keratinocyte suspension. Incubate at 37 °C for 24 h.

3.5 Airlifting

1. Release the SE from the tissue culture plate, and incubate at 37 °C until it contracts, shrinking by two-thirds of its diameter (*see* **Notes 21** and **22**).
2. Airlift the contracted SE by placing it into a 6-well tissue culture plate on the metal grid covered by a previously prepared nylon mesh (*see* **Note 23**).
3. Add 5 mL of the organotypic culture medium, just to the level of the nylon mesh, leaving the epithelial cells exposed to the air (*see* **Note 24**).
4. Renew only half of the medium, by removing 2.5 mL and adding 2.5 mL of the organotypic culture medium every 2 days for a further 10-day organotypic culture (*see* **Notes 25** and **26**).

3.6 Preparation of Frozen Sections

1. Place the sample together with the nylon mesh on a glass object.
2. Wipe off excess medium around the sections using a paper towel, but do not touch the sample.

3. Cut away the uncovered part of the nylon mesh using a scalpel and tweezers.
4. Embed the sample in a tissue freezing medium using Cryomold, and freeze it with liquid nitrogen rapidly by touching only the bottom of the Cryomold for approximately 1 min (*see* **Notes 27–30**).
5. Cut frozen sections at 4–6 μm. Dry the sections at room temperature overnight then fix with ethanol or methanol if necessary. Store at −80 °C until needed.

3.7 Immunohistochemistry

1. Draw a circle around the sections using a PAP pen to create a hydrophobic barrier (*see* **Note 31**).
2. Immerse sections in Tris-buffer for 10 min. Do not allow the sections to become dry for the remaining procedure.
3. Block the endogenous peroxidase activity using peroxidase blocking reagent for 10 min, and wash in Tris-buffer for 2× 5 min (*see* **Note 32**).
4. Block endogenous biotin incubating the section with 0.001% avidin D (w/v) in PBS for 10 min, and wash in Tris-buffer for 2× 5 min.
5. Block the remaining biotin binding sites on the avidin incubating the section with 0.001% biotin for 10 min, and wash in Tris-buffer for 2× 5 min.
6. Block endogenous Fc receptor with the previously prepared Fc receptor block solution for 15 min.
7. Incubate the sections for 60 min at room temperature or overnight at 4 °C with diluted primary antibodies following the instruction of providers. Wash in Tris-buffer for 2× 5 min (*see* **Note 33**).
8. Incubate the sections for 10 min with biotinylated secondary antibodies, and wash in Tris-buffer for 2× 5 min.
9. Incubate the sections for 10 min with streptavidin-HRP, and wash in Tris-buffer for 2× 5 min.
10. Incubate the sections for 15 min with AEC-Substrate, and wash in Tris-buffer for 2× 5 min.
11. Counterstain for 2 min with hematoxylin.
12. Wash under running tap-water for 10 min and rinse with distilled water.
13. Add one to two drops of mounting medium, and place a cover glass over the section (*see* **Note 34**).

4 Notes

1. It is recommended to store the nylon mesh in ethanol at least for 1 day to prevent fungal contamination.
2. Because melanocytes are also CD117-positive, the epidermis should be completely removed. The dissected epidermis is used for the isolation of keratinocytes.
3. Even for experienced individuals, this procedure should take at least 30 min for 15 g of skin tissue.
4. Put a small part of dermis aside for the subsequent isolation of fibroblasts.
5. The volume of the dispersion medium depends on the weight of the skin tissue, approximately 10 mL per gram of skin tissue.
6. A 500 mL sterile PBS bottle is suited for the incubation in a shaking water bath. Parafilm is also used to further seal the bottle to prevent contamination. The bottle is fixed on the bottom of the bath. The shaking frequency should be as high as possible but without spilling the water.
7. In instances where substantial portions of the dermis remain, the dermis can be cut into small pieces again, and the incubation with the used dispersion medium can be repeated.
8. We recommend using a 50 mL centrifuge tube until the blocking and magnetic labeling step and thereafter a 15 mL tube. Carry out the cell wash step by filling tubes to maximal volumes with PBS or MACS buffer, respectively.
9. Add 500 μL of MACS buffer, in case the number of cells is less than 10^8.
10. Our laboratory uses an autoMACS (Miltenyi, Germany), and we select the autoMACS software program Possel-DS.
11. Fibroblasts can also be obtained from the negative fraction. Depending on the source of the skin, however, other dermal cell types may be present in the negative fraction; wash the cells twice with PBS−, and suspend the cells in fibroblast medium. Place the suspension into a 175 cm^2 cell culture flask at 36 °C with 5% CO_2. Renew the culture medium, and continue to cultivate adherent fibroblasts.
12. The source of the tissue material is de-epithelialized dermis from the previous step (*see* Subheading 3).
13. Fibroblasts will migrate from the dermal tissue to the surface of the culture dishes.
14. The source of the tissue material is the dissociated epidermis from the previous step (*see* Subheading 3). A small piece means an approximate shape of 1 × 5 mm. If the skin tissue is cut as

described above, small epidermal pieces can be harvested by removing the epidermal layer from the skin tissue.

15. Mix 140 μL of collagen type 1 and 10 mL of 0.03 N acetic acid. Coat the surface of a cell culture flask with the collagen solution evenly, and leave it at room temperature for 30 min. Wash with PBS− for 3× 5 min.
16. Because of the contraction of the mixture, it is necessary to keep the stated time. Considering the contraction and the loss, one volume should be between 70 and 80 μL for one sample.
17. Try to pipet directly onto the bottom of the wells, not onto the wall. The collagen-Matrigel-mixture is very sticky and easily solidifies at room temperature.
18. It is important not to allow the formation of air bubbles. So, pipet only in one stroke, and do not touch it again.
19. Ascorbic acid in the phase seems to improve the viability of MCs.
20. Only keratinocytes of the second passage are used for the organotypic co-culture.
21. The SE is separated from the 24-well culture plate physically, starting with a 10 μL Eppendorf pipet tip at the side wall of the well and making a gentle stroke along the side wall. It can be repeated if the gel is not released completely.
22. Duration depends on the collagen density and the number of cells that DE contains. In the given condition, it takes approximately 3 h. The contraction is essential for the subsequent development of the epidermal structure.
23. For the airlifting of the contracted SE, a special-made thin stainless iron plate of 0.5 × 7.0 cm with a right-angle bend at 0.8 cm point from one end is used. The SE is spooned up using the bent part and put on the metal grid covered by a prepared nylon mesh.
24. The epidermal layer should be kept free from the medium to the end of the organotypic culture. Otherwise, the epidermal layer will not be developed well.
25. Mix the medium using a 1 mL pipet before removing the medium (4–5 strokes).
26. Add BrdU at a final concentration of 50 μM by the last medium renewal.
27. The preparation of paraffin sections may pose problems with the organotypic culture because of its fragility, e.g., a structure deformation by the fixation and separation of the epidermal layer from the DE during treatment.

28. The sample should be completely surrounded with the tissue-freezing medium, without air bubbles lying in the middle of the medium level.
29. Fill a small Dewar flask with liquid nitrogen. Freeze the sample with liquid nitrogen as rapidly as possible by touching only the bottom of the Cryomold for approximately 1 min—an edge of the Cryomold is fixed with a 30-cm wire for easy handling. Do not allow the freezing medium to touch the liquid nitrogen because of the formation of bubbles.
30. Other freezing techniques, e.g., placing in a cryostat-microtome, leads to structure deformation. Also, the rapid freezing with liquid nitrogen makes the tissue freezing medium more solid, which is crucial for easy cutting work.
31. The circle should be as small as possible to save on reagents. However, do not touch the sections. Add two drops (approximately 100 μL) of reagents at each step.
32. Shake slides vigorously twice to remove excess water or reagents.
33. Antibodies are diluted in the previously prepared antibody diluting solution. The following primary antibodies, dilutions, and incubation periods are used (*see* Figs. 1, 2, 3, 4, 5, and 6): anti-cytokeratin 10, 1:500, overnight at 4 °C; anti-transglutaminase, 1:100, for 60 min at room temperature; anti-ß4 integrin, 1:200, for 60 min at room temperature; anti-tryptase, 1:1000, for 60 min at room temperature; anti-

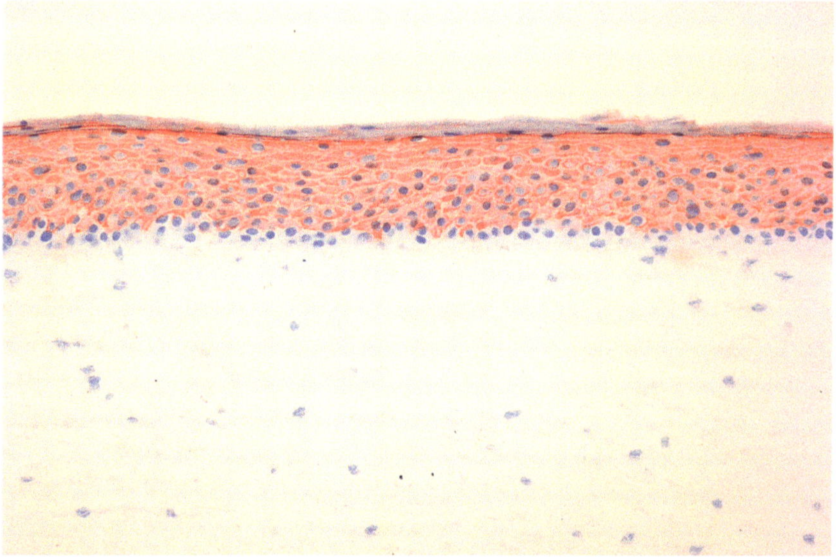

Fig. 1 Immunostaining in skin equivalents: AEC staining of skin equivalents using rabbit or mouse Ab directed against human Cytokeratin 10

chymase, 1:500, overnight at 4 °C; and anti-BrdU, 1:100, for 60 min at room temperature.

34. Shake slides vigorously many times to remove excess water, and leave the slide for some minutes to dry up the water on the slide. This allows long-term preservation of the section.

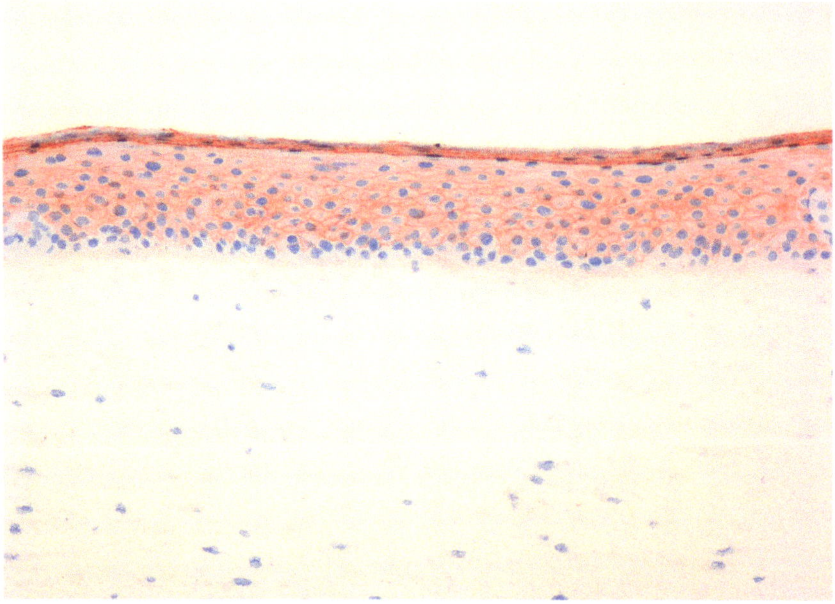

Fig. 2 Immunostaining in skin equivalents: AEC staining of skin equivalents using rabbit or mouse Ab directed against human Transglutaminase

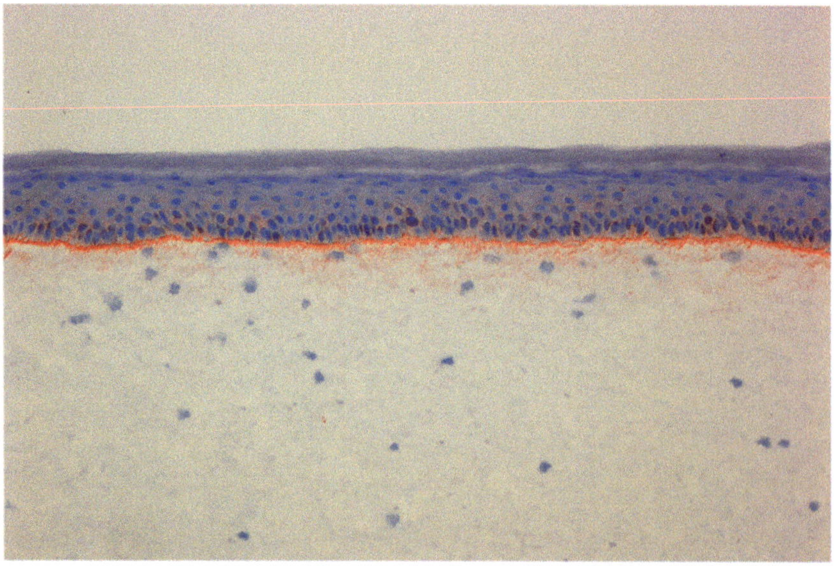

Fig. 3 Immunostaining in skin equivalents: AEC staining of skin equivalents using rabbit or mouse Ab directed against human Integrin ß4

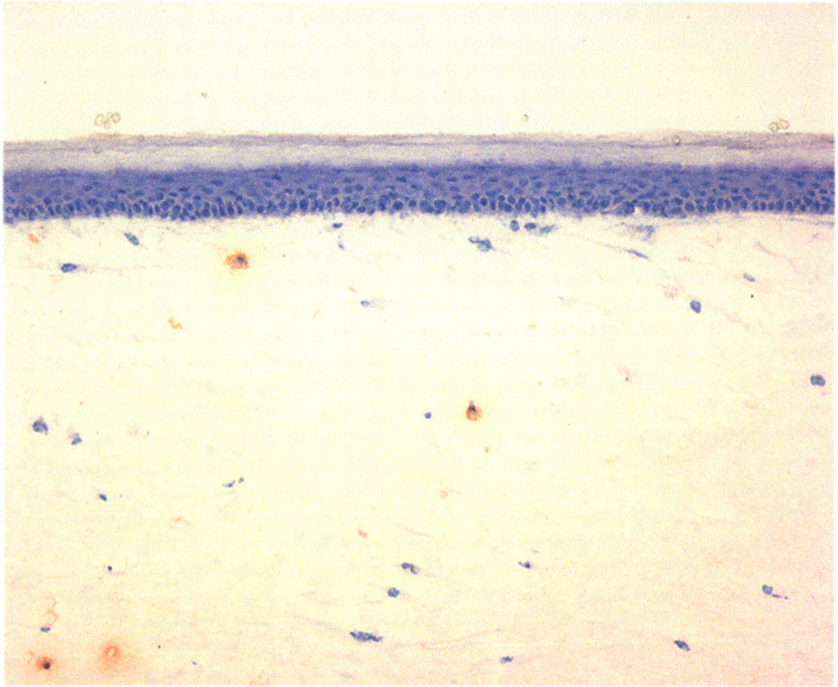

Fig. 4 Immunostaining in skin equivalents: AEC staining of skin equivalents using rabbit or mouse Ab directed against human Tryptase

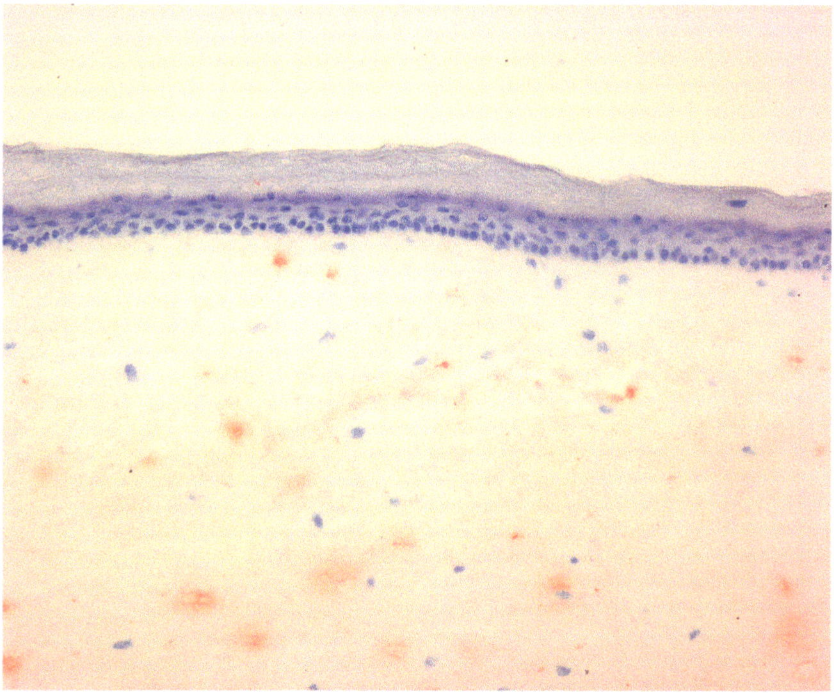

Fig. 5 Immunostaining in skin equivalents: AEC staining of skin equivalents using rabbit or mouse Ab directed against human Chymase

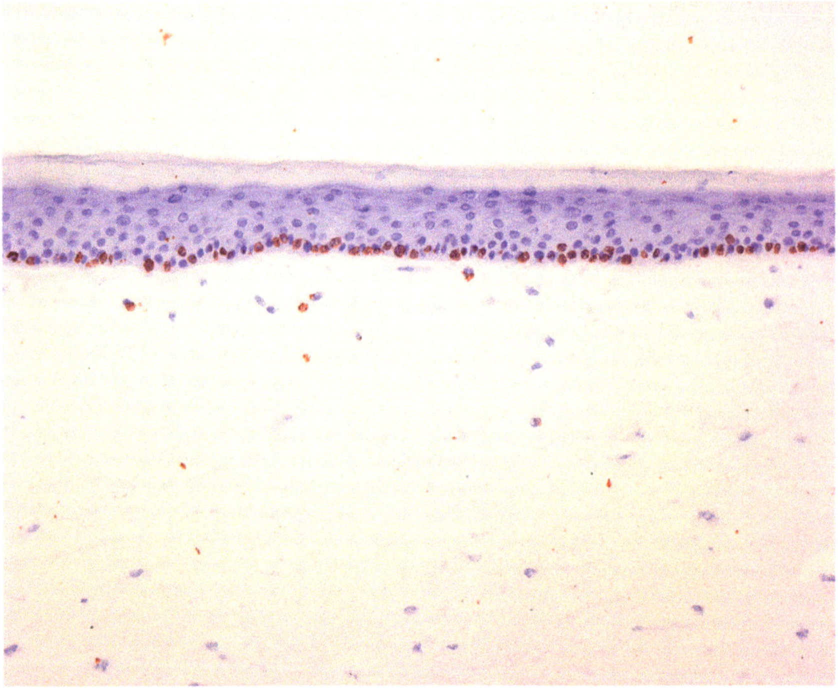

Fig. 6 Immunostaining in skin equivalents: AEC staining of skin equivalents using rabbit or mouse Ab directed against human BrdU

References

1. Bell E, Ehrlich HP, Buttle DJ, Nakatsuji T (1981) Living tissue formed in vitro and accepted as skin-equivalent tissue of full thickness. Science 211:1052–1054
2. Parenteau NL, Nolte CM, Bilbo P et al (1991) Epidermis generated in vitro: practical considerations and applications. J Cell Biochem 45:245–251
3. Fusenig NE (1994) Epithelial-mesenchymal interactions regulate keratinocyte growth and differentiation in vitro. In: Leigh IM, Lane EB, Watt FM (eds) The keratinocytes handbook. Cambridge University Press, Cambridge
4. Asselineau D, Bernard BA, Bailly D, Darmon M (1989) Retinoic acid improves epidermal morphogenesis. Dev Biol 133:322–335
5. Lenoir MC, Bernard BA (1990) Architecture of reconstructed epidermis on collagen lattices varies according to the method used: a comparative study. Skin Pharmacol 3:97–106
6. Parenteau NL, Bilbo P, Nolte CJ et al (1992) The organotypical culture of human skin keratinocytes and fibroblasts to achieve form and function. Cytotechnology 9:163–171
7. Igarashi M, Irwin CR, Locke M, Mackenzie IC (2003) Construction of large area organotypical cultures of oral mucosa and skin. J Oral Pathol Med 32:422–430
8. Maas-Szabowski N, Stark HJ, Fusenig NE (2000) Keratinocyte growth regulation in defined organotypic cultures through IL-1-induced keratinocyte growth factor expression in resting fibroblasts. J Invest Dermatol 114:1075–1084
9. Smola H, Stark HJ, Thiekotter G et al (1998) Dynamics of basement membrane formation by keratinocyte–fibroblast interactions in organotypic skin culture. Exp Cell Res 239:399–410
10. Marionnet C, Pierrard C, Vioux-Chagnoleau C et al (2006) Interactions between fibroblasts and keratinocytes in morphogenesis of dermal epidermal junction in a model of reconstructed skin. J Invest Dermatol 126:971–979
11. Ruoss SJ, Hartmann T, Caughey GH (1991) Mast cell tryptase is a mitogen for cultured fibroblasts. J Clin Invest 88:493–499

12. Nadel JA (1991) Biology of mast cell tryptase and chymase. Ann N Y Acad Sci 629:319–331
13. Algermissen B, Hermes B, Feldmann-Böddeker I et al (1999) Mast cell chymase and tryptase during tissue turnover—analysis on in vitro mitogenesis of fibroblasts and keratinocytes and alterations in cutaneous scars. Exp Dermatol 8:193–198
14. Grabbe J, Welker P, Dippel E, Czarnetzki BM (1994) Stem cell factor, a novel cutaneous growth factor for mast cells and melanocytes. Arch Dermatol Res 287:78–84
15. Metcalfe DD, Baram D, Mekori YA (1997) Mast cells. Physiol Rev 77:1033–1079
16. Artuc M, Hermes B, Steckelings MU et al (1999) Mast cells and their mediators in wound healing—active participants or innocent bystanders? Exp Dermatol 8:1–16
17. Henz BM, Maurer M, Lippert U et al (2001) Mast cells as initiators of immunity and host defense. Exp Dermatol 10:1–10
18. Meng H, Marchese MJ, Garlick JA et al (1995) Mast cells induce T-cell adhesion to human fibroblasts by regulating intercellular adhesion molecule-1 and vascular cell adhesion molecule-1 expression. J Invest Dermatol 105:789–796
19. Gruber BL, Kew RR, Jelask A et al (1997) Human mast cells activate fibroblasts. Tryptase is a fibrogenic factor stimulating collagen messenger ribonucleic acid synthesis and fibroblast chemotaxis. J Immunol 158:2310–2317
20. Gailit J, Marchese MJ, Kew RR, Gruber BL (2001) The differentiation and function of myofibroblasts is regulated by mast cell mediators. J Invest Dermatol 117:1113–1119
21. Artuc M, Steckelings UM, Grützkau A et al (2002) A long-term coculture model for the study of mast cell-keratinocyte interactions. J Invest Dermatol 119:411–415
22. Lawrence ID, Warner JA, Cohan VL et al (1987) Purification and characterization of human skin mast cells. Evidence for human mast cell heterogeneity. J Immunol 139:3062–3069
23. Grützkau A, Henz BM, Kirchhof L et al (2000) Alpha-melanocyte stimulating hormone acts as a selective inducer of secretory functions in human mast cells. Biochem Biophys Res Commun 278:14–19
24. Guhl S, Artuc M, Neou A et al (2011) Long-term cultured human skin mast cells are suitable for pharmacological studies of anti-allergic drugs due to high responsiveness to FcεRI cross-linking. Biosci Biotechnol Biochem 75:382–384
25. Normand J, Karasek MA (1995) A method for the isolation and serial propagation of keratinocytes, endothelial cells, and fibroblasts from a single punch biopsy of human skin. In Vitro Cell Dev Biol Anim 31:447–455
26. Stark HJ, Baur M, Breitkreutz D et al (1999) Organotypic keratinocyte cocultures in defined medium with regular epidermal morphogenesis and differentiation. J Invest Dermatol 112:681–691

Chapter 9

The Absolute Basophil Count

Elena Borzova

Abstract

The absolute basophil count (cells/L) can be determined by manual counting of peripheral blood smears or using cell counting chambers as well as by automated hematology analyzers and fluorescence flow cytometry. Manual basophil counting of peripheral blood smears is currently regarded as the reference method, although the limitations of this method (distribution, observer, and statistical errors) are widely recognized. Automated hematology analyzers offer an advantage of larger numbers of counted cells and high throughput but are characterized by inconsistent analytical performance for basophil enumeration. Flow cytometric enumeration of circulating basophils using panels of monoclonal antibodies is being developed as novel candidate reference method for the absolute basophil count in peripheral blood. Basophil counting using fluorescence flow cytometry is characterized by high precision and statistical superiority. Emerging innovative technologies for absolute cell counts include imaging flow cytometry, mass cytometry, and on-chip blood counting, but their analytical performance for absolute basophil counts is yet to be established. Here, we describe various techniques for absolute basophil counting in peripheral blood including manual basophil counts in smears and hemocytometers and flow cytometric methodologies using double-platform, bead-based, and volumetric approaches.

Key words Basophils, Absolute basophil count, Metachromatic staining, Basopenia, Basophilia, Hemocytometer, Flow cytometry, Peripheral blood smear, Leukocyte differential

1 Introduction

The absolute basophil count is defined as the number of basophils per liter of peripheral blood. Traditionally, basophil counts were presented as a percentage of white blood cells in the manual leukocyte differential. Nowadays, the absolute cell count is recommended for clinical information [1]. In research settings, absolute basophil counts can be used to determine basophil concentrations in samples used for experiments.

The basophil is a granulocyte that occurs at one of the lowest frequencies in the peripheral blood (0.5–1% of circulating leukocytes), with an individual variation in basophil counts being a complex polygenic trait [2–5]. Therefore, precise determination of the normal range and the reference intervals for circulating

Bernhard F. Gibbs and Franco H. Falcone (eds.), *Basophils and Mast Cells: Methods and Protocols*, Methods in Molecular Biology, vol. 2163, https://doi.org/10.1007/978-1-0716-0696-4_9, © Springer Science+Business Media, LLC, part of Springer Nature 2020

basophils can be challenging. Basophil numbers were reported in the range of 20–80 cells/μL (0.02–0.08 × 10^9 cells/L) using microscopic examination of peripheral blood smears and cell-counting chamber techniques [2].

Traditionally, basophil counts greater than 50 cells per μL of peripheral blood (0.05 × 10^9 cells/L) were considered as basophilia, whereas basophil counts of less than 20 cells per μL (0.02 × 10^9 cells/L) were regarded as basopenia [6, 7]. Currently, absolute basophilia in the peripheral blood is defined as absolute basophil counts exceeding 100 cells per microliter of peripheral blood [8]. The term hyperbasophilia is used to define a persistent peripheral blood basophil count of equal or higher than 1000 basophils per microliter [8] (*see* **Note 1**).

Current guidelines (Clinical and Laboratory Standards Institute (CLSI) [9] recommend the statistical approach for defining the reference intervals as the central 95% distribution between lower and upper reference limits (2.5% above and below the interval) for all tests including the absolute basophil count. Published data regarding the reference intervals for absolute basophil counts is limited. A recent study of flow cytometric enumeration of basophils reported the reference interval for absolute basophil counts of 0.014–0.087 × 10^9/L [10]. A cutoff of more than 290 basophils per μL of peripheral blood (0.29 × 10^9 cells/L) was reported for screening for basophilia [11]. More studies are needed to establish and validate the reference intervals for peripheral blood basophils for clinical use.

Manual basophil counting using peripheral blood smears is still considered as the reference method [12]. However, the limitations of this method for cells at a frequency lower than 5%, such as basophils, are well recognized by the current guidelines [12]. In practice, a 100- or 200-cell manual differential count is used for basophil enumeration, although a 200-cell differential count by 2 observers on 2 slides from the same blood sample is recommended for reference studies by the H20-A2 guidelines from CLSI [12]. Basophil counting using manual microscopy of peripheral blood slides is characterized by insufficient precision due to slide distributional and observer errors as well as poor statistical reliability [1, 12–14]. Optimization measures, such as improving the quality of metachromatic Romanowsky dyes and the introduction of blood dispensers, automated slide makers, and automated digital microscopy for differential counting, have reduced distributional and observer errors. Nevertheless, inherent statistical sampling error of manual basophil counting of peripheral blood smears prompted a strong need for an improved reference method [12, 13].

Cell-counting chambers (hemocytometers) are still widely used in laboratories. Several cell-counting chambers are currently available in research practice including the improved Neubauer modified, Fuchs-Rosenthal, Thoma chambers, etc. The protocol using Kimura stain allows basophil enumeration in peripheral blood using cell-counting chambers [15]. Recent technological advances include automated cell counting for chamber slides although this technology is yet to be applied for basophil enumeration. In general, the use of cell counting chambers for manual basophil counts is limited by the statistical sampling error at low number of counted cells [16].

Automated basophil counts can be performed by automated hematology analyzers based on optical light scattering or electrical impedance-based technology [17]. Automated basophil counting offers the advantages of better statistical reliability due to counting a larger number of cells and higher throughput than manual microscopy. For hematology analyzers using differential cell lysis for basophil detection, the phenomenon of spurious pseudobasophilia can occur due to misclassification of plasma cells or abnormal lysis-resistant cells in some diseases including multiple myeloma and chronic leukemia [18]. In our experience, analytical performance of automated hematology analyzers for enumerating basophils appeared to be suboptimal and inconsistent in comparison with fluorescence flow cytometry [10] which is in keeping with findings of other research groups [14, 19].

Flow cytometry-based enumeration of peripheral blood basophils links multicolor immunophenotyping and absolute cell-counting technologies [20]. Several gating strategies are being evaluated for flow cytometric basophil counting [10, 14]. For example, basophil identification can be employed as a part of the monoclonal antibody panels for the five- or eight-part leukocyte differential [21, 22] or can be based on the use of basophil-specific markers for enumeration of basophil granulocytes only [10].

In flow cytometry, human basophils can be identified based on the following phenotypic characteristics: they are in the upper part of the lymphocyte scatter gate (toward the monocytes) and are CD45-medium (positive but clearly lower than other cells in the lymphocyte gate), lin-negative (CD2, CD3, CD14, CD16, CD19, CD20, CD56; however, a small fraction can be weakly positive for CD16), HLA-DR-negative (only a small fraction is weakly positive), CCR3-positive, CRTH2-positive, CD123-positive (at high levels), FcεRI-/IgE-positive (depending on serum IgE level), and CD203c-variable (depending on the activation state).

The markers best suited for basophil enumeration are discussed below:

CCR3: The only other cell type strongly positive for CCR3 are eosinophils which have very distinct light scatter characteristics. Although T cells can be CCR3-positive in tissues, CCR3-positive T cells are extremely rare in peripheral blood even under pathological conditions. Some authors argue that CCR3-positive T cells may co-localize with basophils in SSC^{low} gate [23]. Additionally, CCR3 expression may alter on basophil activation [23]. Although CCR3 can be used to identify basophils by a single marker, a combination with other markers (e.g., CD123) may result in a better correlation with automated complete blood count [24].

CRTH2: In addition to basophils and eosinophils, Th2-type T cells are CRTH2-positive. Thus, for the identification of basophils, a CD3 marker must be included to gate out Th2 cells. Thus, the combination of T-cell markers with CRTH2 is very useful if enumeration of basophils, eosinophils, and T-cell subsets is required.

CD123: The only other cell type expressing IL-3 receptors at such high levels as basophils are plasmacytoid dendritic cells which are strongly positive for HLA-DR. Thus, CD123 in combination with an HLA-DR marker is a useful gating strategy for human basophils. There is a controversy in the field as to whether the expression of CD1 may change following a basophil activation, thereby affecting the gating accuracy [23, 25].

FcεRI/IgE: Although used in some studies, staining of IgE or IgE-receptors does not allow the unequivocal identification of basophils in individuals with low-serum IgE concentrations or in patients treated with omalizumab [26, 27] (*see* **Notes 2** and **3**).

CD203c: CD203c is expressed on basophils and circulating mast cell progenitors in the peripheral blood. Basophil identification based on CD203c can be difficult because its expression varies from nearly negative to clearly positive depending on the activation state of these cells [23]. However, the combination of CD203c with another basophil marker such as CCR3 can be useful to study in vivo basophil activation, since CD203c surface expression is enhanced by a variety of IgE-dependent and IgE-independent stimuli.

Methods for basophil quantification by flow cytometry include single- or double-platform methodologies [20]. "Double-platform" methods allow basophil enumeration based on multiplying the percentage of basophils among leukocytes, as detected by flow cytometry, and absolute white blood cell count performed by an automated hematology analyzer [10, 28]. Absolute cell counting using microbead technology or volumetric assessment is referred to as "single-platform" method. The performance of single-platform counting techniques is characterized by lesser variation than that of a double-platform counting method [20, 28].

Flow cytometry-based absolute basophil counting offers advantages of greater precision, analysis of large number of cells, and high throughput and high specificity and, therefore, holds a promise of becoming a new candidate reference method instead of manual counting in peripheral blood smears [21]. At present, flow cytometric methods are being increasingly used for comparisons between methods for enumeration of circulating basophils [10, 14] although poor correlations were noted between absolute basophil counts derived by flow cytometry and manual microscopy [21, 22,

29]. Furthermore, disease-related changes in basophil immunophenotype (e.g., HLA-DR upregulation) [30] may interfere with the performance of this method. Imaging flow cytometry (ImageStream® technology) and mass cytometry (CyTOF, Fluidigm) have emerged as innovative approaches for cell classification and identification [31, 32], but their performance for basophil enumeration is yet to be established. Novel HematoFlow methodology (Beckman Coulter) combines the capabilities of automated hematology analyzer, flow cytometry, and Cytodiff™ staining panel for extended leukocyte differential and cellular analysis [24, 33]. This enables the prospect of basophil counting based on combined morphological and flow cytometric parameters in the future. Future perspectives for absolute basophil counting could be linked to the development of microfluidic lab-on-a-chip technology for the complete blood cell count and further development of digital applications for cell counting [34, 35].

From a clinical perspective, abnormal basophil counts were reported in various clinical conditions including allergic diseases, chronic spontaneous urticaria, and endocrine and myeloproliferative disorders [2, 6–8, 36]. Therefore, there is a potential for integrating absolute basophil counts in clinical decision-making for screening, diagnosis, and monitoring of these diseases in the future. In clinical practice, understanding the strengths and the limitations of counting methods for peripheral blood basophils can be important for accurate interpretation of the blood test results and for an informed choice of analytical procedures for absolute basophil count. Better morphological and immunophenotypical definition of peripheral blood basophils would enhance the clinical value of basophils as biomarkers in routine clinical practice. Further advances in the knowledge of basophil biology in health and disease will contribute to the development and validation of basophil-related biomarkers for clinical use.

2 Materials

2.1 Manual Basophil Counting

2.1.1 Hemocytometer Method

1. Optical microscope with ×40 objective.
2. Hemocytometer: improved Neubauer counting chamber or modified Fuchs-Rosenthal counting chamber.
3. Thick glass coverslip.
4. 1.5 mL Microcentrifuge tubes.
5. Micropipette calibrated with a maximum capacity of 200 μL.
6. Whatman Grade 41 filter paper—110 mm diameter.
7. Kimura stain [12, 27]:

(a) Prepare a stock toluidine blue solution by dissolving 0.05 g of Toluidine blue (Fluka), in 50 mL of 1.8% NaCl solution and 11 mL 96% ethanol. Dilute with distilled water to a final volume of 100 mL.

(b) Prepare a stock solution (0.03%) of light-green yellowish SF (Fluka) in distilled water.

(c) Prepare a saturated solution of saponin in 50% ethanol with continuous stirring until orange coloration of the solution emerges [37].

(d) 67 mM Sodium phosphate buffer, pH 6.4.

(e) Prepare Kimura stain by mixing 11 mL of 0.05% toluidine blue solution, 0.8 mL of 0.03% light-green solution, 0.5 mL of saturated saponin solution, and 5 mL of 67 mM phosphate buffer (pH 6.4). Prepare Kimura stain freshly each week. Store at room temperature protected from the light by aluminum foil.

2.1.2 Peripheral Blood Smears

1. Absolute methanol.
2. May-Grünwald-Giemsa stain:
   (a) Prepare Staining Solution I: 0.3 g May-Grünwald powder in 100 mL absolute methanol, and filter before use [13]. The stain should be stored in a closed container at room temperature for 24 h (do not freeze Romanowsky dyes [13]). Handle the stain with care, and avoid skin contact in view of potential carcinogenicity [13].

   (b) Prepare Staining Solution II (Giemsa stain): Dissolve 1 g Giemsa stain powder in 66 mL glycerol, and heat to 56 °C for 90–120 min. Add 66 mL of absolute methanol, mix thoroughly, and filter before use [13]. The stain should be stored in a closed container at room temperature [13]. Prepare fresh twice a day if needed.

   (c) Sörensen's buffer at pH 6.8 (*see* below).
3. Wright's stain [13]:
   (a) Dissolve 0.3 g of Wright's stain powder in 100 mL of methanol, and filter before use. The stain should be stored tightly sealed in a closed container at a room temperature for 24 h (*see* **Note 4**). Do not freeze. Handle Wright stain with care, and avoid skin contact because of potential carcinogenicity.

   (b) Sörensen's buffer solution at pH 6.4: Dissolve 6.63 g anhydrous KH_2PO_4 and 2.56 g anhydrous Na_2HPO_4 in 1 L distilled water.
4. Frosted glass microscope slides (75 × 25 mm).
5. Manual staining rack or glass Coplin staining jars.
6. Light microscope with ×10 and ×40 objectives.

2.2 Automated Basophil Counting Using Flow Cytometry

1. Vacutainer with K_2EDTA or sodium citrate (*see* **Note 5**).
2. Directly conjugated monoclonal antibodies:
   (a) Monoclonal antibody panel 1:
      5 μL CD123-PE (BD Biosciences).
      10 μL HLA-DR-FITC (BD Biosciences).
   (b) Monoclonal antibody panel 2:
      10 μL anti-CCR3-PE (R&D).
3. Lysing Solution (BD Biosciences).
4. CellWash® (BD Biosciences).
5. CellFix® (BD Biosciences).
6. 5 mL FACS polypropylene tubes.
7. Cell counting microbeads (e.g., BD TRUCount Beads (BD Biosciences), CountBright™ Absolute Counting Beads (Invitrogen), or Cytocount™(Dako), etc.).
8. High-performance flow cytometer (optimal for multicolor immunophenotyping).
9. Automated hematology analyzer.

3 Methods

3.1 Manual Absolute Basophil Counts in Whole Blood Stained with Kimura Stain Using Cell-Counting Chambers

1. Mix freshly drawn anticoagulated venous blood with Kimura stain in a proportion 1:10 by gentle pipetting [15]. Avoid vortexing (*see* **Note 6**).
2. Adjust a thick coverslip to the counting chamber until Newton's rings (concentric alternating dark and light rings pattern) appear.
3. Load each chamber of an improved Neubauer hemocytometer with 10 μL of diluted blood by a capillary tube or a micropipette.
4. Leave the loaded counting chamber on a bench for 1–2 min before cell counting.
5. Examine the counting chamber under the microscope. Focus the microscope on the grid of the counting chamber at low power magnification. Perform basophil count in four to five large squares of the counting chamber under 40× high dry objective and the blue (daylight) filter. Count cells touching upper and left borders of the counting chamber but not cells touching lower and right borders.

   To obtain basophil counts per mL using the improved Neubauer chamber, multiply average cell count per square by the dilution factor and 10^4.

3.2 Manual Absolute Basophil Counting Procedure Using a Peripheral Blood Smear

3.2.1 Prepare a Peripheral Blood Smear

1. Mix anticoagulated blood in the tube before preparing a smear. Prepare a peripheral blood smear as quickly as possible after blood collection (preferably within 2 h) [38].
2. Use a glass capillary pipette or a blood dispenser for vacutainers to place a 2–3 mm drop of blood on a glass microscope slide approximately 1 cm from the frosted end.
3. Use another glass microscope slide as a spreader slide. Position the spreader slide at an angle of 45° in front of the drop of blood. Pull the spreader slide back toward the drop of blood at an angle of 45°. Allow the blood to spread along the edge of the spreader slide by capillary action.
4. Push the spreader slide forward at an angle around 30° with a swift smooth motion to prepare the smear (*see* **Note** 7).
5. Allow the smear to air dry at room temperature for about 10 min. Fix the air-dried smear by flooding with absolute methanol for 2 min. Remove methanol by tilting the slide. Warning: methanol is highly toxic even in small amounts.
6. Ensure that peripheral blood smear occupies 1/2–3/4 of the slide length [38], is wedge shaped, and has a smooth rounded feathered end. The blood smear should not touch any slide edges and should have a sufficient area for morphological evaluation [10] where red blood cells close to each other but do not touch or overlap [39].

3.2.2 Staining a Peripheral Blood Smear with May-Grünwald-Giemsa Stain

1. Immerse the slide in May-Grünwald staining solution I (diluted with an equal volume of Sörensen's buffer, pH = 6.8) for 5 min in a Coplin jar or a staining rack.
2. Transfer the slide from a jar with May-Grünwald staining solution I to a jar with staining solution II (Giemsa stain) without washing.
3. Cover the slide with Giemsa staining solution II (diluted with nine parts Sörensen's buffer, pH = 6.8) for 10–15 min.
4. Remove the slides from the jar, and rinse the slide with Sörensen buffer (pH 6.8).
5. Wash with deionized water.
6. Place the slide into a jar with water for 2–5 min.
7. Allow the slides to air-dry in a tilted position. Do not blot dry the slide.
8. Mount the slide with a cover glass if necessary (*see* **Note 8**).

3.2.3 Staining a Peripheral Blood Smear with Wright Stain

1. Place the slide with a fixed peripheral blood smear on the horizontal staining rack.
2. Flood the slide with Wright stain for 2 min. Do not rinse off Wright stain from the slide.

3. Add an equal volume of the Sörensen's buffer (pH = 6.4) to Wright stain on the slide.
4. Mix the stain solution and the buffer gently without touching the smear on the slide, and observe a metallic sheen on the surface of the staining solution mixture. Make sure that the staining solution mixture does not run off the slide.
5. Leave for 3 min.
6. Rinse the slide with deionized water for 30 s.
7. Allow the slide to air-dry in a tilted position; do not blot-dry.
8. Mount the slide with a cover glass if required.

3.2.4 Evaluation of Stained Peripheral Blood Smears

1. Examine the slide at a low magnification (10× objective).
2. For optimal differential staining for Romanowsky dyes, the nuclei of white blood cells should be stained in bluish purple [39, 40]:
   Erythrocytes—pale to moderate pink.
   Neutrophils—pale pink cytoplasm with reddish to light purple granules.
   Eosinophils—blue cytoplasm with orange-red granules.
   Basophils—deep purple or dark violet granules.
   Lymphocytes—pale blue cytoplasm.
   Monocytes—bluish gray cytoplasm.
   Platelets—violet to purple granules in light-blue cytoplasm.
3. Check the lateral edges for leukocyte distribution and basophil clumping (*see* **Note 9**).
4. Choose the examination area of the blood smear where red blood cells are close to each other and do not overlap (in at least 50%).
5. Review the slide under higher magnification (40× objective).
6. Obtain an estimate of the white blood cell count (cells/mL) by counting the total number of leukocytes in ten 40× objective fields. For WBC estimate (WBC/mL), multiply the average number of counted WBC per field by 2000 [41] (*see* **Note 10**).
7. Examine the smear under high oil magnification (100× objective) for WBC differentiation. Check the distribution of red blood cells of 200–250 per 100× oil immersion field [41]. Perform enumeration of at least 100 consecutive leukocytes; usually the 200 cell differential is used. (For validation studies, 2200-cell differentials on the same slide by 2 observers are optimal.) For differential cell counting, use a "battlement track" technique to ensure that each cell is counted only once [39] (*see* **Note 11**). Basophil counts are expressed as the percentage of the total white blood cell count.

3.3 Absolute Basophil Counts in Whole Blood Using Flow Cytometry Labeling

1. Obtain EDTA-anticoagulated venous blood.
2. Incubate 100 μL of whole peripheral blood with the combinations of monoclonal autoantibodies for 15 min at room temperature.
3. Add 5 μL CD123-PE, 10 μL HLA-DR-FITC (BD Biosciences), or 10 μL anti-CCR3-PE (R&D Systems).
4. Lyse erythrocytes with Lysing Solution® (BD Biosciences) as per manufacturer's recommendations.
5. Wash with CELLWASH® (BD Biosciences) as per manufacturer's specifications (*see* **Note 12**).
6. Fix the sample with CellFIX® (BD Biosciences) as per manufacturer's recommendations if the sample analysis is delayed for 2–4 h.
7. Acquire data of 50,000 events with a flow cytometer.
8. Analyze using flow cytometric analysis software.
9. Carry out an absolute basophil count using a double-platform method, a volumetric method, or cell-counting microbeads [17] (*see* **Note 13**).

3.3.1 Double-Platform Counting Method

1. Obtain the percentage of gated basophils in the sample by flow cytometric immunophenotyping.
2. Obtain an absolute white blood cell count using an automated hematology analyzer.
3. Calculate an absolute basophil count by multiplying the percentage of gated basophils and an absolute white blood cell count and dividing by 100.

3.3.2 Single-Platform Microbead-Based Method

1. Add cell counting microbeads (CountBright™ Absolute Counting Beads (Molecular Probes), Flowcount beads (Beckman Coulter), Flow Cytometry Absolute Count Standard™ (Bangs Laboratories, Inc.), etc.) to the sample, or add the sample to the test tubes with reference fluorescent microbeads (TRUCount™ tubes, BD Biosciences) as per manufacturer's specification. Use the same pipette and precision reverse pipetting for dispensing blood sample and cell counting beads.
2. Add the beads to the sample just before data acquisition by flow cytometry. Mix carefully the sample and beads by gentle pipetting.
3. Record the known concentration of the beads.
4. Add the same amount of the beads to a control buffer sample of the same volume. Acquire the data to place a gate for a bead population in the control sample. Use this gate to obtain the bead population in the blood sample.

5. Acquire the data in the whole blood sample. Make sure that at least 1000 events are acquired in the gate for cell counting beads.
6. Obtain the percentage of the gated cell-counting beads in the sample. Obtain the number of events in the gate for basophils in the sample.
7. Calculate the absolute basophil count (cells/μL) in the sample by dividing the number of events in the gate for basophils by the number of bead events, and multiply by the known concentration of cell-counting beads.

3.3.3 Single-Platform Volumetric Method

Some of the flow cytometers (Accuri™ C6 (BD Biosciences), Guava™ easyCyte™ (Merck Millipore), etc.*) allow a precise measurement of the sample volume and thereby volumetric cell counting.*

1. Pipette the blood sample in the test tube. Use "wet tip" reverse pipetting [20] to ensure accurate pipetting. Obtain the sample volume (*see* **Note 14**).
2. Acquire the data in the sample. Obtain the number of events in the gate for basophils.
3. Calculate the absolute basophil count as a ratio of the number of events in the gate for basophils to the sample volume (*see* **Note 15**).

4 Notes

1. Persistent basophilia (at least 8 weeks) may be one of the early signs of myeloproliferative disorders [8, 42, 43]. Unexplained hyperbasophilia in peripheral blood should trigger a detailed hematological workup [8, 42, 44, 45]. Firstly, patients undergo noninvasive investigations including testing for *BCR-ABL1* and *JAK2 V617F*, allergy testing, serum tryptase, etc. [8]. Secondly, if no underlying disease is detected, invasive investigations can be carried out such as bone marrow investigations and cytogenetic and molecular studies [8]. If lymphoma is suspected, radiologic studies, computed tomography, T-cell receptor, and Ig receptor rearrangement analysis are warranted [8]. In patients with myelodisplastic syndromes, the presence of basophilia (>250 cells/μL) and eosinophilia (>350 cells/μ L) was shown by multivariate analysis to have a prognostic significance for reduced survival [46].
2. In patients with allergic diseases, basophil numbers can be elevated but may rapidly decline in acute and severe clinical manifestations (asthma [15], anaphylaxis [7]). Total absence of basophils and eosinophils was reported in peripheral blood, bone marrow, and the skin of a patient with chronic urticaria

and vitiligo [47]. Basophil counts were also noted to be inversely related to the severity of chronic spontaneous urticaria as examined by metachromatic staining [29]. It is unknown as to whether these observations reflect a decrease in basophil counts in severe manifestations of these diseases, whether metachromatic staining is affected by basophil degranulation [19] or whether basopenia is a result of migration of basophils from the circulation into the skin [26]. Therefore, flow cytometric enumeration of circulating basophils may be preferable in disorders associated with basophil degranulation. Furthermore, serial assessments of absolute basophil counts in these patients may be helpful to monitor the course of the disease.

3. Basophil counts, as measured by flow cytometry, are shown to increase during the treatment with omalizumab in patients chronic spontaneous urticaria [26]. Additionally, the FcεRI receptor density on basophils is rapidly declined following treatment with omalizumab [27].
4. Romanowsky dyes are characterized by inherent instability and high precipitate formation [48]. On standing, certain components of Romanowsky dyes can be oxidized leading to variation in their composition including azure B content and its ratio to eosin [38, 48]. Also, minimal standing time is crucial to avoid precipitation in Romanowsky dyes and onto slides during staining [38, 48].
5. The choice of anticoagulant can be important for absolute basophil count. Dry K_2EDTA is recommended as anticoagulant of choice for full blood cell counts by current guidelines. This solution is spray-dried by the manufacturers on the interior surface of the blood collection tubes. For liquid anticoagulants, such as K_3EDTA or sodium citrate, the dilution factor should be taken into account. Heparin is not recommended to avoid pre-analytical activation of platelets and artifacts in the staining of cells [49].
6. Basophil identification can be affected by sample handling techniques, fixation procedures, and exposure to water-based diluents. Pre-analytical sample handling should involve gentle pipetting and avoid vortex mixing. Exposure to water-based diluents affects basophil identification by metachromatic staining and should be avoided before fixation in view of high water solubility of their granules [6]. Basophils tend to form clumps [7]; therefore, use gentle pipetting to mix the sample before data acquisition to prevent cellular aggregates.
7. In slide preparation, it is important that the drop of blood is pulled by the spreader slide rather than pushed ahead. The smear should be gradually spreading from a thicker base (closer to the frosted end of the slide) to a thinner feathered end.

8. Slide mounting with a cover glass can be considered if there is a need for a slide review by several examiners or long-term storage [13]. Slide mounting and coverslipping can protect the slides from mechanical damage or dye deterioration and reduce the biohazard when handling the slides [13].
9. In wedged slides, basophils tend to be unevenly distributed toward the feathered end of the slide [7, 13]. This may result in basophil underestimation by manual microscopy of peripheral blood smears compared to flow cytometric enumeration [13]. The use of automated slide makers may improve cell distribution on the slide [1].
10. The WBC estimate is used for internal quality control only but not for clinical use [41].
11. On microscopic examination of the peripheral blood slides, it is important to make sure that each cell is counted only once. For this purpose, a "battlement track" technique is used [50]. Using this approach, cells in the smear area selected for the morphological examination are counted in three to five fields between the edges of the smear back and forth in zigzag pattern until the necessary number of cells has been counted [50]. Following "battlement track" pattern ensures that cells distributed toward the edges of the slide will be included in the cell count thereby accounting for distributional error [50].
12. In sample preparation for flow cytometric assays, lyse-and-wash techniques are mostly used for double-platform counting method, especially for surface markers with low-level expression [28]. Lyse-and-wash technique offers the advantage of lower fluorescence background but introduces cell losses that may affect the absolute cell count [28]. Using lyse-and-wash sample preparation leads to the loss of the original relationship between absolute cell count and the initial sample volume thereby affecting the precision of cell count [28].

    The lyse-no-wash technique is widely used for single-platform cell counting methods and is characterized by minimal sample manipulation [28]. Sample preparation using the lyse-no-wash technique enables absolute cell counting in relation to the initial sample volume by correcting for known dilution factors [28]. However, lyse-no-wash techniques have the drawbacks of higher background fluorescence and cell debris contamination [28]. Both techniques may have limitations due to the use of lysing solutions which may affect basophil scatter characteristics or their surface marker expression [28].
13. Flow cytometric enumeration of circulating basophils using monoclonal antibodies can be particularly helpful for patients with basophil counts at the lower reference limit. Basophil

counts at the lower detection limit by manual and automated counting is characterized by increasing imprecision.

14. For single-platform methodologies, accurate pipetting should be ensured by using "wet-tip" reverse pipetting. This technique is achieved by pressing the micropipette plunger to the second stop, then placing the tip in the sample and aspirating the sample and then dispensing the sample by pressing the plunger to the first stop [20]. Repeat this technique with the same pipette tip at least twice before dispensing the sample ("wet-tip" pipetting). Careful and regular calibration of the micropipettes is essential for precise pipetting [20].

15. In clinical practice, the proportional (percentage) leukocyte differential is being replaced by absolute cell count as more informative and of greater clinical significance than the cell percentage [1, 12]. In clinical settings, the repeated differential count at the interval of less than a week on the same patient was proven not to be clinically necessary in most patients except certain clinical situations such as bleeding, severe infections, and monitoring chemotherapy [1].

References

1. Pierre RV (2002) Peripheral blood film review. The demise of the eyecount leukocyte differential. Clin Lab Med 22:279–297
2. Galli SJ, Metcalfe DD, Dvorak AM (2010) Basophils and mast cells and their disorders. In: Beutler E, Lichtman MA, Coller BS, Kipps TJ, Seligsohn U (eds) Williams' Hematology, Edn 8. McGraw-Hill, New York, pp 915–932
3. Falcone FH, Haas H, Gibbs BF (2000) The human basophil: a new appreciation of its role in immune responses. Blood 96:4028–4038
4. Tajuddin SM, Schick UM, Eicher JD et al (2016) Large-scale exome-wide Associatoin analysis identifies loci for white blood cell traits and pleiotropy with immune-mediated diseases. Am J Hum Genet 99(1):22–39
5. Jain D, Hodonsky CJ, Schick UM et al (2017) Genome-wide association of white blood cell counts in Hispanic/Latino Americans: the Hispanic community health study/study of Latinos. Hum Mol Genet 26(6):1193–1204
6. Parwaresch MR (1976) The human blood basophil. Morphology, origin, kinetics, function and pathology. Springer-Verlag, Berlin/Heidelberg, p 235
7. Shelley WB, Parnes HM (1965) The absolute basophil count. JAMA 192:368–370
8. Valent P, Sotlar K, Blatt K et al (2017) Proposed diagnostic criteria and classification of basophilic leukemias and related disorders. Leukemia 31:788–797
9. Horowitz GL, Altaie S, Boyd JC et al (2008) Clinical and laboratory standards institute (CLSI). Defining, establishing and verifying reference intervals in the clinical laboratory; approved guideline—third edition. CLSI document EP28-A3. Clinical and Laboratory Standards Institute, Wayne, P.A.
10. Ducrest S, Meier F, Tschopp C, Pavlovic R, Dahinden CA (2005) Flowcytometric analysis of basophil counts in human blood and inaccuracy of hematology analyzers. Allergy 60:1446–1450
11. Mitre E, Nutman T (2003) Lack of basophilia in human parasitic infections. Am J Trop Med Hyg 69:87–91
12. Koepke JA et al (2007) Clinical and laboratory standards institute (CLSI). Reference leukocyte (WBC) differential count (proportional) and evaluation of instrument methods; approved standard—second edition. CLSI document H20-A2. Clinical and Laboratory Standards Institute, Wayne, PA
13. Houwen B (2002) Blood film preparation and staining procedures. Lab Hematol 6:1–7
14. Amundsen EK, Henriksson CE, Holthe MR, Urdal P (2012) Is the blood basophil count sufficiently precise, accurate and specific? Am J Clin Pathol 137:86–92

15. Kimura I, Moritani Y, Tanizaki Y (1973) Basophils in bronchial asthma with reference to reagin-type allergy. Clin Allergy 3:195–202
16. Falcone FH (2004) Basophils and immunity to parasites: an update. Rev Fr Allergol 44:14–22
17. Buttarello M, Plebani M (2008) Automated blood cell counts. State of the art. Am J Clin Pathol 130:104–116
18. Hur M, Lee YK, Lee KM et al (2004) Pseudobasophilia as an erroneous white blood cell differential count with a discrepancy between automated cell counters: report of two cases. Clin Lab Haem 26:287–290
19. Lesesve J-F, Benbih M, Lecompte T (2005) Accurate basophil counting: not an easy goal! Clin Lab Haem 27:143–144
20. Storie I (2006) Absolute cell counting. In: Pursley SW et al (eds) Guide to flow cytometry. Dako, Philadelphia, pp 107–111
21. Roussel M, Davis BH, Fest T, Wood BL (2012) International Council for Standardization in Hematology (ICSH). Toward a reference method for leukocyte differential counts in blood: comparison of three flow cytometric candidate methods. Cytometry A 81:973–982
22. Cherian S, Levin G, Lo WY et al (2010) Evaluation of an 8-colour flow cytometric reference method for white blood cell differential enumeration. Cytometry B Clin Cytom 78:319–328
23. Chirumbolo S (2016) Commentary; the expression of CD123 can decrease with basophil activation: implications for the gating strategy of the basophil activation test. Front Immunol 7:260
24. Kim Z, Choi BS, Kim JK, Won DI (2016) Basophil markers for identification and activation in the indirect basophil activation test by flow cytometry for diagnosis of autoimmune urticaria. Ann Lab Med 36(1):28–35
25. Santos AF, Becares N, Stephens A et al (2016) The expression of CD123 can decrease with basophil activation: implications for the gating strategy of the basophil activation test. Clin Transl Allergy 6:11
26. Saini SS, Omachi TA, Trzaskoma B et al (2017) Effect of omalizumab on blood basophil counts in patients with chronic idiopathic/spontaneous urticaria. J Invest Dermatol 137:958–961
27. Jörg L, Pecaric-Petkovic T, Reichenbach S et al (2018) Double-blind placebo-controlled trial of the effect of omalizumab on basophils in chronic urticaria patients. Clin Exp Allergy 48:196–204
28. Brando B, Barnett D, Janossy G et al (2000) Cytofluorometric methods for assessing absolute numbers of cell subsets in blood. European working group on clinical cell analysis. Cytometry 42:327–346
29. Grattan CE, Dawn G, Gibbs S, Francis DM (2003) Blood basophil numbers in chronic ordinary urticaria and healthy controls: diurnal variation, influence of loratadine and prednisolone and relationship to disease activity. Clin Exp Allergy 33:337–341
30. Charles N, Hardwick D, Daugas E, Illei GG, Rivera J (2010) Basophils and the T helper 2 environment can promote the development of lupus nephritis. Nat Med 16:701–707
31. Zuba-Surma EK, Mariusz ZR (2011) Analytical capabilities of the ImageStream cytometer. Meth Cell Biol 102:207–230
32. Tordesillas L, Rahman AH, Hartmann BM, Sampson HA, Berin MC (2016) Mass cytometry profiling the response of basophils and the complete peripheral blood compartment to peanut. J Allergy Clin Immunol 138 (6):1741–1744
33. Kim JE, Kim BR, Woo KS, Han JY (2012) Evaluation of the leukocyte differential on a new automated flow cytometry hematology analyzer. Int J Lab Hematol 34(5):547–550
34. Heikali D, Di Carlo D (2010) A niche for microfluidics in portable hematology analyzers. J Assoc Lab Autom 15:319–328
35. Thurman AC, Davis JL, Jan M et al (2015) Development and validation of an app-based cell counter for use in the clinical laboratory setting. J Pathol Inform 6:2
36. Denburg JA, Wilson WE, Bienenstock J (1982) Basophil production in myeloproliferative disorders: increases during acute blastic transformation of chronic myeloid leukemia. Blood 60:113–120
37. Sanz M-J, Jose PJ, Williams TJ (2000) Measurement of eosinophil accumulation in vivo. In: Proudfoot AEI, Wells TNC, Power CA (eds) Meth Mol Biol (138): Chemokine Protocols. Humana Press, Totowa, New Jersey, pp 275–283
38. Woronzoff-Dashkoff KK (2002) The Wright-Giemsa stain: secrets revealed. Clin Lab Med 22:15–23
39. Barth D (2012) Approach to peripheral blood film assessment for pathologists. Semin Diagn Pathol 29:31–48
40. Bain BJ (2007) Blood sampling and blood film preparation and examination. In: Blood cells: a practical guide, 4th edn. Blackwell Publishing, Hoboken, New Jersey, pp 1–20
41. Maedel LB, Doig K (2008) Examination of the peripheral blood smear and correlation with the complete blood count. In: Rodak BF,

Fritsma GA, Doig K (eds) Hematology: clinical principles and applications, 3rd edn. Sauders Elsevier, Philadelphia, pp 175–190
42. Nau RC, Hoagland HC (1971) A myeloproliferative disorder manifested by persistent basophilia, granulocytic leukemia and erythroleukemic phases. Cancer 3:662–665
43. Howden B (2001) The differential cell count. Lab Hematol 7:89–100
44. George TI (2012) Malignant or benign leukocytosis. Hematology Am Soc Hematol Educ Program 2012:475–484
45. Tang G, Woods LJ, Wang SA et al (2009) Chronic basophilic leukemia: a rare form of chronic myeloproliferative neoplasm. Hum Pathol 40:1194–1199
46. Wimazal F, Germing U, Kundi M, Noesslinger T, Blum S, Geissler P, Baumgartner C, Pfeilstoecker M, Valent P (2010) Evaluation of the prognostic significance of eosinophilia and basophilia in a larger cohort of patients with myelodysplastic syndromes. Cancer 116:2372–2381
47. Juhlin L, Venge P (1988) Total absence of eosinophils in a patient with chronic urticaria and vitiligo. Eur J Hematol 40(4):368–370
48. Horobin RW (2011) How Romanowsky stains work and why they remain valuable—including a proposed universal Romanowsky staining mechanism and a rational troubleshooting scheme. Biotech Histochem 86:36–51
49. Macey M, Azam U, McCarthy D, Webb L et al (2002) Evaluation of the anticoagulants EDTA and citrate, theophylline, adenosine and dipyridamole (CTAD) for assessing platelet activation on the ADVIA 120 hematology system. Clin Chem 48:891–899
50. Tefferi A, Hanson CA, Inwards DJ (2005) How to interpret and pursue an abnormal complete blood cell count in adults. Mayo Clin Proc 80:923–936

Part III

Basophil/Mast Cell Lines

Chapter 10

Mast Cell and Basophil Cell Lines: A Compendium

Egle Passante

Abstract

Mast cells and basophils play a crucial role during type I hypersensitivity reactions. However, despite efforts to elucidate their role in the pathogenesis of allergy and inflammation, our understanding of MC and basophil biology is still relatively scarce. The practical difficulty in obtaining a sufficient number of purified primary cells from biological samples has slowed down the process of reaching a full understanding of the physiological role of these functionally similar cell types. The establishment of several immortalized cell lines has been a useful tool to establish and perform sophisticated laboratory protocols that are impractical using primary cells. Continuous cell lines have been extensively used to investigate allergen/IgE-mediated cell activation, to elucidate the degranulation dynamics, to investigate structural and functional properties of the high-affinity receptor (FcεRI), and to test cell-stabilizing compounds. In this chapter, we review the most widely used and better-characterized MC and basophil cell lines, highlighting their advantages and drawbacks. It must be pointed out, however, that while cell lines represent a useful in vitro tool due to their easy manipulability and reduced culture costs, they often show aberrant characteristics which are not fully representative of primary cell physiology; results obtained with such cells therefore must be interpreted with due care.

Key words Mast cells, Basophils, Cell lines, IgE receptors (FcεRI), Humanized rat basophilic leukemia cells, RBL, LAD-2, HMC-1, KU812, LUVA, ROSA

1 Mast Cells and Basophils, an Introduction

Mast cell (MC) and basophil cell lines are widely used as a tool for diagnostic assays to evaluate the biological activity of allergen extracts and to assess quality control protocols and batch analysis control of allergen preparations [1–3]. Their primary cell counterparts are, in fact, key effectors in allergic diseases and inflammatory conditions such as hypersensitivity reactions, asthma, and atopic dermatitis. Primary MCs are also key players in autoimmune diseases such as Crohn's disease, inflammatory bowel disease, and ulcerative colitis [4–6]. Mast cells' physiological roles are numerous; not only do they regulate inflammation, but they also regulate wound healing, fibrosis and tissue remodeling, innate and adaptive immune responses, and angiogenesis [7, 8].

Bernhard F. Gibbs and Franco H. Falcone (eds.), *Basophils and Mast Cells: Methods and Protocols*, Methods in Molecular Biology, vol. 2163, https://doi.org/10.1007/978-1-0716-0696-4_10, © Springer Science+Business Media, LLC, part of Springer Nature 2020

Both MCs and basophils release a large number of preformed and newly synthesized pro-inflammatory mediators following cross-linking of IgE-bound FcεRIs by multivalent allergens [9]. When cross-linked, FcεRIs activate an intracellular cascade that leads to the exocytosis of granules and the production of de novo mediators. Preformed mediators include histamine, heparin, tryptases, and chymases, while newly synthesized factors include prostaglandins, leukotrienes, and cytokines such as TNF-α, IL-1, IL-4, and IL-10 [7–9]. To note traces of TNF-α have been found associated with MC intracellular granules and released quickly following MC stimulation; however the bulk of the produced TNF-α derives from de novo synthesis [8]. Mukai and co-workers in their recent review compiled an exhaustive list of mediators produced by MC lineages and subsets [8]; however, the authors suggest that, while it is feasible assessing whether a particular lineage or subset is a source of a specific mediator, it is still unclear which role such mediators may play in physiological or pathological conditions [8]. Most of the studies investigating MCs as a source of cytokines and chemokines have been performed in in vitro settings, and therefore further studies are needed to ascertain the dynamics and physiological importance of mediator release by different MCs subsets in vivo.

Mast cells can also release chemotactic factors that can promote the recruitment, proliferation, survival, and activation of other figurate elements of the immune system. For example, CCL3 promotes tissue migration of macrophages and neutrophils [10], and IL-16 can recruit $CD4^+$ T cells [11]. Mukai and co-workers have recently reported a comprehensive list of MC-derived chemokines [8].

In the light of their anatomical distribution, mediator content, and susceptibility to a vast array of stimuli, it does not surprise that MC and basophil roles are not limited to the initiation of allergic responses but that they play more complex roles in adaptive and innate immune responses [12].

Mast cells can be found in the proximity of structures at the interface of the body with the external environment (i.e., lungs, intestine, blood vessels, and nerves), while basophils are short lived, circulating cells that originate and mature in the bone marrow and infiltrate the site of infection/inflammation following activation of pro-inflammatory pathways. Mast cells and basophils can also be activated directly by pathogens through a family of pattern recognition receptors called "toll-like receptors" (TLRs) [12, 13] and are the first activators of inflammatory response, not only releasing cytokines and chemoattractants for other immune cells such as neutrophils and macrophages but also phagocytizing the invading pathogens and presenting antigens to T cells [14]. The role of basophils during host defense is less well known, but they are thought to sustain the inflammatory process with the release of immunoregulatory cytokines such as IL-4, IL-13, and TNF-α [15].

Mast cell and basophil research has often been hindered by practical difficulties in obtaining large amounts of homogeneous cells from biological samples. Mast cells can be obtained either by enzymatic digestion of sample tissues (*see* Chapter 3) followed by density gradient centrifugation or from circulating progenitors following in vitro differentiation (*see* Chapters 4 and 5). Basophils, on the other hand, represent less than 1% of the white cell population and their isolation and purification can consist of expensive, elaborate, and tedious multiple-step protocols (*see* Chapter 2). Such purification protocols often suffer from practical pitfalls resulting in cells with low viability and purity. Furthermore, the limited yield of basophils and MCs following such isolation procedures involves the necessity of large amounts of blood/tissue with consequent ethical and practical drawbacks. In the light of this, it is not surprising that efforts have been made to generate immortalized cell lines that could offer a practical and flexible tool to investigate several aspects of MC and basophil biology such as activation, signaling, and survival. In the early 1970s, the establishment of the first rat basophilic MC line (RBL-2H3) signaled a turning point providing scientists with a suitable tool to develop a solid and reliable experimental system for routine experiments and innovative protocols [16–18]. The availability of an easy-to-cultivate basophilic MC line allowed the development of a systematic approach to characterize fundamental aspects of MC/basophil molecular biology [17]. Since then, several MC lines have been established, all showing different physicochemical properties, expression levels of surface markers and receptors, as well as intracellular contents.

It must be stressed, however, that immortalized cells, despite facilitating several practical aspects of research, show abnormalities and aberrant characteristics. Furthermore, misidentification, contamination issues, and intra-laboratory reproducibility due to different culture conditions are limitations that afflict the use of cell lines. Results obtained from such studies need therefore to be interpreted with the necessary care.

This chapter provides a general overview of the development of the most commonly used MC and basophil lineages from their inception, and it goes on to describe their major applications in experimental settings. This review is intended as a guide only and it is far from being a comprehensive description of cytochemical properties of each lineage.

1.1 Human Mast Cell Lines

Despite efforts to establish long-term human MC cultures, at present there is a limited number of immortalized lineages. For more than 10 years, the only established human MC line was the so-called human MC line-1 (HMC-1) [18]. HMC-1 has been widely characterized and exploited to investigate MC degranulation and the anti-inflammatory properties of countless compounds, but it has also been used to investigate the molecular dynamics of several intracellular pathways involved in MC activation and survival. In the early 2000s, Kirshenbaum and co-workers isolated two MC lineages (LAD-1 and LAD-2) that more closely resemble the primary culture of CD34^{+}-derived human MCs [19]. Culturing conditions for these cells however have prohibitive costs and therefore LAD-1 and LAD-2 often are not the first choice for routine experiments. In 2011 Laidlaw [20] and co-workers published results on a newly established human MC line (LUVA) that spontaneously arose during culture of peripheral blood CD34^{+} cells from a healthy donor. LUVA cells have been reported to degranulate after IgE receptor cross-linkage [20] and present a phenotype of fully mature MCs. More recently, Saleh and co-workers have added two new cell lines to the arsenal of lineages that researchers can exploit. The authors reported to have generated both an FcεRI-positive, stem cell factor (SCF)-dependent human lineage (ROSA$^{kit\ wt}$) and an SCF-independent subclone (ROSA$^{kit\ D816V}$). These two new cell lines represent a flexible and useful tool to investigate aberrations of the MC biology due to a constitutive activation of the KIT receptor. In the next sections, we summarize the major features of the most used human cell lines.

1.2 HMC-1 and HMC-1 Sublines

The HMC-1 cell line was originally established from the peripheral blood of a patient with MC leukemia [19–22]. Despite the fact that HMC-1 cells' phenotype is typical of immature MCs, HMC-1 cells show, even though at low levels, several hallmarks that characterize mastocytes [22, 23]. It has been demonstrated that HMC-1 cells express many of the typical MC surface antigens like CD2, CD18, CD25, CD29, CD40, and CD54; HMC-1 cells also show intracellular metachromatically staining granules containing histamine, β-tryptase, and heparin. On the other hand, HMC-1 cells lack functional FcεRI, α-tryptase, and chymases [21]. Another key MC surface marker is the so-called KIT receptor (*alias* CD117). KIT is a type III tyrosine kinase receptor encoded by the proto-oncogene *c-kit*. By binding of SCF, KIT activates an intracellular pathway leading to the activation of a variety of transcription factors with a vital role during cell differentiation, proliferation, and survival [23]. KIT is expressed not only on MC but also on some hematopoietic stem cells, germ cells, and melanocytes [24].

HMC-1 cells carry an isoform of KIT constitutively active due to two point mutations in the intracellular domain of the receptor [25]. Such mutations consist of (a) the substitution of the aspartic acid (D) in position 816 for Valine (V) (D816V) and (b) the substitution of Valine (V) in position 560 for glycine (G) (V560G) [25]. At present, two HMC-1 sublines are available (HMC1.1 and HMC1.2) that differ in the specific harbored CD117 (KIT) mutation. In particular both HMC-1.1 and HMC-1.2 cell lines harbor the V560G mutation, while HMC-1.2 only shows the V560G mutation [22, 26]. The presence of the D^{V}

mutation represents a key feature of HMC-1.2, since this mutation is often present in patients affected by adult sporadic mastocytosis and myeloproliferative disease [25, 27, 28]. HMC-1.1 and HMC-1.2 show different phenotypical characteristics such as proliferation rate and other growth characteristics. HMC-1 cells therefore can represent a useful model to better understand the effects of a constitutively active KIT on MC regulation, proliferation, and survival and to test the effect of KIT inhibitors. The presence of a constitutively active KIT receptor makes this cell line independent of the presence of SCF in the culture media, which considerably reduces laboratory costs.

As mentioned above, HMC-1 cells lack surface expression of the high-affinity receptor FcεRI and of the low-affinity FcεRII. Initial studies reported that mRNA content of the α- and β-chain of FcεRI is negligible, while mRNA for the γ-subunit is present at a high amount [21]. However, more recent evidence suggests that the FcεRIβ subunit is expressed at low levels by these cells [29, 30]. Consistently with the lack of functional FcεRI, HMC-1 cells fail to degranulate after IgE/antigen sensitization, and this represents a crucial limitation to the applicability of this cell line in allergy/inflammation-related studies. The HMC-1 cell lines however have been employed for degranulation studies following non-immunological stimulation with the calcium ionophore A23187, compound 48/80, and phorbol esters. However, the physiological meaning of such results is limited and the interpretation of these data needs to consider that elicited pathways might not be relevant under physiological conditions, especially in the context of allergen/IgE-mediated activation.

Thus, despite its limitations, HMC-1 cells show undoubted advantages in that they are relatively easy-to-cultivate cell lineages and a great number of homogeneous cells can be obtained in a limited period. HMC-1 cells' growth is independent of the presence of growth factors and this reduces subsequent handling costs. In the light of this, it is not surprising that these cell lineages have been widely used in laboratories worldwide. To extend the applicability of this versatile MC line, Xia and co-workers [30] have generated an HMC-1 subline that stably expresses the human FcεRIα subunit. This new cell line (HMCα) has been proven to bind to IgE and to release cytokines (e.g., IL-8, IL-6) following antigen stimulation. However, HMCα fails to degranulate following immunological activation. The array of cytokines released from this transformed HMC-1 subline accurately reflects the panel of cytokines normally released by human MCs during IgE/antigen challenge. The reason why HMCα cells fail to degranulate, despite their ability to bind to IgE/antigen, is not clear; a hypothesis suggests that the ratio between specific forms of FcεRI ($\alpha\gamma_2$ vs. $\alpha\beta\gamma_2$) might play a role. HMCα certainly represents a step ahead toward a more representative model of human tissue MCs.

To assess the validity of using HMC-1.2 cells in laboratory settings to investigate MC diseases, Haenisch and co-workers [31] have: (a) compared the transcriptome data of HMC-1.2 cells with those of normal human resting lung and tonsillar MCs and (b) compared the HMC-1.2 transcriptome with that of leukocytes and mononuclear cells from patients harboring the D816V mutation. Data suggest that HMC-1.2 cells show increased expression of genes involved in cell cycle, apoptosis, and neoplastic processes. These findings in HMC-1.2 cells are in line with transcriptome data from leukocytes from peripheral blood and mononuclear cells from bone marrow aspirates of patients harboring the D816V mutation. Due to this similarity in gene expression, it is believed that this cell lineage offers a strong and valid tool to study the biology of MCs in relation to the D816V mutation [31].

1.3 LAD Cell Lines

In 2003 Kirshenbaum and co-workers characterized two novel human MC lineages, namely, LAD-1 and LAD-2 [19]. These cell lines originated from bone marrow aspirates of a male patient affected by MC sarcoma/leukemia. LAD-2 cell line can be currently obtained subject to a Material Transfer Agreement from the NIH/NIAID/Laboratory of Allergic Disease, Bethesda (USA). LAD-1 cells have not been distributed to the public [32] and literature search provided a paucity of information. The LAD-2 cell line has been used worldwide to investigate several aspects of MC biology such as degranulation, signaling pathways, and genetic markers [33]. Unlike HMC-1 cells, LAD cells do not express the activating mutation of Kit and therefore their growth is dependent on the presence of SCF in the culture medium (10 ng/mL). The exact mechanism that confers immortality to the LAD-2 cell line is still unknown. Kirshenbaum and co-workers have hypothesized that increased levels of telomerase coupled with changes in spectral karyotyping might account for the acquired immortality. This characteristic makes LAD cells more similar to primary cultured MCs. The LAD-2 cell line expresses ultrastructural features of mature MCs such as CD4, CD9, CD13, CD14, CD22, CD31, CD32, CD45, CD63, CD64, CD71, CD103, CD117, CD132, CD203c, and CXCR4 (CD184); CCR5 (CD195); and intracytoplasmic histamine, tryptase, and chymase. LAD cells also express functional FcεRI and FcγRI and they can degranulate following IgE-mediated receptor cross-linking. LAD-2 cells can be used as a tool to investigate IgE-FcεRI binding and subsequent downstream events, MC signaling pathways, and the effect of novel drugs. Some groups, however, have reported a decreased ability of LAD cells to release cytokines compared to other lineages and primary MCs [34]. LAD cells represent one of the most powerful tools in MC research but, due to the high cost of rhSCF, LAD cells are often not the first choice as a MC model. Additionally, LAD-2 cells also show a slow doubling time (approximately 2 weeks) which limits their applicability when large numbers of cells are required. In a 10-year retrospective analysis, Kirshenbaum and co-workers have reported minimal phenotypical variation of this cell line from what was originally described in 2003. The authors compared surface markers, KIT expression, β-hexosaminidase release, and karyotyping of a newly thawed and cultured batch with data originally obtained in 2003; the only reported variation from the original (2003) cell lineage was a different karyotyping. The 2013 batch was found to be 47,XXY +2, −21 while the original was 48,XXY [33]. Recent data report single nucleotide polymorphisms of the KIT receptor at amino acid position aa529 and aa531, a deletion polymorphism at S715aa, and the presence of the GNNK variation. However, the absence of the activation mutation D816V was confirmed [33]. More recently, Kirshenbaum and co-workers have characterized LAD-1 and compared their characteristics with LAD-2 cells. LAD-1 cells showed a doubling time of 4 weeks and expressed CD133, CCR3 (CD193), CCR5 (CD195), CXCR4 (CD184), and IL3R (CD123). LAD-1 cells showed an increased FcεRI/CD117 and released beta-hexosaminidase following IgE stimulation [32].

1.4 LUVA Cell Line

Established in 2011, the LUVA lineage is a MC line derived from a culture of non-transformed hematopoietic progenitor cells. LUVA cells have been extensively reviewed elsewhere [20] and therefore their main features are only briefly summarized here. LUVA cells spontaneously differentiated from $CD34^+$-enriched mononuclear cells from a donor with aspirin-exacerbated respiratory disease but without any clonal MC disorder. It has been reported that LUVA cells can degranulate following FcεRI cross-linking as proved by a modest release of β-hexosaminidase [20]. Cross-linking Fc receptors also evoked the secretion of prostaglandin D2 (PGD2) and thromboxane A2 (TXA2). In addition to the expression of normal levels of FcεRI, LUVA cells also constitutively express FcγRI and FcγRII. Metachromatic intracellular granules resulted positive for tryptase, chymase, cathepsin G, and carboxypeptidase A3 (CPA3). Although LUVA cells do not show KIT mutations, they can survive and proliferate in culture without the presence of SCF in the culture media; intriguingly, exogenous addition of SCF to the culture media evoked a dose-dependent increase of phosphorylation of the KIT receptor and consequent increase of cell proliferation. The LUVA cell line represents a new precious tool for future studies, as it is a fully functional MC line without some of the aberrant characteristics that limit the applicability of other MC lineages.

1.5 $ROSA^{KIT\ WT}$ Cell Line, $ROSA^{KIT\ D816V}$, and $ROSA^{KIT\ D816V\text{-}Gluc}$ Subclones

The lineages described so far (HMC-1, LAD, and LUVA) offer useful tools for the investigation of several aspects of MC biology. However, none of these cell lines express the D816V KIT mutation alone or simultaneously express a functional FcεRI and a D816V-mutated KIT receptor. This limitation is of particular importance in the context of studies aimed to elucidate the biology of systemic mastocytosis (SM). Patients affected by SM suffer from the effect of aberrant MC mediator release together with abnormal MC proliferation due to the presence of the D816V mutation. Considering that the most severely affected SM patients are those who present coexisting allergy that implies FcεRI and that 90% of the SM patients harbor a D816V, a cell lineage expressing a functional FcεRI and a D816V-mutated KIT receptor would represent an ideal tool to dissect the molecular pathways of this debilitating disease. Furthermore, a possibility to compare results obtained with D816V-mutated lineages with subclones harboring the wild-type KIT receptor would be ideal. Saleh and co-workers report to have established both an SCF-dependent MC line expressing a normal KIT receptor (KIT WT) along with a functional FcεRI and an SCF-independent subclone of such cell line harboring the D816V KIT [26]. These cell lines called $ROSA^{KIT\ WT}$ and $ROSA^{KIT\ D816V}$, respectively, might represent an important tool to directly compare signaling pathways activated by normal or mutant forms of KIT (*see* Table 1).

The $ROSA^{KIT\ WT}$ cell line was derived in 2009 from $CD34^{+}$ cells of normal human umbilical cord blood. The phenotypic characterization has shown the expression of typical MC surface markers such as CD32, CD44, CD50, CD66a, CD116, and CD203c. Furthermore, $ROSA^{KIT\ WT}$ cells express a fully functional IgE receptor and a normal KIT receptor (CD117). Due to the presence of a functional FcεRI, following immunological stimulation with human monoclonal IgE/anti-human IgE, $ROSA^{KIT\ WT}$ cells can undergo degranulation with a substantial release of β-hexosaminidase (up to 38% of the total enzyme content) and histamine. The release of TNF-α and PGD2 can also be elicited by immunological stimulation. Calcium ionophore A23187 induced cell degranulation with the release of up to 80% of the β-hexosaminidase content.

The $ROSA^{KIT\ WT}$ cell line can be maintained in culture indefinitely without loss of clonal stability; its short doubling time is approximatively 24 h in the presence of rhSCF (80 ng/mL). This feature represents an advantage compared to LAD cells that, as previously described, show a doubling time of approximately 2 weeks.

$ROSA^{KIT\ D816V}$ subclone was obtained following transfection of $ROSA^{WT}$ cell line with a lentiviral vector encoding KIT D816V. The introduction of the D816V mutation makes the KIT receptor constitutively active. This results in the ability of the $ROSA^{KIT\ D816V}$ cells to proliferate in the absence of rhSCF with a doubling time of 24 h. Interestingly NGF, IL-3, and GM-CSF could partially prevent $ROSA^{KIT\ WT}$ death due to SCF deprivation. Like the parental cell lineage, $ROSA^{KIT\ D816V}$ displays a functional IgE

Table 1
Comparative table for functional FcεRI, KIT receptor mutations, and SCF dependence of human MC lines

	Functional FcεRI	KIT receptor	SCF dependence	Additional information
HMC 1.1	No	V^{560G}	No	
HMC 1.2	No	D^{816V} and V^{560G}	No	
LUVA	Yes	No	No	
LAD-2	Yes	No activating mutations	Yes	
$ROSA^{KIT\ WT}$	Yes (high levels)	No activating mutations	Yes	Engraft NSG mice; BM, lung, and spleen
$ROSA^{KIT}$ D816V	Yes (high levels)	D^{81} $D^{816V\ 6V}$	No	Engraft NSG mice (high rate); BM, lung, and spleen
$ROSA^{KIT}$ D816V-Gluc	Yes	D^{816V}	No	Engraft NSG mice (high rate); BM, lung, and spleen Express *G. princeps* luciferase

receptor and showed a strong positivity for KIT (which levels are higher than those expressed by the $ROSA^{KIT\ WT}$ lineage). Surprisingly immunological stimulation of the subclone $ROSA^{KIT\ D816V}$ did not elicit the release of β-hexosaminidase and histamine, and this cell lineage showed an overall decreased responsiveness to IgE-dependent stimuli despite the presence of a functional IgE receptor and activation of several signaling pathways.

The morphology of the two cell types differs in several aspects: $ROSA^{KIT\ WT}$ cells appear to be rounded, with a high nuclear-to-cytoplasm ratio, while the $ROSA^{KIT\ D816V}$ subclone shows a spindle-shaped morphology. Both ROSA cell lines resulted positive for tryptase, FcεRIα, and KIT but $ROSA^{KIT\ D816V}$ were negative for chymase. The expression level of KIT was higher for $ROSA^{KIT\ D816V}$ cells compared with $ROSA^{KIT\ WT}$ cells.

Interestingly the two cell lines can engraft NSG mice even though with different rates ($ROSA^{KIT\ D816V}$ engraft NSG mice more efficiently than $ROSA^{KIT\ WT)}$. This result is of importance because it provides an animal model that can mimic human SM. More recently Bibi and co-workers [35] have generated a variant of the $ROSA^{KIT\ D816V}$ cell line expressing *Gaussia princeps* luciferase (Gluc). The new cell line, $ROSA^{KIT\ D816V\text{-}Gluc}$, allows simultaneous quantification of the engraftment and in vivo imaging by measuring luciferase activity. Importantly $ROSA^{KIT\ D816V\text{-}Gluc}$ did not show alteration of morphology, *KIT* D^{816V} expression, and therefore SCF independence [35].

Table 2
Synoptic table of the most important mast cell/basophil cell lines of human and rodent origin

Cell line	Cell type	Origin	FcεRI surface expression	FcεRI chain expressed	Growth factor dependence	Suggested culture conditions
HMC-1 HMC-α	Mast cells Mast cells	Human Human	– +	β[a], γ α[b], β, γ	N N	IMDM +10% FBS + 4 mM L-glutamine + α-thioglycerol
LAD-1 LAD-2	Mast cells Mast cells	Human Human	+ +	Not known α, β, γ	Y Y	StemPro-34 SFM, 2 mM L-glutamine +100 ng/mL SCF
LUVA	Mast cells	Human	+	α, expression of β- and γ-chain not known	N	StemPro-34SFMat with 50% fresh medium weekly
ROSA$^{kit}_{WT}$	Mast cells	Human	+	Not known	Y	IMDM + 10% FCS + 2 mM L-glutamine + 80 ng/mL murine SCF
ROSA$^{kit}_{D816V}$	Mast cells	Human	+	Not known	N	IMDM + 10% FCS + 2 mM L-glutamine
NCL-2	Mast cells	Mouse	+	Not known	N	MEMα + 10% FCS + 2 mM L-glutamine, 50 μM 2-mercaptoethanol
RBL-2H3	Mast cells/basophils	Rat	+	α, β, γ	N	α-MEM + 10% FBS
RBL-hEIa-2B121	Mast cells/basophils	Rat/humanized	+	α	N	80% MEM + 20% RPMI 4 mM L-glutamine + 5% FCS
RBL-30/25	Mast cells/basophils	Rat/humanized	+	α	N	
RBL SX-38	Mast cells/basophils	Rat/humanized	+	α, β, γ	N	
RBL-703/21	Mast cells/basophils	Rat/humanized	+	α	N	RPMI 1640 + 10% FBS + 2 mM L-glutamine
KU812	Basophils	Human	+[c]	α, β[d], γ	N	RPMI 1640 + 10% FBS + 2 mM L-glutamine
LAMA-84	Basophils	Human	+	α, β	N	RPMI 1640 + 10% FBS

[a]Initial study published by Nilsson and co-workers showed negligible amount of the β-subunit of FcεRI [21]; however more recent data have shown the presence of low amounts of FcεRIβ in HMC-1 [29, 64]
[b]Cells stably transfected with the full-length cDNA of the IgE-binding α-chain of the human Fcε [30]; HMC-α fails to show a degranulative response following immunological activation despite the ability to bind hIgE/antigen
[c]A study from Hara and co-workers reports that KU812 lacks FcεRI but cells can express it upon stimulation with IL-4 [65]
[d]The β-chain shows an unexpected low expression in KU812 clones, suggesting the possibility that both $\alpha\gamma_2$ and $\alpha\beta\gamma_2$ isoforms of the FcεRI are present

Saleh et al. report to have submitted patent applications for the two cell lines (Patent WO/2013/064639) in 2013. Both ROSA$^{KIT\ WT}$ and ROSA$^{KIT\ D816V}$ cell lineages are currently deposited with the National Collection of Cultures of Microorganisms (CNCM, Institut Pasteur Paris) under the numbers CNCM I-4551 and CNCM I4552, respectively [26].

1.6 KU812

KU812 is an immature pre-basophilic cell line obtained from a male patient affected by chronic myeloid leukemia, and it shows phenotypic features of basophil precursors [36]. Characteristic of this cell line is the presence of scarce metachromatic granules containing low amounts of histamine and tryptase. KU812 cells express low levels of FcεRI and no FcεRII [37, 38]. Originally, it was reported that KU812 cells do not show surface expression of FcεRI despite the presence of mRNA transcripts for the α-chain of the receptor [38–40]. However, Magnusson and co-workers [41] reported that KU812 cells do express, even though at low densities, FcεRI on their surface as proven by positive direct IgE-FITC binding. Moreover, the same authors demonstrated that IgE binds to FcεRI with high affinity and specificity toward the IgE isotype. It has been reported that under specific conditions KU812 cells can be "directed" toward basophilic differentiation [42]. KU812 cells show the ability to spontaneously differentiate to basophil-like cells under serum-free conditions; differentiation however can be also achieved following several physiological and nonphysiological stimuli. In particular, IL-3, IL-4, IL-6, TNF-α, and hydrocortisone have been suggested not only to stimulate basophil-like maturation but also to enhance cell surface expression of FcεRI [38, 40, 43, 44]. Other stimuli that can induce KU812 maturation are conditioned media from the human T-cell line Mo, sodium butyrate, and dimethyl sulfoxide [45, 46]. KU812 spontaneous maturation could be due to an autocrine effect of self-produced IL-6. During differentiation it is possible to observe a morphological maturation accompanied by an increase in granulation, histamine content, and cell growth inhibition [42]. Interestingly, it has been shown that KU812 cells maintain the ability to differentiate toward other elements of the erythroid lineage. Such features, however, are lost after multiple culturing passages [38, 47]. KU812 represents the only human basophilic cell line and it has been used as a model to both study cell commitment toward erythron id or basophilic differentiation and investigate antigen-dependent degranulation of basophils.

2 Rodent Cell Lines

Due to the lack of a fully mature, easy-to-cultivate human basophil or MC line, for many years almost the entire MC/basophil research landscape exploited cell lines of rodent origin. However, it must be stressed that MCs/basophil originating from different species present important differences in their physiological properties. Therefore, between different animal species, MC/basophil-associated diseases show marked differences in their clinical picture, and it is not a surprise if data obtained from murine cells cannot always be extrapolated to human MC/basophil biology. Furthermore, findings achieved with cell lines of animal origin (murine, rat, canine, equine, etc.) should always be confirmed with cells of human origin. The distinctive differences between MCs of murine and human origin have already been extensively reviewed by Bischoff [9]. One of the major functional differences between rodent and human MCs is the different protease contents. Murine MCs express several tryptases and chymases (e.g., mast cellP, mast cellP14), with different specificities, while human MCs express just three tryptases (α, β, and γ) and one chymase [9]. Expression levels of IL-4, IL-5, and TNF also show marked differences. Responsiveness to IL-3 differs too; murine MC growth and survival depend on IL-3 while human MCs respond poorly to IL-3 [48, 49]. In the next sections, we summarize the major features of the most exploited rodent cell lines.

2.1 RBL Cell Lines

RBL cell lines were established following serial limited dilution of cells originating from a single rat basophilic leukemia [50, 51]. Over the years, several RBL cell lines have been created, each with different histochemical and ultrastructural features. The most widely used among the RBL cell lines is the histamine-releasing subclone RBL-2H3 obtained by Kulczycki in 1979 [50]. This cell line has been extensively reviewed elsewhere [18, 52] and here we report just a brief description of its main characteristics. RBL-2H3 cells present features of both mucosal MCs and basophils and respond with degranulation to both immunological and non-immunological stimuli in a manner that closely resembles the degranulation dynamic of primary MCs and basophils [17]. RBL-2H3 cells present on their surface functional FcεRI and, due to their manipulability, homogeneity, and fast growth rate, have been widely used for studies of IgE-mediated degranulation, to test novel MC stabilizers and to investigate the structural physicochemical properties of FcεRs [53, 54]. In more recent years, the use of the RBL-2H3 cell line has been extended to diagnostic and therapeutic applications following the establishment of chimeric sublines expressing human FcεRI subunits (*see* Chapter 11). FcεRI presents a tetrameric structure composed by one α-, one β-, and two γ-subunits (αβγ2) and binds IgE with high affinity ($1 \times 10^{10}\ M^{-1}$) [54]. The α-subunit has the only function of binding the ligand through its extracellular domain. The β-chain is essential for the receptor's translocation to the membrane and provides receptor stability. The β-chain can also amplify the activation signal that subsequently will be transduced by the γ-chain [55]. The β-chain is not functionally essential and therefore a trimeric αγ2 structure, despite presenting a reduced stability, is the minimum requirement for a functional Fc receptor. Chimeric cell lines have been used as a tool to determine the biologic activity of allergen extracts. Native RBL-2H3 cells express all three subunits (α, β, γ) of the high-affinity receptor of IgE. However, rodent FcεRs do not recognize human IgE. Transfection of the RBL-2H3 cell line with human alpha/ beta/gamma subunits has yielded a panel of cell lines (*see* Table 1) that show the ability to degranulate after cross-linking IgE-bound FcεRI. Chimeric RBL cells show different levels of degranulation and different IgE-binding capacities; this might probably be due to the diverse expression levels and ratio of the subunits as well as the different surface expressions of the receptors. RBL SX-38 cells express all three α-, β-, and γ-subunits while RBL-hEIa-2B121 and RBL-30/25 cells only express the α-subunit of human FcεRI. It has been shown, however, that despite the absence of the human γ-subunit, the signal transduction mechanisms in RBL-hEIa-2B121 and RBL-30/25 cells are intact [56]. The differential expression of FcεRI subunits provides a reason for the different binding capacities of these cells toward human IgE, with RBL SX-38 having the highest binding capacity followed by RBL-30/25 and then RBLhEIa-2B121 (99.9%, 97%, and 83%, respectively) [2]. Like the native RBL cells, all the chimeric clones show reduced ability to bind purified IgE after approximately 2 weeks in culture. The IgE-binding capacity of these cell lines is mirrored by their ability to degranulate after allergen stimulation leading to 94% of mediator release for RBL SX-38 followed by RBL-30/25 (84%) and RBL-hEIa-2B121 (54%) [2, 57]. Chimera RBL-2H3 cell lines show the same advantages as native RBL-2H3 in terms of high speed of growth, simple cell culture techniques, and affordable maintenance costs; this makes RBL cells a good research instrument. RBL chimeras have been used to ascertain the allergenicity of antigens to humans in commercialized products and diagnostic allergen preparations [58] and to assess the activity of serum IgE from allergic patients [57, 59]. Chimeric RBL-2H3 cells have been generated also following transfection with canine and equine FcεRI chains [51].

2.2 NCL-2 Cell Line

NCL-2 cell line has been isolated in 2014 [60] from the long-term culture of BMMC from Cinnamon coat-colored Nishiki (NC)/Kuj mice (an inbred strain of Japanese fancy mice). Interestingly, NC/Kuj mice are widely used as mouse model of atopic dermatitis due to their ability to spontaneously develop eczema with increased serum concentration of IgE.

Being derived by BMMC, NCL-2 cells are classified as mucosal MCs. However, during prolonged time in culture, their phenotype might switch to CTMC-like due to the suppression of the production of MCP-1 following exposure to SCF (and IL-13) [60]. NCL-2 cells are non-tumoral, immortalized nonadherent cells that have been reported to grow without IL-13 or SCF [61]. FcεRI and KIT have been found to be expressed at a cellular surface level. The KIT receptor is not constitutively active in this cell line and this is reflected by a slow growth (doubling time has been estimated to be 30.1 h) and an *indolent tumor growth* in vivo if compared with P815 mastocytoma cells. However, the authors warn that this cell lineage could acquire a de novo mutation of KIT in long-term culture, a characteristic that NCL-2 cells have in common with BMMC from NC/Kuj mice.

Following immunological sensitization with anti-DNP-IgE and stimulation with DNP-HSA, NCL-2 cells released histamine and produced LTB4 in a dose-dependent manner. Non-immunological stimuli like ionomycin and ATP could evoke degranulation as well [60]. NCL-2 cells can also release β-hexosaminidase, tryptase, IL-6, and TNF-α [62].

NCL-2 cells have not been widely used in scientific research so far, however, they have been shown to change the pattern of proliferation when cultured on polymer-based biomaterials organized in honeycomb-like films (HCFs) [61]. Choi and co-workers observed that when pore size of the HCF decreased at a dimension smaller than the cellular dimension, NCL-2 cells divided and formed multinuclear cells along the rim of the pore to then enter the HCF holes [61]. Differences in proliferation, morphological, and secretory characteristics between NCL-2 cells and RBL-2H3 cells following culture on HCFs have been reported by Choi and co-workers in 2016 [62]. To note the same group that isolated NCL-2 reported to have generated another NCL lineage (NCL-1) with an activating Kit mutation (D814F) on a single allele [60].

3 Concluding Remarks

Mast cells and basophils are key players during allergy, inflammation, as well as innate immunity [63]. Mast cell and basophil research, however, has often been impeded by the lack of readily available model systems. At present in fact, there is a limited number of MCs and basophil cell lines, and such cultures present huge variability in terms of phenotype and functionality. Over the years, several cell lineages have been established, all bearing some, but not all, the typical MC and/or basophil hallmarks. The HMC-1 cell line is the most widely used but presents the limitation of a not fully mature phenotype with the lack of surface expression of FcεRI. On the other hand, the LAD-2 cell line offers an experimental system model closer to the normal MC physiology, but the prolonged doubling time and exorbitant maintenance costs are not ideal for routine experiments. However, despite the high running costs of LAD-2 cells, 10 years after establishment, over 300 material transfer agreements and 40 licensing agreements were granted to researches worldwide proving the research potential of this lineage.

The most recent addition to the human lineage arsenal is the ROSA family of cell lines: $ROSA^{KIT\ WT}$ cell line, $ROSA^{KIT\ D816V}$, and $ROSA^{KIT\ D816V\text{-}Gluc}$ cell lineages. Together these cell lines offer the possibility to investigate signaling of MC harboring the D816V mutation along with a functional FcεRI and directly compare the results with parent cells expressing a wild-type KIT receptor. Thanks to the ability of the ROSA cell to engraft NSG mice, a new in vivo model of SM is now available.

Transfection of rat basophilic leukemia cells with human FcεRI subunits has led to the generation of chimeric cell lines that have been used in both research and diagnostics (*see* Chapters 11 and 12). Cell lines have been proven useful to study specific molecular aspects of MCs and basophils, clarifying their role in allergy and inflammation, acquired immunity, and other pathological conditions. However, several crucial pitfalls make this research tool inadequate for a complete understanding of molecular and physiological cell dynamics. The validity of results obtained using cultured cells is often a matter of debate, as immortalized cell lines present abnormal characteristic that may not reflect primary cell physiology. Another important pitfall of the use of cell lines is the intra-laboratory reproducibility of experimental results. Therefore, there is a need to standardize laboratory protocols among research groups worldwide.

Information regarding experimental procedures must be shared and made readily available among investigators in order to achieve well-defined and constant experimental conditions. This will minimize not only the interlaboratory variations but also the costs of protocol development and optimization. Another aspect that must be considered is the potential risk of contamination and misidentification of cell lines. As a result, cell lines should only be procured from certified institutions or trusted sources. An excellent example is given by RBL cells that have been mainly regarded as MCs despite showing a phenotype in between MCs and basophils [17]. The routine use of RBL cells to investigate MC behavior has produced a large amount of published material, but also ambiguous and contradictory data [58].

References

1. Vogel L, Lüttkopf D, Hatahet L et al (2005) Development of a functional in vitro assay as a novel tool for the standardization of allergen extracts in the human system. Allergy 60:1021–1028
2. Ladics GS, van Bilsen JHM, Brouwer HMH et al (2008) Assessment of three human FcεRI-transfected RBL cell-lines for identifying IgE induced degranulation utilizing peanut-allergic patient sera and peanut protein extract. Regul Toxicol Pharmacol 51:288–294
3. Lin J, Renault N, Haas H et al (2007) A novel tool for the detection of allergic sensitization combining protein microarrays with human basophils. Clin Exp Allergy 37:1854–1862
4. Gelbmann CM, Mestermann S, Gross V et al (1999) Strictures in Crohn's disease are characterised by an accumulation of mast cells colocalised with laminin but not with fibronectin or vitronectin. Gut 45:210–217
5. Raithel M, Winterkamp S, Pacurar A et al (2001) Release of mast cell tryptase from human colorectal mucosa in inflammatory bowel disease. Scand J Gastroenterol 36:174–179
6. Dvorak AM, Monahan RA, Osage JE et al (1980) Crohn's disease: transmission electron microscopic studies. II. Immunologic inflammatory response. Alterations of mast cells, basophils, eosinophils, and the microvasculature. Hum Pathol 11:606–619
7. Wilcock A, Bahri R, Bulfone-Paus S et al (2019) Mast cell disorders: from infancy to maturity. Allergy 74(1):53–63
8. Mukai K, Tsai M, Saito H et al (2018) Mast cells as sources of cytokines, chemokines, and growth factors. Immunol Rev 282(1):121–150
9. Bischoff SC (2007) Role of mast cells in allergic and non-allergic immune responses: comparison of human and murine data. Nat Rev Immunol 7:93–104
10. da Silva EZM, Jamur MC, Oliver C (2014) Mast cell function. J Histochem Cytochem 62:698–738
11. Mekori YA, Metcalfe DD (1999) Mast cell-T cell interactions. J Allergy Clin Immunol 104:517–523
12. Supajatura V, Ushio H, Nakao A et al (2002) Differential responses of mast cell toll-like receptors 2 and 4 in allergy and innate immunity. J Clin Invest 109:1351–1359
13. Puxeddu I, Piliponsky A, Bachelet I et al (2003) Mast cells in allergy and beyond. Int J Biochem Cell Biol 35:1601–1607
14. Dawicki W, Marshall JS (2007) New and emerging roles for mast cells in host defence. Curr Opin Immunol 19:31–38
15. Brunner T, Heusser CH, Dahinden CA (1993) Human peripheral blood basophils primed by interleukin 3 (IL-3) produce IL-4 in response to immunoglobulin E receptor stimulation. J Exp Med 177:605–611
16. Eccleston E, Leonard BJ, Lowe JS et al (1973) Basophilic leukaemia in the albino rat and a demonstration of the basopoietin. Nat New Biol 244:73–76
17. Passante E, Frankish N (2009) The RBL-2H3 cell line: its provenance and suitability as a model for the mast cell. Inflamm Res 58:737–745
18. Butterfield JH, Weiler D, Dewald G et al (1988) Establishment of an immature mast cell line from a patient with mast cell leukemia. Leuk Res 12:345–355
19. Kirshenbaum AS, Akin C, Wu Y et al (2003) Characterization of novel stem cell factor responsive human mast cell lines LAD 1 and 2 established from a patient with mast cell sarcoma/leukemia; activation following aggregation of FcepsilonRI or FcgammaRI. Leuk Res 27:677–682
20. Laidlaw TM, Steinke JW, Tiñana AM et al (2011) Characterization of a novel human mast cell line that responds to stem cell factor and expresses functional FcεRI. J Allergy Clin Immunol 127:815–822.e5
21. Nilsson G, Blom T, Kusche-Gullberg M et al (1994) Phenotypic characterization of the human mast-cell line HMC-1. Scand J Immunol 39:489–498
22. Sundström M, Vliagoftis H, Karlberg P et al (2003) Functional and phenotypic studies of two variants of a human mast cell line with a distinct set of mutations in the c-kit proto-oncogene. Immunology 108:89–97
23. Edling CE, Hallberg B (2007) C-kit-a hematopoietic cell essential receptor tyrosine kinase. Int J Biochem Cell Biol 39(11):1995–1998
24. Miettinen M, Lasota J (2005) KIT (CD117): a review on expression in normal and neoplastic tissues, and mutations and their clinicopathologic correlation. Appl Immunohistochem Mol Morphol 13(3):205–220
25. Furitsu T, Tsujimura T, Tono T et al (1993) Identification of mutations in the coding sequence of the proto-oncogene c-kit in a human mast cell leukemia cell line causing ligand-independent activation of c-kit product. J Clin Invest 92:1736–1744

26. Saleh R, Wedeh G, Herrmann H et al (2014) A new human mast cell line expressing a functional IgE receptor converts to tumorigenic growth by KIT D816V transfection. Blood 124:111–120
27. Worobec AS, Semere T, Nagata H et al (1998) Clinical correlates of the presence of the Asp816Val c-kit mutation in the peripheral blood mononuclear cells of patients with mastocytosis. Cancer 83:2120–2129
28. Longley BJ, Metcalfe DD, Tharp M et al (1999) Activating and dominant inactivating c-KIT catalytic domain mutations in distinct clinical forms of human mastocytosis. Proc Natl Acad Sci U S A 96:1609–1614
29. Cruse G, Kaur D, Leyland M et al (2010) A novel FcεRIβ-chain truncation regulates human mast cell proliferation and survival. FASEB J 24:4047–4057
30. Xia YC, Sun S, Kuek LE et al (2011) Human mast cell line-1 (HMC-1) cells transfected with FcεRIα are sensitive to IgE/antigen-mediated stimulation demonstrating selectivity towards cytokine production. Int Immunopharmacol 11:1002–1011
31. Haenisch B, Herms S, Molderings GJ (2013) The transcriptome of the human mast cell leukemia cells HMC-1.2: an approach to identify specific changes in the gene expression profile in kit D816V systemic mastocytosis. Immunol Res 56:155–162
32. Kirshenbaum AS, Yin Y, Sundstrom JB et al (2019) Characterization and comparison of human LAD1 mast cells with human LAD2 mast cells. J Allergy Clin Immunol 143:AB180
33. Kirshenbaum AS, Petrik A, Walsh R et al (2014) A ten-year retrospective analysis of the distribution, use and phenotypic characteristics of the LAD2 human mast cell line. Int Arch Allergy Immunol 164:265–270
34. Guhl S, Babina M, Neou A et al (2010) Mast cell lines HMC-1 and LAD2 in comparison with mature human skin mast cells - drastically reduced levels of tryptase and chymase in mast cell lines. Exp Dermatol 19(9):845–847
35. Bibi S, Zhang Y, Hugonin C et al (2016) A new humanized in vivo model of KIT D816V+ advanced systemic mastocytosis monitored using a secreted luciferase. Oncotarget 7:82985–83000
36. Kishi K (1985) A new leukemia cell line with Philadelphia chromosome characterized as basophil precursors. Leuk Res 9:381–390
37. Yamashita M, Ichikawa A, Katakura Y et al (2001) Induction of basophilic and eosinophilic differentiation in the human leukemic cell line KU812. Cytotechnology 36:179–186
38. Fukuda T, Kishi K, Ohnishi Y et al (1987) Bipotential cell differentiation of KU-812: evidence of a hybrid cell line that differentiates into basophils and macrophage-like cells. Blood 70:612–619
39. Fischkoff SA, Kishi K, Benjamin WR et al (1987) Induction of differentiation of the human leukemia cell line, KU812. Leuk Res 11:1105–1113
40. Blom T, Huang R, Aveskogh M et al (1992) Phenotypic characterization of KU812, a cell line identified as an immature human basophilic leukocyte. Eur J Immunol 22:2025–2032
41. Magnusson CG, Håård J, Matsson P et al (1995) Demonstration of specific high-affinity fc epsilon-receptors on the human basophil-like leukemia cell line KU812 by flow cytometry. Allergy 50:72–77
42. Almlof I, Nilsson K, Johansson V et al (1988) Induction of basophilic differentiation in the human basophilic cell line KU812. Scand J Immunol 28:293–300
43. Nilsson G, Carlsson M, Jones I et al (1994) TNF-alpha and IL-6 induce differentiation in the human basophilic leukaemia cell line KU812. Immunology 81:73–78
44. Hutt-Taylor SR, Harnish D, Richardson M et al (1988) Sodium butyrate and a T lymphocyte cell line-derived differentiation factor induce basophilic differentiation of the human promyelocytic leukemia cell line HL-60. Blood 71:209–215
45. Valent P, Besemer J, Kishi K et al (1990) IL-3 promotes basophilic differentiation of KU812 cells through high affinity binding sites. J Immunol 145:1885–1889
46. Nakazawa M, Mitjavila MT, Debili N et al (1989) KU 812: a pluripotent human cell line with spontaneous erythroid terminal maturation. Blood 73:2003–2013
47. Valent P, Besemer J, Sillaber C et al (1990) Failure to detect IL-3-binding sites on human mast cells. J Immunol 145:3432–3437
48. Razin E, Ihle JN, Seldin D et al (1984) Interleukin 3: a differentiation and growth factor for the mouse mast cell that contains chondroitin sulfate E proteoglycan. J Immunol 132:1479–1486
49. Dearman RJ, Skinner RA, Deakin N et al (2005) Evaluation of an in vitro method for the measurement of specific IgE antibody responses: the rat basophilic leukemia (RBL) cell assay. Toxicology 206:195–205
50. Kulczycki A, Isersky C, Metzger H (1974) The interaction of IgE with rat basophilic leukemia cells. I. Evidence for specific binding of IgE. J Exp Med 139:600–616

51. Rashid A, Sadroddiny E, Ye HT et al (2012) Review: Diagnostic and therapeutic applications of rat basophilic leukemia cells. Mol Immunol 52(3-4):224–228
52. Holowka D, Hartmann H, Kanellopoulos J et al (1980) Association of the receptor for immunoglobulin E with an endogenous polypeptide on rat basophilic leukemia cells. J Recept Res 1:41–68
53. Goetze A, Kanellopoulos J, Rice D et al (1981) Enzymatic cleavage products of the alpha subunit of the receptor for immunoglobulin E. Biochemistry 20:6341–6349
54. Kraft S, Kinet J-P (2007) New developments in FcepsilonRI regulation, function and inhibition. Nat Rev Immunol 7:365–378
55. Ra C, Jouvin MH, Kinet JP (1989) Complete structure of the mouse mast cell receptor for IgE (fc epsilon RI) and surface expression of chimeric receptors (rat-mouse-human) on transfected cells. J Biol Chem 264:15323–15327
56. Gilfillan AM, Kado-Fong H, Wiggan GA et al (1992) Conservation of signal transduction mechanisms via the human fc epsilon RI alpha after transfection into a rat mast cell line, RBL 2H3. J Immunol 149:2445–2451
57. Takagi K, Nakamura R, Teshima R et al (2003) Application of human fc epsilon RI alpha-chain-transfected RBL-2H3 cells for estimation of active serum IgE. Biol Pharm Bull 26:252–255
58. Passante E, Ehrhardt C, Sheridan H et al (2009) RBL-2H3 cells are an imprecise model for mast cell mediator release. Inflamm Res 58:611–618
59. Hoffmann A, Jamin A, Foetisch K et al (1999) Determination of the allergenic activity of birch pollen and apple prick test solutions by measurement of β-hexosaminidase release from RBL-2H3 cells. Comparison with classical methods in allergen standardization. Allergy 54:446–454
60. Hiragun T, Yanase Y, Okabe T et al (2014) Establishment of a mast cell line, NCL-2, without kit mutation, derived from NC mouse bone marrow. FEBS Open Bio 4:342–346
61. Choi H, Tanaka M, Hiragun T et al (2014) Non-tumor mast cells cultured in vitro on a honeycomb-like structured film proliferate with multinucleated formation. Nanomedicine 10:313–319
62. Sugimoto K (2016) Difference in cell proliferation and spontaneous mediator release between two mast cell lines, NCL-2 and RBL-2H3 on honey comb-like structured film. Nanotechnol Nanomed Nanobiotechnol 3:1–7
63. Galli SJ, Nakae S, Tsai M (2005) Mast cells in the development of adaptive immune responses. Nat Immunol 6:135–142
64. Akizawa Y, Nishiyama C, Hasegawa M et al (2003) Regulation of human FcepsilonRI beta chain gene expression by Oct-1. Int Immunol 15:549–556
65. Hara T, Yamada K, Tachibana H (1998) Basophilic differentiation of the human leukemia cell line KU812 upon treatment with interleukin-4. Biochem Biophys Res Commun 247:542–548

Chapter 11

Use of Humanized RBL Reporter Systems for the Detection of Allergen-Specific IgE Sensitization in Human Serum

Daniel Wan, Xiaowei Wang, Ryosuke Nakamura, Marcos J. C. Alcocer, and Franco H. Falcone

Abstract

Determination of allergen-specific immunoglobulin E (IgE) levels in human blood samples is an important diagnostic technology for the assessment of allergic sensitization. The presence of specific IgE in human serum samples can be measured by sensitizing humanized rat basophil leukemia (RBL) cell lines with diluted serum and measuring cellular activation after challenge with the suspected allergens. This has been traditionally performed by measuring the levels of β-hexosaminidase released upon RBL degranulation. Here, we describe the use of two recently developed humanized RBL reporter cell lines, which offer higher sensitivity and are amenable to high-throughput scale experiments.

Key words Reporter system, Luciferase, DsRed, RBL, IgE, Allergy, RS-ATL8, NFAT-DsRed

1 Introduction

Immunoglobulin E (IgE) is an antibody isotype, which is at the center of the pathology of allergic disease. Many of its effects are mediated via the tetrameric high-affinity IgE receptor FcεRI. Because of the key role of IgE, the diagnosis of type I allergy has been traditionally focused on measuring the levels of allergen-specific (sIgE) and total IgE in the serum or plasma of allergic individuals. This is routinely performed in vitro by measuring IgE binding to solid-phase bound allergens and detection via enzymatically labeled secondary antibodies and fluorescent substrates, using devices available in automated or semi-automated formats from various companies [1]. More recently, new methods have taken advantage of protein array technology (ISAC: Immuno Solid-phase Allergen Chip) [2], allowing simultaneous testing of more than 100 allergens using a small blood sample. Methods based on allergen-specific IgE measurements have high sensitivity, but their clinical relevance is not always clear, as they report the ability of IgE

Bernhard F. Gibbs and Franco H. Falcone (eds.), *Basophils and Mast Cells: Methods and Protocols*, Methods in Molecular Biology, vol. 2163, https://doi.org/10.1007/978-1-0716-0696-4_11,

to bind allergen but not to induce an allergic reaction. As a result, IgE tests are frequently accompanied by clinically more relevant skin prick tests (SPTs) and a detailed medical history. However, SPTs also have a range of disadvantages, since the use of certain medications can interfere with the test results resulting in some patients being excluded from this type of test. The residual hazard also requires these tests to be carried out in specialized units. Another more recent type of allergy test is based on the assessment of basophil activation in whole blood (basophil activation tests; BATs) using flow cytometry (*see* Chapters 16–19). These tests work by monitoring the upregulation of CD63 and/or CD203c on the surface of basophils after incubation of heparinized whole blood with the suspected allergen. There is currently no universally accepted standard protocol for BATs, and these tests need to be performed on the day of blood sampling (whereas serum samples can be stored for a long period). As a result, this test is still not widely used clinically, but only performed in specialized, research-active laboratories.

Another technique has been available for about two decades but, due to its perceived limitations, has not found a widespread use. This is the use of humanized rat basophilic leukemia (RBL) cells for monitoring specific allergen sensitization in patient serum. The RBL cell lines need to be humanized with at least the human α-chain of FcεRI [3], as human IgE does not bind to the rodent IgE receptor [4]. A critical issue with this cell line has been the cytotoxicity of certain human sera for the rat cells, requiring high dilutions and resulting in insufficient sensitization and subsequently negative tests. However, a new generation of reporter cell lines is now replacing the previous humanized RBL cell lines taking advantage of sensitive reporter genes such as luciferase or fluorescent proteins [5–8]. This enables the detection of allergen-specific IgE even when using serum dilutions as high as 1:100 [8–10]. In this chapter, we describe the use of two such humanized RBL reporter systems, RS-ATL8 and NFAT-DsRed, for the detection of allergen-specific IgE in human serum samples. Both methods pave the way for high-throughput measurements, and the latter can also be used in conjunction with allergen arrays [9].

2 Materials

All reagents and supplies should be sterile. Follow local waste disposal regulations.

2.1 Reagents

1. RBL medium: minimum essential medium (MEM), 10% v/v heat-inactivated fetal bovine serum, 100 U/mL penicillin, 100 μg/mL streptomycin, 2 mM L-glutamine (if not already present in MEM). Store at 4 °C.

2. RS-ATL8 and NFAT-DsRed reporter cell lines, available from the authors through a Material Transfer Agreement (MTA).
3. G418 sulfate (Geneticin).
4. Blasticidin S (for NFAT-DsRed cells).
5. Hygromycin B (for RS-ATL8 cells).
6. DPBS without Ca^{2+} or Mg^{2+}.
7. Trypsin-EDTA or cell scrapers.
8. Patient sera.
9. Allergens.
10. Polyclonal Anti-human IgE, e.g. goat anti-human.
11. Lysis buffer: 1% v/v Triton X-100 in DPBS (for fluorescence assay). Store at 4 °C.
12. ONE-Glo Luciferase Assay System (Promega), for luminescence assay. Follow local regulations regarding the handling of suspected carcinogens. Prepare and store as described by the manufacturer. Warm to room temperature before use.

2.2 Supplies

1. Graduated pipettes and pipette tips.
2. 25 cm^2/75 cm^2 cell culture flasks.
3. 50 mL centrifuge tubes.
4. Clear 96-well plates.
5. Black 96-well plates (for fluorescence assay).
6. White 96-well plates (for luminescence assay).

2.3 Equipment

1. Pipettors for graduated pipettes/tips.
2. Water bath.
3. Class II microbiological safety cabinet.
4. Cell culture incubator: humidified incubator set to 37 °C and 5% CO_2.
5. Heat block.
6. Hemocytometer.
7. Centrifuge capable of holding 50 mL tubes.
8. Plate reader for fluorescence or luminescence.

3 Method

3.1 General Cell Culture

Work in a class II microbiological safety cabinet using sterile techniques.

1. If starting from frozen stocks, partially defrost a vial of cells by warming in hand, and then add 1 mL pre-warmed RBL medium to fully defrost.

2. Transfer cells to a 25 cm^2 flask containing 8 mL RBL medium, and place in a cell culture incubator overnight.
3. When the cells are confluent in the 25 cm^2 flask (this may take several days, depending on the viability of frozen cells), transfer cells to a 75 cm^2 flask by detaching adherent cells using trypsin-EDTA (*see* **steps 2–11** of Subheading 3.3).
4. Maintain cells in 10 mL RBL medium supplemented with 1 mg/mL G418 sulfate and 20 μg/mL blasticidin S (NFAT-DsRed) or 500 μg/mL G418 sulfate and 200 μg/mL hygromycin B (RS-ATL8), in a 75 cm^2 flask, in the cell culture incubator.
5. Passage cells when confluent using trypsin-EDTA (*see* **steps 2–10** of Subheading 3.3). Replacing the flask with 20% cells will result in confluence after 3 days. Replace with 10% cells for confluence (*see* **Note 1**) after 4 days (refresh the medium after 2 or 3 days). Top up the level of liquid in the flask to 10 mL.
6. Freeze stocks of cells in larger batches, i.e., from several flasks. Centrifuge cells at 300 × *g* for 5 min; resuspend in RBL medium with 10% v/v DMSO to obtain a cell concentration of 5×10^6 cells/mL. Work quickly since DMSO is toxic to cells at room temperature. Aliquot 1 mL cells/cryovial, transferring cryovials to a Mr. Frosty™ (Nalgene) container filled with isopropanol, and place at −80 °C for several hours or overnight. After freezing, transfer the cryovials to a liquid nitrogen Dewar flask for long-term storage. Defrost a vial to check the viability of the batch.

3.2 (Optional) Treatment of Serum Before Sensitization

1. If necessary, heat human serum in an Eppendorf tube at 56 °C for 5 min (*see* **Note 2**) on a heat block and then place it on ice.

3.3 Sensitization

Cells are sensitized for 16–20 h by incubating them in RBL medium with the desired dilution of previously heat-inactivated serum (*see* **Note 3**).

1. Pre-warm RBL medium to 37 °C in a water bath.
2. Remove medium from the 75 cm^2 flask of confluent cells (*see* **Note 4**).
3. To the flask, add 10 mL DPBS without Ca^{2+} or Mg^{2+} (*see* **Note 5**).
4. Tilt the flask to wash the cells with DPBS.
5. Discard the DPBS in the flask.
6. Add 2 mL trypsin-EDTA to the flask.
7. Tilt the flask to ensure that the cells are exposed to the trypsin-EDTA solution.

8. Incubate the flask for 5–15 min at 37 °C in a cell culture incubator, until clumps of cells are visibly detached (*see* **Note 6**).
9. Add 8 mL pre-warmed RBL medium to the flask and resuspend the cell suspension (*see* **Note 7**).
10. Transfer the cells to a 50 mL tube. Some cells can be left behind in the flask for passaging (*see* Subheading 3.1).
11. Count a sample of cells in a hemocytometer and calculate the total cell number.
12. Pellet the cells by centrifuging at 300 × *g* for 5 min.
13. Resuspend the cells to a concentration of 1×10^6 cells/mL using RBL medium.
14. Sensitize cells using appropriate concentration of serum (*see* **Note 2**).
15. Pipette 50 μL cells/well of a 96-well plate.
16. Incubate the plate overnight in the cell culture incubator.

3.4 Stimulation

1. Check the plate under a light microscope to see if the cells are in a healthy state.
2. Prepare appropriate dilutions of allergens or other stimuli in RBL medium. Optimal concentrations may vary for each allergen and will need to be determined experimentally. Use an appropriate amount of anti-human IgE as a positive control (*see* **Note 8**).
3. Remove the medium from the cells in the 96-well plate.
4. Add 50 μL stimuli to appropriate wells of the plate. Do this quickly so the cells don't dry out when the medium is removed. Smaller steps, e.g., 3–6 wells at a time, can be done instead.
5. Incubate the plate in the cell culture incubator for 24 h (fluorescence assay with NFAT-DsRed) or 3 h (luciferase assay with RS-ATL8).

3.5 Measurement of Activation

3.5.1 Fluorescence Assay with NFAT-DsRed

1. Wash the cells once with 50 μL DPBS, remove DPBS, and add 50 μL 1% Triton X-100 in DPBS to lyse the cells.
2. Tap the plate gently to mix, then transfer the lysate in each well to a black 96-well plate, and measure fluorescence in a plate reader (*see* Fig. 1).

3.5.2 Luciferase Assay (EXiLE Test, See **Note 9**)

1. Allow the luciferase assay reagent and cells to come to room temperature before mixing. The assay is temperature sensitive and this method allows more reproducible results.
2. Add 50 μL ONE-Glo Luciferase Assay Substrate (Promega) directly to stimulated cells (no need to remove the medium containing stimulus).

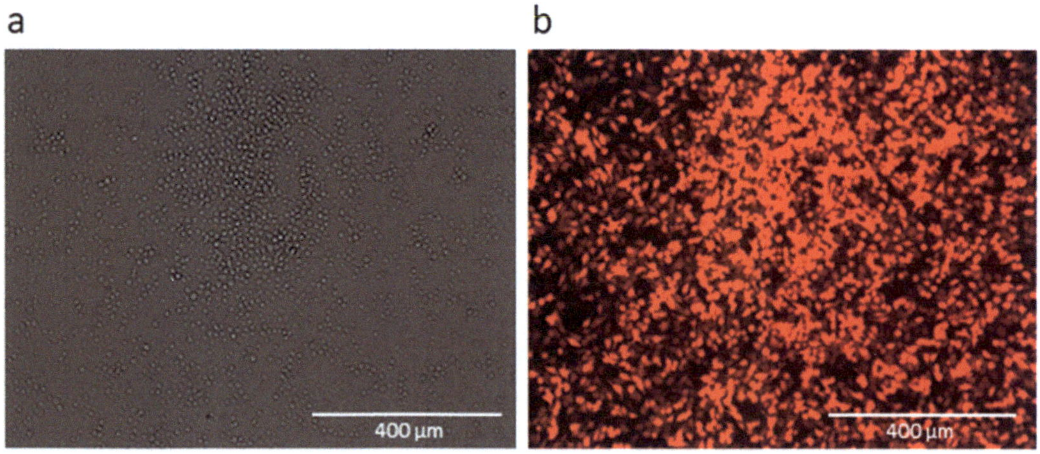

Fig. 1 Fluorescence microscopy image of NFAT-DsRed reporter cell line sensitized with monoclonal IgE without stimulation (**a**) and 24 h after stimulation with anti-IgE antibody (**b**). Note that not all cells appear to have been activated by this treatment and that the intensity of fluorescence varies between individual cells

3. Transfer the 100 µL mixture in each well to a white 96-well plate, and measure luminescence in a plate reader (*see* **Notes 10–12**).

4 Notes

1. A 1/20 split (5% cells) is not recommended for routine passaging as this can cause "colonies" to form and unhealthy cells in the middle of the cell clusters.
2. Human sera are frequently cytotoxic for RBL cells (*see* Fig. 2). This is in part due to the presence of activated complement, which is traditionally inactivated by heating at 56 °C for 30 min. This prolonged treatment however also induces irreversible conformational changes in IgE molecules, ablating its ability to bind to FcεRI. A shorter heat inactivation will be sufficient to inactivate complement while preserving the integrity of IgE. We have had good results with this protocol, even with 20% serum concentration (1:5). Not all sera, however, require heat inactivation. In most cases, a 1:50 or 1:100 dilution will be sufficient to remove cytotoxicity without any heat treatment. Thus the heat treatment is described as optional.
3. We routinely start the sensitization protocol at 3 p.m. and the stimulation at 9 a.m. the following day.
4. Once fully confluent, a 75 cm^2 flask contains enough cells to fill two 96-well plates.

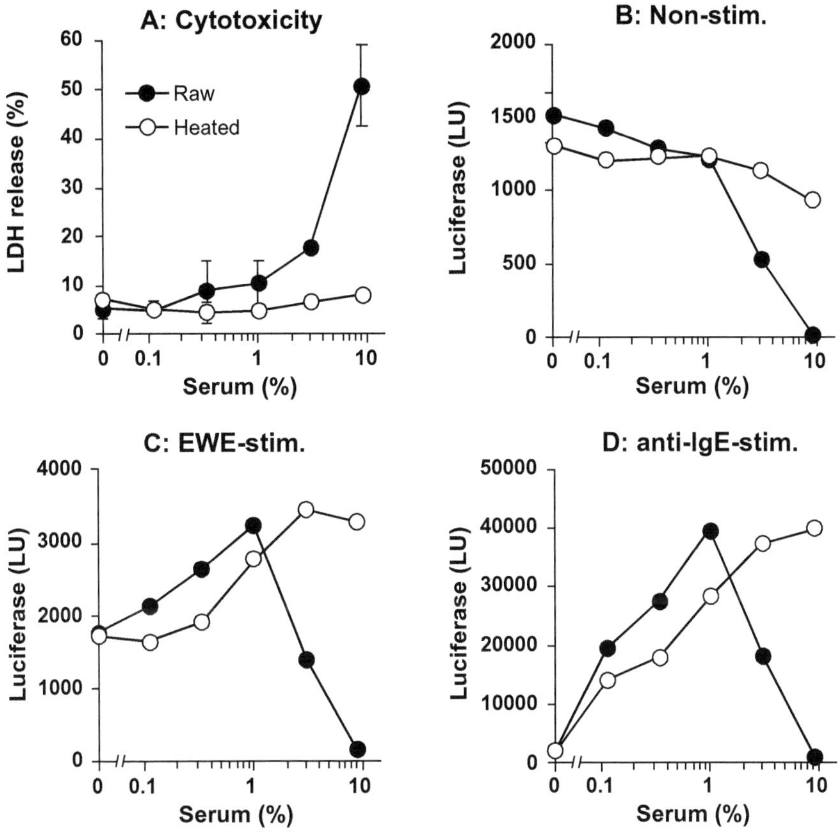

Fig. 2 Cytotoxicity of human serum on the RS-ATL8 cells. A serum of an egg white-allergic patient (7 years, male, total IgE 1388 IU/mL egg white sIgE 12.8 U_A/mL) was heated at 56 °C for 30 min. The RS-ATL8 cells were sensitized overnight with the serum at various concentrations. (**a**) Cytotoxicity of the sera on the cells as revealed by LDH release assay. EXiLE response after 3 h treatment of (**b**) control medium, (**c**) egg white extract (EWE), and (**d**) anti-human IgE

5. Cell scrapers can be used to detach the cells instead of trypsin-EDTA. Simply refresh the RBL medium and scrape the cells. There is no need for a DPBS wash.
6. You can speed up the detachment process by repeatedly knocking the bottom of the flask with your knuckles, holding it tight with the other hand. Be careful, however, as the use of too much force can split the flask.
7. Make sure any visible clumps are broken up, by pipetting up and down. We routinely use a 5 mL serological pipette for this purpose, as these have a narrower tapered end. Make sure you do not create bubbles when pipetting up and down by always leaving some medium in the pipette when expelling the liquid and not taking up any air when aspiring it.

8. The optimal amount of anti-IgE will depend on the type (species, clonality, affinity, etc.) used and needs to be determined empirically. Optimum concentrations can range between 100 ng/mL and 2 μg/mL which therefore need to be determined by testing the anti-IgE antibody over a wide range of concentrations (e.g., at least 5 orders of magnitude). This should result in a bell-shaped stimulation curve, with a clear sub-optimal and supraoptimal stimulation range.
9. This test is now known as EXiLE test, for Ig*E* cross-linking (=*X*) *i*nduced *L*uciferase *E*xpression [8].
10. In our lab, we have tested different types of multiwell plates. Clear plates are not recommended for final scans due to lower sensitivity than black or white plates. However, if there is no option but to use clear plates, light detection from below (bottom scans) will give higher sensitivity than top scans. Black plates give the greatest sensitivity for fluorescence assays, and white plates give the greatest sensitivity for luminescence assays. For opaque plates, for obvious reasons, light detection (and fluorescence excitation) must be from above. We found that opaque-walled, clear-bottomed plates do not greatly increase sensitivity. However, different plate readers may give different results. If you have the resources, it may be worthwhile to repeat these tests for the plate readers in your own lab.
11. If cells are sensitized in black/white plates, then there is no need to transfer cell lysates from clear to opaque plates; the entire experiment can be carried out on a single plate. However, we find it essential to examine the cells after an overnight incubation to check for cell death or contamination. A clear-bottomed plate is required for light microscopes.
12. Bubbles can be created by an incorrect pipetting technique or sample evaporation and may interfere with plate readings. Large bubbles can be removed by stabbing with a dry pipette tip (a wet pipette tip cannot burst bubbles). If there are many small bubbles throughout the plate, most can be removed by briefly centrifuging the plate.

Acknowledgments

This work was supported by the COST Action BM1007 Mast Cells and Basophils—Targets for Innovative Therapies.

References

1. Ricci G, Capelli M, Miniero R et al (2003) A comparison of different allergometric tests, skin prick test, Pharmacia UniCAPR and ADVIA CentaurR, for diagnosis of allergic diseases in children. Allergy 58:38–45
2. Sastre J (2010) Molecular diagnosis in allergy. Clin Exp Allergy 40:1442–1460
3. Taudou G, Varin-Blank N, Jouin H et al (1993) Expression of the alpha chain of human Fc epsilon RI in transfected rat basophilic leukemia cells: functional activation after sensitization with human mite-specific IgE. Int Arch Allergy Immunol 100:344–350
4. Miller L, Blank U, Metzger H, Kinet JP (1989) Expression of high-affinity binding of human immunoglobulin E by transfected cells. Science 244:334–337
5. Falcone FH, Wan D, Barwary N, Sagi-Eisenberg R (2018) RBL cells as models for in vitro studies of mast cells and basophils. Immunol Rev 282(1):47–57. https://doi.org/10.1111/imr.12628
6. Ali EA, Nakamura R, Falcone FH (2017) Use of humanized RS-ATL8 reporter system for detection of allergen-specific IgE sensitization in human food allergy. Methods Mol Biol 1592:147–161. https://doi.org/10.1007/978-1-4939-6925-8_12
7. Falcone FH, Alcocer MJ, Okamoto-Uchida Y, Nakamura R (2015) Use of humanized rat basophilic leukemia reporter cell lines as a diagnostic tool for detection of allergen-specific IgE in allergic patients: time for a reappraisal? Curr Allergy Asthma Rep 15(11):67. https://doi.org/10.1007/s11882-015-0568-3
8. Nakamura R, Uchida Y, Higuchi M et al (2010) A convenient and sensitive allergy test: IgE crosslinking-induced luciferase expression in cultured mast cells. Allergy 65:1266–1273
9. Wang X, Cato P, Lin H-C et al (2014) Optimisation and use of humanised RBL NF-AT-GFP and NF-AT-DsRed reporter cell lines suitable for high-throughput scale detection of allergic sensitisation in array format and identification of the ECM-integrin interaction as critical factor. Mol Biotechnol 56 (2):136–146. https://doi.org/10.1007/s12033-013-9689-x
10. Lin J, Renault N, Haas H et al (2007) A novel tool for the detection of allergic sensitization combining protein microarrays with human basophils. Clin Exp Allergy 37:1854–1862

Chapter 12

Use of Humanized Fluorescent Reporter Cell Line RBL NFAT-DsRed for the Detection of Allergen-Specific IgE in Patient Sera Using Allergen Microarrays

Marina Kalli, Marcos J. C. Alcocer, Andrew J. Blok, and Franco H. Falcone

Abstract

The presence of allergen-specific IgE (sIgE) in human sera can be determined by measuring the binding of sIgE to solid phase-bound preparations containing the allergens to be tested. These can be complex extracts, purified or recombinant allergens, or peptides. Older methods, such as the IgE CAP test, only allow sIgE measurements to multiple allergens in individual measurements. Newer technologies such as the ImmunoCAP® ISAC test allows semiquantitative testing of sIgE to over a hundred allergens on a protein array. Allergen arrays have higher numerical power, allowing testing to many allergens at the same time, using only a small amount of serum. We have previously demonstrated how allergen arrays can be used in combination with purified peripheral blood basophils, introducing a clinically relevant readout. Here, we describe a protocol and materials that allow the testing of sIgE with multiple allergens in array format, using a humanized fluorescent IgE reporter system (RBL NFAT-DsRed).

Key words Fluorescent reporter cell line, RBL, IgE, Allergen, Microarrays, NFAT-DsRed

1 Introduction

The introduction of allergen arrays and component-resolved diagnostics (CRD) is widely considered a milestone in the improvement of diagnostic techniques for IgE-mediated allergy [1]. CRDs take advantage of the ability to assess the binding of allergen-specific IgE (sIgE) to a large number of allergens (purified or recombinant proteins, peptides) printed in array (multiplex) format on suitable surfaces. CRD now plays an increasingly important role in assessing IgE-mediated cross-reactivities in food allergy, hymenoptera venom allergy, and pollinosis.

CRD in multiplex format, in particular when used for diagnosing IgE-mediated food allergy, is able to avoid some issues associated with the more established diagnostic methods, such as oral food challenge (OFC), considered the gold standard in its double-blind placebo-controlled food challenge form (DBPCFC), as well

Bernhard F. Gibbs and Franco H. Falcone (eds.), *Basophils and Mast Cells: Methods and Protocols*, Methods in Molecular Biology, vol. 2163, https://doi.org/10.1007/978-1-0716-0696-4_12, © Springer Science+Business Media, LLC, part of Springer Nature 2020

as skin prick testing (SPT). CRDs can enable the distinction between primary sensitizations to "true" allergens and "mere" IgE cross-reactivities [2]. To some extent, CRDs are also thought to enable prediction of the severity of reactions and potential for dangerous anaphylactic reactions. This is because, in the case of food allergens, sensitizing proteins/epitopes, which are not destroyed by heat or are resistant to digestive enzymes, are more likely to cause symptoms than heat-labile, easily digested proteins [3]. This information becomes assessable through the use of CRD, where sIgE to individual molecules, rather than using complex extracts, is measured. While this would a priori appear to tilt the balance in favor of only using purified or recombinant allergens for diagnosis, some arguments can be made in support of the use of complex extracts. For example, extracts can be useful in cases where the sensitizing allergens may not yet be known or are not yet available in recombinant form, as in recently more globally introduced food sources, such as goji berries [4], tapioca [5], or even edible insects [6]. Furthermore, the safety coverage achieved by using only a few selected major allergens for assessment, bearing in mind the size of the genomes of commonly used food crops such as wheat (*Triticum aestivum*; 14.5 Gbp), soy (*Glycine max*; 1.115 Gbp) [7], or canola [8] (*Brassica napus*; 1.13 Gbp) [9], can be questioned. However, using complex extracts, in addition to the known problematic issues such as stability or difficult standardization, potentially brings back one of the issues already solvable by CRD, the presence of cross-reactive carbohydrate determinants (CCDs) in food extracts. These are carbohydrate epitopes characterized by their ability to bind IgE, but inability to cause clinical symptoms [10]. Therefore, to take full advantage of using complex extracts as an additional way of assessing sensitization to food, a method introducing a biologically (i.e., ultimately clinically) relevant readout is desirable, as this can distinguish between unproductive IgE binding (not leading to cellular activation, therefore clinically irrelevant) and cross-linking sIgE/allergen combinations.

Our group has shown proof of principle that allergen arrays can be combined with live peripheral blood basophils for a biologically relevant assessment of sensitization to food or any other allergen [11]. However, due to the cost and other difficulties with purifying sufficient amounts of basophils from peripheral blood (*see* Chapter 2) of patients, this solution is unlikely to find widespread use in a clinical facility. Therefore, using a low-cost reporter cell line, which could be grown in practically unlimited amounts and is able to indicate FcεRI-mediated activation by allergens, would make this technology more amenable to a widespread use. However, no detailed protocols for the use of IgE reporter cell lines with allergen arrays are available. Critical parameters in this technology, which combines protein arrays with living cells, are the chemical structure of the surface used for printing the allergen arrays and details of the protocol, in order to avoid cell toxicity during incubation and cell losses during the washing steps.

Here, we present a protocol, which describes how sensitization to food can be assessed using extracts, purified or recombinant allergens, printed onto allergen arrays, in combination with the recently developed, humanized fluorescent IgE reporter cell line, NFAT-DsRed [12, 13].

2 Materials

2.1 Reagents

1. RBL medium: Eagle's Minimum Essential Medium (EMEM), 10% v/v heat-inactivated fetal bovine serum, 100 U/mL penicillin, 100 μg/mL streptomycin, and 2 mM L-glutamine. Store at 4 °C.
2. RBL-703/21 NFATp-DsRed-Express2 cell line, available from the authors on the basis of an MTA.
3. G418 sulfate (Geneticin).
4. Blasticidin S (for NFAT-DsRed cells).
5. DPBS without Ca^{2+} or Mg^{2+}.
6. Trypsin–EDTA or cell scrapers.
7. Patient sera.
8. Allergens.
9. Anti-human IgE.
10. 0.05% (v/v) Tween 20.
11. 1 mg/mL bovine serum albumin (BSA).
12. 10 μg/mL Fibronectin solution in DPBS.

2.2 Supplies

1. Graduated pipettes and pipette tips.
2. 25 cm^2/75 cm^2 cell culture flasks.
3. 50 mL centrifuge tubes.
4. Clear six-well tissue culture-treated plates.
5. 35 mm culture dishes.
6. 16-well ProPlate® Multi-well Chambers (Grace Bio Labs, US, Cat. No 244864).
7. ProPlate® Clips (Grace Bio Labs, US, Cat. No 204830).
8. Amine-coated slides (Surmodics, US, Cat. No. DN01-0025).
9. Parafilm M.

2.3 Equipment

1. Pipettors for graduated pipettes/tips.
2. Water bath.
3. Class II microbiological safety cabinet.

4. Cell culture incubator: humidified incubator set to 37 °C and 5% CO_2.
5. Hemocytometer.
6. Centrifuge capable of holding 50 mL tubes.
7. GenePix® 4000B Microarray Scanner (Molecular Devices, US) or a comparable suitable microarray scanner.

3 Methods

3.1 General Cell Culture

Work in a Class II microbiological safety cabinet using sterile techniques.

1. If starting from frozen stocks, partially defrost a vial of cells by warming in hand, and then add 1 mL pre-warmed RBL medium to fully defrost.
2. Transfer cells to a 25 cm^2 flask containing 8 mL RBL medium, and place in a cell culture incubator overnight.
3. When the cells are confluent in the 25 cm^2 flask, transfer cells to a 75 cm^2 flask by detaching adherent cells using trypsin–EDTA (*see* **steps 1–7** of Subheading 3.3).
4. Maintain cells in 10 mL RBL medium supplemented with 1 mg/mL G418 sulfate and 20 μg/mL blasticidin S (NFAT-DsRed), in a 75 cm^2 flask, in the cell culture incubator.
5. Passage cells when confluent using trypsin–EDTA (*see* **steps 1–8**).
6. Freeze stocks of cells in larger batches. Centrifuge cells at 300 × *g* for 5 min, and resuspend in RBL medium with 10% DMSO to obtain a cell concentration of 5×10^6 cells/mL. Aliquot 1 mL cells/cryovial, transferring cryovials to a Mr. Frosty™ (Nalgene) container filled with isopropanol, and place in −40 °C for 1 h and then to −80 °C overnight. After freezing, transfer the cryovials to a liquid nitrogen Dewar flask for long-term storage. Defrost a vial to check the viability of the batch.

3.2 Treatment of Serum Before Sensitization

1. Heat aliquot of human serum at 56 °C for 5 min.

3.3 Sensitization

Cells are sensitized for 16–20 h by incubating them in RBL medium with the desired dilution of previously heat-inactivated serum.

1. Pre-warm RBL medium to 37 °C in a water bath.
2. Remove medium from 75 cm^2 flask of confluent cells.

3. Wash the flask once with 10 mL DPBS without Ca^{2+} or Mg^{2+}.
4. Aspirate and discard DPBS.
5. Add 2 mL trypsin–EDTA to the flask and tilt the flask from one side to another to ensure that the entire surface has been covered with the solution.
6. Incubate the flask for approximately 15 min at 37 °C in a cell culture incubator, until clumps of cells are visibly detached.
7. Add 8 mL pre-warmed RBL medium to the flask and gently resuspend the cells.
8. Transfer the cells in a sterile 50 mL tube.
9. Count the cells using a hemocytometer and calculate the total cell number.
10. Pellet the cells by centrifuging at 300 × *g* for 5 min.
11. Resuspend the cells to a concentration of 3×10^5 cells/mL using RBL medium.
12. Sensitize cells using appropriate concentration of serum (*see* **Note 1**).
13. Plate 3 mL cells/well of a 6-well plate.
14. Incubate the plate overnight in the cell culture incubator.

3.4 Protein Blocking

1. Sterilize allergen printed surfaces using ultraviolet light in a Class II biosafety cabinet at 30,000 J/cm^2 for 30 min (*see* **Note 2**).
2. Carefully adjust the slides to the 16-well multi-well chambers (*see* **Note 3**).
3. Wash the surfaces 3× with 100 μL 0.05% v/v Tween 20 in DPBS (*see* **Note 4**).
4. Add 100 μL (1 mg/mL) BSA diluted in DPBS in each well. Incubate at 37 °C for 1 h.
5. Wash 3× with 100 μL 0.05% v/v Tween 20 in DPBS.
6. Wash 1× with sterile DPBS.
7. Add 100 μL (10 μg) fibronectin, diluted in DPBS, in each well of the multi-well chamber. Incubate at 37 °C for 3 h.
8. Wash 3× with sterile DPBS.

3.5 Stimulation

1. Collect sensitized cells using a sterile cell scraper (30 mm) (*see* **Note 5**).
2. Transfer 200 μL cell suspension in each well of the 16-well multi-well chambers (*see* **Note 6**).
3. Incubate for 24 h in a 37 °C/5% CO_2 humidified cell incubator.

3.6 Measurement of Activation

1. Gently remove the slides from each multi-well chamber (*see* **Note 7**).
2. Scan the slides using a GenePix® 4000B Microarray Scanner. Excitation laser light was set at 532 nm = standard green (PMT gain: 700), matching pre-installed emission filter standard green (558 nm to 593 nm; 575DF35575).
3. Using the GenePix Pro 6.1 software, circles can be drawn around every spot of interest and then automatically adjusted to size for every spot using the align features in the selected block mode function of the software. This allows the software to calculate the spot-specific fluorescence intensity. Care should be taken that the features are aligned correctly in order for accurate results to be obtained.
4. Once acquisition is completed, the image is interrogated using the GenePix Pro 6.1 software (Molecular Devices, US). Fluorescence intensity from every spot is then determined (Fig. 1).

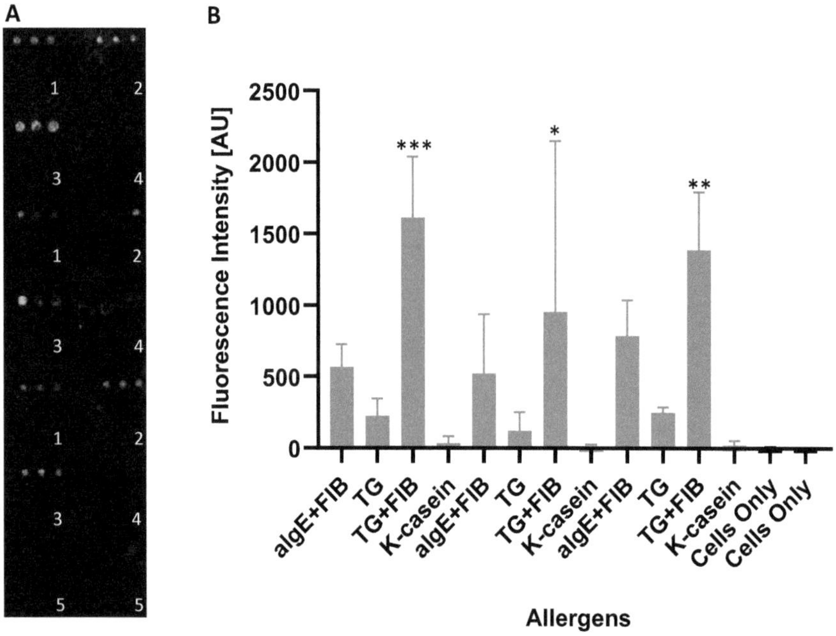

Fig. 1 (**a**) Cell binding of RBL 703/21 NFAT-DsRed to immobilized proteins printed in triplicate (**1**. anti-human IgE+10% fibronectin, **2**. timothy grass (TG) pollen extract, **3**. TG + 10% fibronectin, **4**. k-casein, and **5**. cells only) on amine-coated slides. (**b**) Fluorescence intensity values of individually activated protein spots on amine-coated slides. Data analysis was performed using one-way ANOVA (Dunnett's multiple-comparison test, $n = 3$, $* = p < 0.05$) compared to the negative control (cells only). RBL cells were sensitized with the serum of a grass pollen allergic patient overnight and incubated with the slides (8.4×10^5 cells/slide) for 24 h. Images of each protein spot were scanned using a GenePix 4000B Microarray Scanner at 532 nm laser wavelength

3.7 Data Analysis

1. Data is exported from the scanner using the GenePix Pro 6.1 software as a .txt file. Data analysis can then be performed using e.g. Microsoft Excel.
2. The fluorescence intensity of each allergen spot is determined via the subtraction of B_{532} values from the F_{532} value determined using the GenePix Pro 6.1 software. Each measurement was performed in triplicate (*see* **Note 8**).
3. Perform statistical analysis using GraphPad Prism 8 or comparable software.

4 Notes

1. Human sera are frequently cytotoxic for RBL cells. The presence of activated complement can be partially responsible, which can be traditionally inactivated by heating at 56 °C for 30 min. However, this prolonged treatment can result in irreversible conformational changes in IgE molecules, affecting its ability to bind to FcεRI. Thus, a shorter heat inactivation of just 5 min will be sufficient to inactivate complement while preserving the integrity of IgE. In our hands, even 20% serum concentration (1:5 dilution) has proven adequate to produce good results. In most cases, a 1:50 or a 1:100 dilution will be sufficient to remove cytotoxicity without any heat treatment. Beware, however, that increased serum dilution may reduce the sensitivity of the essay below the threshold needed for activation.
2. We advise sterilizing the slides, as well as the 16-well multi-well chambers, in a Class II biosafety cabinet in sterile petri dishes. After sterilization, securely close the petri dishes and seal with Parafilm. These can be kept at room temperature until needed and should be opened only while working in a sterile safety cabinet.
3. Carefully adjust the amine-coated slides onto the sterilized multi-well chambers. Do not apply too much pressure, as this might result in breaking the slides.
4. All washing and blocking steps are performed in a Class II biosafety cabinet. Open the lid of the sterile petri dish again in the safety cabinet. In new sterile petri dishes, pour the washing and blocking solution. Aspirate 100 μL using a multichannel pipette and slowly release the content with the tips of the pipette touching only the walls of the multi-well chamber. Aspirate the content by touching with the tips the bottom edges of every well to repeat the washes.
5. Gently scrape the cells from the six-well plate without using too much force and/or pressure. To make sure the cells have

detached from the surface, use a 1 mL pipette and gently resuspend them in pre-warmed EMEM cell medium.

6. Add the 200 μL cell suspension gently and slowly into the wells of the multi-well chamber. Excessive force might damage the cells and reduce cell viability.
7. Care should be taken when removing slides from the multi-well chambers as there is a reasonable chance that the slide may break. Try to remove one clip of the multi-well chamber at a time, gently applying some force with a sharp material at the edges. Allow air to move within the edges before lifting it up. Always wear nitrile gloves during this step.
8. In F_{532} Mean—B_{532}, the F and B parameters stand for feature pixel intensity and background intensity, respectively, where the feature is the printed spot and the background is the area immediately around the printed spot.

Acknowledgments

The work described in this chapter was fully funded by the Rosetrees Trust and Stoneygate Trust (JS15/M562) award to F.H.F. and M.J.C.A.

References

1. Alessandri C et al (2017) Diagnosing allergic sensitizations in the third millennium: why clinicians should know allergen molecule structures. Clin Transl Allergy 7:21
2. Muraro A et al (2014) EAACI food allergy and anaphylaxis guidelines: diagnosis and management of food allergy. Allergy 69:1008–1025
3. Weinberger T, Sicherer S (2018) Current perspectives on tree nut allergy: a review. J Asthma Allergy 11:41–51
4. Carnés J et al (2013) Recently introduced foods as new allergenic sources: sensitisation to Goji berries (*Lycium barbarum*). Food Chem 137:130–135
5. Santos KS et al (2013) Novel allergens from ancient foods: Man e 5 from manioc (*Manihot esculenta* Crantz) cross reacts with Hev b 5 from latex. Mol Nutr Food Res 57:1100–1109
6. Pali-Schöll I et al (2019) Edible insects: cross-recognition of IgE from crustacean- and house dust mite allergic patients, and reduction of allergenicity by food processing. World Allergy Organ J 12:100006
7. Schmutz J et al (2010) Genome sequence of the palaeopolyploid soybean. Nature 463:178–183
8. Fiocchi A, Dahdah L, Riccardi C, Mazzina O, Fierro V (2016) Precautionary labelling of cross-reactive foods: the case of rapeseed. Asthma Res Pract 2:13
9. Chalhoub B et al (2014) Early allopolyploid evolution in the post-neolithic Brassica napus oilseed genome. Science 345:950–953
10. Homann A, Schramm G, Jappe U (2017) Glycans and glycan-specific IgE in clinical and molecular allergology: sensitization, diagnostics, and clinical symptoms. J Allergy Clin Immunol 140:356–368
11. Lin J et al (2007) A novel tool for the detection of allergic sensitization combining protein microarrays with human basophils. Clin Exp Allergy 37:1854–1862
12. Falcone FH, Alcocer MJC, Okamoto-Uchida-Y, Nakamura R (2015) Use of humanized rat basophilic leukemia reporter cell lines as a diagnostic tool for detection of allergen-specific IgE in allergic patients: time for a reappraisal? Curr Allergy Asthma Rep 15:67
13. Ali EA et al (2019) Characterization of human FcεRIα chain expression and gene copy number in humanized rat basophilic leukaemia (RBL) reporter cell lines. PLoS One 14: e0221034

Chapter 13

NPY-mRFP Rat Basophilic Leukemia (RBL) Reporter: A Novel, Fast Reporter of Basophil/Mast Cell Degranulation

Nafal J. S. Barwary, Daniel Wan, and Franco H. Falcone

Abstract

Humanized rat basophilic leukemia (RBL) reporter cell lines are increasingly used for the detection of allergen-specific IgE and other purposes, such as the detection of allergens and standardization of allergen preparations. Existing reporter systems have many strengths and advantages but can be expensive or require longer incubation times. The new NPY-mRFP reporter cell line addresses such problems, as it requires neither expensive substrates nor overnight incubation for detection of activation. The fusion of Neuropeptide Y (NPY) with monomeric Red Fluorescent Protein (mRFP) results in localization of the fluorescent protein in granules. As NPY-mRFP is preformed in granules, the reporter system activation can be assessed using fluorescence measurements after as soon as 45–60 min, as described in this chapter, without the need to add any substrates.

Key words Reporter system, NPY-mRFP, RBL, IgE, Degranulation, Allergy

1 Introduction

Humanized rat basophilic leukemia (RBL) reporter systems are increasingly replacing the older generation of humanized RBL cell lines, due to their ease of use and higher sensitivity [1, 2]. The two most currently used humanized RBL reporter systems are RS-ATL8 [3] and NFAT-DsRed [4]. Both systems report the translocation of NFAT from the cytosol to the nucleus, which is known to occur as a consequence of Ca^{2+} influx after IgE-/FcεRI-dependent activation [5]. Therefore, the main application of humanized RBL reporter systems or the older non-humanized, non-reporter systems, has been the detection of allergen-specific IgE, e.g., in food allergy [6–9], but also the evaluation of the potential allergenicity of parasite antigens [10], as well as the detection of allergens or standardization of allergen preparations [11–13].

As previously described [2], the humanized reporter systems have a series of advantages over the older RBL cells, which used measurement of beta-hexosaminidase as a readout for activation.

Bernhard F. Gibbs and Franco H. Falcone (eds.), *Basophils and Mast Cells: Methods and Protocols*, Methods in Molecular Biology, vol. 2163, https://doi.org/10.1007/978-1-0716-0696-4_13,

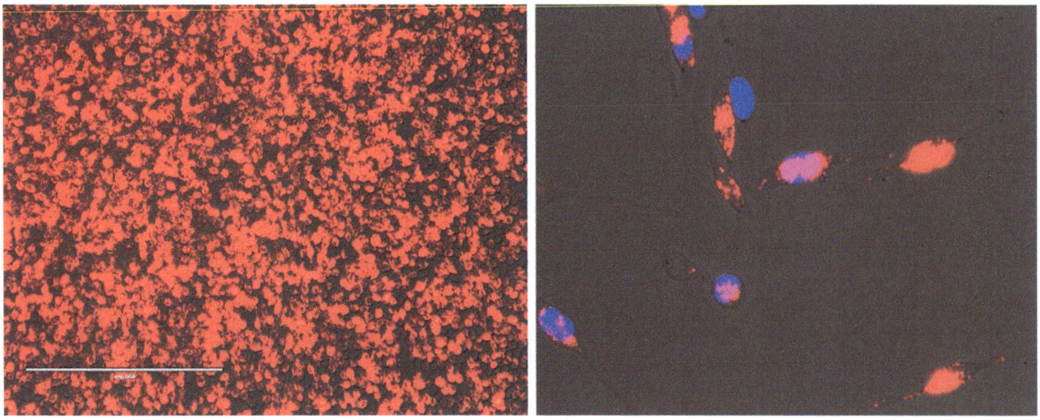

Fig. 1 Fluorescence microscopy images of unstimulated NPY-mRFP RBL cells at lower (100×) and higher (400×) magnification, showing strong red fluorescence localized to the granules. Nuclei were stained with 4′,6-diamidino-2-phenylindole (DAPI) nuclear stain in the right image

However, both available reporter systems also have some disadvantages. The luciferase reporter RS-ATL8 is unsuitable for use in allergen arrays, as the substrate requires cell lysis and uses expensive luciferase substrates, while the NFAT-DsRed reporter needs an overnight incubation before the results can be evaluated (*see* Chapter 12).

To address this need, we created a new RBL reporter system, which is characterized by lower running costs and shorter times of detection. The system exploits the targeting of neuropeptide Y (NPY) to the RBL granules, which results in granular localization of fluorescent proteins fused with it [14] (*see* Fig. 1).

This is explained in more detail in our recent review [1]. There are several key advantages over the NFAT RBL reporters; the system does not require any costly reagents, as degranulation is assessed by measuring the red fluorescent protein contained in the granules; incubation times are short, as degranulation will be terminated within 45–60 min after the addition of the stimulus; and finally, the system reports the degranulation event directly, rather than parameters associated with RBL activation, such as NFAT nuclear translocation. This means that the newly developed NPY-mRFP RBL reporter cell line can also be used for mechanistic studies of degranulation or degranulation inhibitors.

2 Materials

All reagents and supplies used for cell culture should be sterile. Follow local waste disposal regulations.

2.1 Reagents

1. RBL medium: Minimum Essential Medium (MEM), 10% v/v heat-inactivated fetal bovine serum, 100 U/mL penicillin, 100 μg/mL streptomycin, 2 mM L-glutamine (freshly added when making up medium; *see* **Note 1**). Store at 4 °C.
2. Phenol red-free medium. The same as above, but without any phenol red.
3. NPY-mRFP reporter cell lines (humanized or non-humanized), available from the authors through a Material Transfer Agreement (MTA). The reporter cells are available in two versions: NPY-mRFP RBL cells, which do not express the human high-affinity IgE receptor, and *hu*NPY-mRFP RBL, which expresses the human FcεRI alpha chain, enabling them to bind human IgE.
4. G418 sulfate (Geneticin) for humanized and non-humanized NPY-mRFP RBL.
5. Zeocin (for humanized NPY-mRFP RBL only).
6. Dulbecco's phosphate-buffered saline (DPBS) without Ca^{2+} or Mg^{2+}.
7. Trypsin-EDTA or cell scrapers.
8. Patient sera.
9. Allergens.
10. Anti-human IgE.
11. CELLBANKER cryopreservation medium.

2.2 Supplies

1. Graduated pipettes and pipette tips.
2. 25 cm^2/75 cm^2 cell culture flasks.
3. 50 mL centrifuge tubes.
4. Clear 96-well plates.
5. Black 96-well plates (for fluorescence assay).

2.3 Equipment

1. Pipettors for graduated pipettes/tips.
2. Water bath.
3. Class II microbiological safety cabinet.
4. Cell culture incubator: humidified incubator set to 37 °C and 5% CO_2.
5. Heat block.
6. Hemocytometer (e.g., Neubauer modified).
7. Centrifuge capable of holding 50 mL tubes.
8. Plate reader for fluorescence.

3 Method

3.1 General Cell Culture

Work in a class II microbiological safety cabinet using sterile techniques.

1. If starting from frozen stocks, partially defrost a vial of cells by warming in hand, and then add 1 mL pre-warmed RBL medium to fully defrost.
2. Transfer cells to a 25 cm^2 flask containing 9 mL RBL medium, and place in a cell culture incubator overnight.
3. When the cells are confluent in the 25 cm^2 flask (this may take several days, depending on the initial viability of defrosted cells), transfer cells to a 75 cm^2 flask by detaching adherent cells using trypsin-EDTA (*see* **steps 2–11** of Subheading 3.3).
4. Maintain cells in 10 mL RBL medium supplemented with 1 mg/mL G418 sulfate and 200 μg/mL Zeocin (*see* **Note 2**) in a 75 cm^2 flask, in the cell culture incubator.
5. Passage cells when confluent using trypsin-EDTA (*see* **steps 2–10** of Subheading 3.3 below). Replacing the flask with 20% cells will result in confluence after 3 days. Replace with 10% cells for confluence (*see* **Note 3**) after 4 days (refresh the medium after 2 or 3 days). Top up the level of liquid in the flask to 10 mL.
6. Freeze stocks of cells in larger batches, i.e., from several flasks. Centrifuge cells at 300 × *g* for 5 min, resuspend in CELLBANKER (*see* **Note 4**) to obtain a cell concentration of 5×10^6 cells/mL. Aliquot 1 mL cells/cryovial, transferring cryovials to a Mr. Frosty™ (Nalgene) container filled with isopropanol, and place in −80 °C for several hours or overnight. After freezing, transfer the cryovials to a liquid nitrogen Dewar flask for long-term storage. Defrost a vial to check the viability of the batch.

3.2 [Optional] Treatment of Serum Before Sensitization

1. If necessary, heat human serum in an Eppendorf tube at 56 °C for 5 min (*see* **Note 5**) on a heat block and then place it on ice.

3.3 Sensitization

Cells are sensitized for 16–20 h by incubating them in RBL medium with the desired dilution of previously heat-inactivated (if necessary) serum (*see* **Note 6**).

1. Pre-warm RBL medium to 37 °C in a water bath.
2. Remove the medium from 75 cm^2 flask of confluent cells.
3. To the flask, add 10 mL DPBS without Ca^{2+} or Mg^{2+} (*see* **Note 7**).
4. Tilt the flask to wash the cells with DPBS.

5. Discard the DPBS in the flask.
6. Repeat the washing step (**4** and **5**).
7. Add enough trypsin-EDTA to the flask so that the cells are fully covered with a thin layer of fluid (*see* **Note 8**).
8. Incubate the flask for 5–15 min at 37 °C in a cell culture incubator, until clumps of cells are visibly detached (*see* **Note 9**).
9. Add 8 mL pre-warmed RBL medium to the flask and resuspend the cells (*see* **Note 10**).
10. Transfer the cells to a 50 mL tube. Some cells can be left behind in the flask for passaging (*see* Subheading 3.1).
11. Count a sample of cells in a hemocytometer and calculate the total cell number.
12. Pellet the cells by centrifuging at 300 × *g* for 5 min.
13. Resuspend the cells to a concentration of 1×10^6 cells/mL using RBL medium.
14. Sensitize cells using appropriate concentration of serum (*see* **Note 5**).
15. Pipette 150 μL cells/well of a 96-well plate.
16. Incubate the plate overnight in the cell culture incubator.

3.4 Stimulation

1. Check the plate under a light microscope to see if the cells are in a healthy state.
2. Remove the medium from the cells in the 96-well plate.
3. Carefully wash the cells once with phenol red-free cell culture medium (*see* **Note 11**).
4. Replace the medium with 50 μL of phenol red-free cell culture medium.
5. Prepare appropriate dilutions of allergens or other stimuli in phenol red-free medium. Optimal concentrations may vary for each allergen and will need to be determined experimentally. Use an appropriate amount of anti-human IgE as a positive control (*see* **Note 12**).
6. Add 50 μL stimuli to appropriate wells of the plate.
7. Incubate the plate in the cell culture incubator for 45–60 min.

3.5 Fluorescence Assay with NPY-mRFP RBL

1. Transfer supernatants to a black 96-well plate (*see* **Note 13**). The medium only should be used as a blank, in addition to the supernatant of unstimulated cells. This will give an idea of spontaneous degranulation.
2. Measure fluorescence at 530 nm (excitation) and 590 nm (emission).

4 Notes

1. Glutamine is considered a nonessential amino acid, and it may therefore appear counterintuitive to add this to the medium. However, glutamine can become "conditionally" essential under certain conditions, such as in the case of fast-growing cells. L-Glutamine is unstable at physiological pH and is rapidly degraded into NH_4^+ and pyroglutamate (a heterocyclic lactone form of glutamic acid). The degradation of glutamine depends on the type of medium, the pH, and the temperature. Increasing amounts of NH_4^+ can have toxic effects and will affect cell growth and behavior. For these reasons, L-glutamate is added freshly to the medium together with other components such as serum or antibiotics. It is then stored at 4 °C.
2. The NPY-mRFP is a stable transfectant. However, as for all stable transfectants, there is the risk that expression of the transgene is lost over time. This will depend on the transgene copy number, the chromosomal integration sites, and multiple other factors. Thus, keeping the antibiotics used for selection in the medium can help prevent this unwanted loss of expression. It is possible to omit the antibiotics for a short (days) time, e.g., to reduce the cost of the media, but this needs to be carefully weighed against the risks. In this case, G418 is added to prevent loss of the NPY-mRFP transgene, while Zeocin is added to prevent loss of the stably transfected human FcεRIα chain.
3. A 1/20 split (5% cells) is not recommended for routine passaging as this can cause "colonies" to form and unhealthy cells in the middle of the cell clusters.
4. While other commercial and self-made freezing media are possible, we have good experience with CELLBANKER. The manufacturers state that cells can be transferred to a −80 °C freezer directly after resuspending the pelleted cells in CELLBANKER, without the need for the controlled 1 °C/min cooling rate provided by the Mr. Frosty device, and that cells can also be stored at −80 °C instead of liquid nitrogen, but we have not tested this. If using DMSO-based freezing media, work quickly, since 10% v/v DMSO is toxic to cells at room temperature.
5. Human sera are frequently cytotoxic for RBL cells. This is in part due to the presence of activated complement, which is traditionally inactivated by heating at 56 °C for 30 min. This prolonged treatment however also induces irreversible conformational changes in IgE molecules, ablating its ability to bind to FcεRI. A shorter heat inactivation will be sufficient to inactivate complement while preserving the integrity of IgE. We have

had good results with this protocol, even with 20% serum concentration (1:5). Not all sera, however, require heat inactivation. In most cases, a 1:50 or 1:100 dilution will be sufficient to remove cytotoxicity without any heat treatment. Thus the heat treatment is described as optional.

6. The exact duration of the overnight sensitization time can vary, as long as it is performed for a minimum of 8–10 h. We routinely start the sensitization protocol at 3 p.m. and the stimulation at 9 a.m. the following day.
7. Once fully confluent, a 75 cm^2 flask contains enough cells to fill 2 96-well plates.
8. You can use cell scrapers to detach the cells instead of trypsin-EDTA. Simply refresh the RBL medium and scrape the cells. There is no need for a DPBS wash.
9. You can speed up the detachment process by repeatedly knocking the bottom of the flask with your knuckles, holding it tight with the other hand. Be careful, however, as the use of too much force can split the flask.
10. Make sure any visible clumps are broken up, by pipetting up and down. We routinely use a 5 mL serological pipette for this purpose, as these have a narrower tapered end. Make sure you do not create bubbles when pipetting up and down by always leaving some medium in the pipette when expelling the liquid and not taking up any air when aspiring it.
11. Phenol red has an absorption maximum at 550 nm, which is very close to the 530 nm excitation wavelength of mRFP, and would therefore interfere with the measurements.
12. The optimal amount of anti-IgE will depend on the type (species, clonality, affinity, etc.) used and needs to be determined empirically. Optimum concentrations can range between 100 ng/mL and 2 μg/mL which therefore need to be determined by testing the anti-IgE antibody over a wide range of concentrations (e.g., at least 5 orders of magnitude). This should result in a bell-shaped stimulation curve, with a clear sub-optimal and supraoptimal stimulation range.
13. In our lab, we have tested different types of multiwell plates. Black plates give the greatest sensitivity for fluorescence assays. However, their background fluorescence can vary greatly between manufacturers. Always give preference to black plates optimized for low autofluorescence.

References

1. Falcone FH, Wan D, Barwary N, Sagi-Eisenberg R (2018) RBL cells as models for in vitro studies of mast cells and basophils. Immunol Rev 282:47–57
2. Falcone FH, Alcocer MJC, Okamoto-Uchida-Y, Nakamura R (2015) Use of humanized rat basophilic leukemia reporter cell lines as a diagnostic tool for detection of allergen-specific IgE in allergic patients: time for a reappraisal? Curr Allergy Asthma Rep 15:67. https://doi.org/10.1007/s11882-015-0568-3
3. Nakamura R, Uchida Y, Higuchi M et al (2010) A convenient and sensitive allergy test: IgE crosslinking-induced luciferase expression in cultured mast cells. Allergy 65:1266–1273. https://doi.org/10.1111/j.1398-9995.2010.02363.x
4. Wang X, Cato P, Lin H-C et al (2013) Optimisation and use of humanised RBL NF-AT-GFP and NF-AT-DsRed reporter cell lines suitable for high-throughput scale detection of allergic sensitisation in array format and identification of the ECM-integrin interaction as critical factor. Mol Biotechnol 56:136–146. https://doi.org/10.1007/s12033-013-9689-x
5. Hutchinson LE, McCloskey MA (1995) Fc epsilon RI-mediated induction of nuclear factor of activated T-cells. J Biol Chem 270:16333–16338
6. Ali EA, Nakamura R, Falcone FH (2017) Use of humanized RS-ATl8 reporter system for detection of allergen-specific IgE sensitization in human food allergy. Methods Mol Biol 1592:147–161
7. Nakamura R, Nakamura R, Adachi R et al (2013) Evaluation of allergenicity of acid-hydrolyzed wheat protein using an in vitro elicitation test. Int Arch Allergy Immunol 160:259–264. https://doi.org/10.1159/000341671
8. Jarupalee T, Chatchatee P, Komolpis K et al (2018) Detecting allergens from black tiger shrimp Penaeus monodon that can bind and cross-link IgE by ELISA, western blot, and a humanized rat basophilic leukemia reporter cell line RS-ATL8. Allergy, Asthma Immunol Res 10:62–76. https://doi.org/10.4168/aair.2018.10.1.62
9. Lv L, Lin H, Li Z et al (2018) Allergenicity of acrolein-treated shrimp tropomyosin evaluated using RBL-2H3 cell and mouse model. J Sci Food Agric 98:4374–4378. https://doi.org/10.1002/jsfa.8954
10. Wan D, Ludolf F, Alanine DGW et al (2014) Use of humanised rat basophilic leukaemia cell line RS-ATL8 for the assessment of allergenicity of Schistosoma mansoni proteins. PLoS Negl Trop Dis 8:e3124. https://doi.org/10.1371/journal.pntd.0003124
11. Knipping K, van Roest M, Kruijssen L et al (2016) Intra- and inter-laboratory validation of an innovative huFcεRIα-RBL-2H3 degranulation assay for in vitro allergenicity assessment of whey hydrolysates. Toxicol Vitr 33:29–34. https://doi.org/10.1016/j.tiv.2016.02.018
12. Hoffmann A, Jamin A, Foetisch K et al (1999) Determination of the allergenic activity of birch pollen and apple prick test solutions by measurement of β-hexosaminidase release from RBL-2H3 cells. Comparison with classical methods in allergen standardization. Allergy Eur J Allergy Clin Immunol 54:446–454. https://doi.org/10.1034/j.1398-9995.1999.00917.x
13. Kaul S, Lüttkopf D, Kastner B et al (2007) Mediator release assays based on human or murine immunoglobulin E in allergen standardization. Clin Exp Allergy 37:141–150. https://doi.org/10.1111/j.1365-2222.2006.02618.x
14. Lang T, Wacker I, Steyer J et al (1997) Ca1+-triggered peptide secretion in single cells imaged with green fluorescent protein and evanescent-wave microscopy. Neuron 18:857–863. https://doi.org/10.1016/S0896-6273(00)80325-6

Chapter 14

Use of Engineered Nanoparticles (ENPs) for the Study of High-Affinity IgE FcεRI Receptor Engagement and Rat Basophilic Leukemia (RBL) Cell Degranulation

Walla Alelwani, Raed A. Alharbi, Daniel Wan, Driton Vllasaliu, Franco. H. Falcone, and Snow Stolnik

Abstract

Degranulation of mast cells and basophils occurs after the cross-linking of FcεRI receptor-bound IgE by multivalent allergens, resulting in the release of a range of de novo synthesized and preformed mediators of the allergic response. β-Hexosaminidase release is usually measured as a simple readout for degranulation. Furthermore, the rat basophilic leukemia (RBL)-2H3 cell line is commonly used for measuring degranulation, monitoring β-hexosaminidase release. Here, we describe surface-engineered and modified nanoparticles with specific ligands in order to study the signaling and cellular responses of the RBL-2H3 cell line.

Key words Cross-linking, Degranulation, High-affinity receptor, FcεRI, IgE, RBL-2H3, Nanoparticles, Polystyrene

1 Introduction

Activation of mast cells and basophils is induced by clustering of high-affinity receptor (FcεRI)-bound IgE by multivalent antigens, resulting in a signaling cascade, culminating in degranulation and mediator release [1, 2]. This cross-linking of surface receptors will activate the tyrosine kinase Lyn under the regulation of the protein tyrosine phosphatase CD45 [3] and will promote phosphorylation of the ITAM within FcεRIβ and FcεRIγ chains [4]. Following the phosphorylation, the cytoplasmic signals are propagated by the stimulation of Src homology domain 2 (SH2) containing protein such as the spleen protein kinase Syk [5]. Then, several proteins including Src SH2 containing leukocyte-specific phosphoprotein of 76 kDa (SLP76), phospholipase C (PLC) γ1 and γ2, and Vav are phosphorylated by the activated Syk [6]. Cross-linking of FcεRI will also lead to Fyn activation, which led to the phosphorylation of adapter protein Grb2-associated binder-like protein 2 (Gab2).

Bernhard F. Gibbs and Franco H. Falcone (eds.), *Basophils and Mast Cells: Methods and Protocols*, Methods in Molecular Biology, vol. 2163, https://doi.org/10.1007/978-1-0716-0696-4_14,

Consequently, Gab2 will attach to the p85 subunit of phosphatidylinositol 3-kinase enzyme (PIK3) in order to produce phosphatidylinositol-1,4,5 triphosphate (IP3) and recruit pleckstrin homology (PH) domains such as Bruton's tyrosine kinase (Btk) and PLC-γ to the membrane. PLC-γ1 and PLC-γ2 are also phosphorylated. These phosphorylated PLCs catalyze the hydrolysis of plasma membrane PIK3, producing IP_3 and 1,2-diacylglycerol (DAG). These second messengers, IP_3 and DAG, stimulate the release of calcium ions from intercellular store sites and activate protein kinase C (PKC), respectively [7]. The depletion of Ca^{2+} from the endoplasmic reticulum lumen triggers Ca^{2+} influx across the plasma membrane through a specific Ca^{2+} channel, also known as CRAC (for Ca^{2+} release-activated Ca^{2+} channel. This influx enhances a free cytoplasmic Ca^{2+} concentration, which is a very crucial step in further intracellular signaling cascade ultimately resulting in degranulation [8]. Based on previous studies, the early signaling events of FcεRI will stimulate ERK-associated MAPKs (Vav, Raf1, MEK) and GTPases (Rac, Ras, Rho). These processes trigger MAPK pathways, which lead to the degranulation and induction of de novo leukotriene and cytokine synthesis [9, 10]. Additionally, degranulation releases preformed inflammatory mediators, including histamine [11], serotonin [3], β-hexosaminidase [12], and arachidonic acid [3]. This mechanism has been studied extensively for several decades using rat basophilic leukemia (RBL) cells as a well-established model [13].

The interaction of basophils or mast cells with nanomaterials, whether FcεRI receptor dependent or independent, is increasingly under focus. Recent work explored the potential of nanoparticles (NPs) to induce or inhibit allergic reactions. For example, ZnO, but not TiO_2, NPs were found to inhibit the degranulation of RBL-2H3 cells or bone marrow-derived mast cells (BMMCs); in fact, TiO_2 NPs induced degranulation [14]. A similar anti-allergic effect was described for subnanometer-sized C_{60} fullerene particles using human skin or lung mast cells and peripheral blood basophils; preincubation with these particles led to the strong inhibition of Lyn and Syk, inhibiting early signal transduction events [15]. In contrast, 5 nm silver NPs were able to induce RBL-2H3 degranulation, while 100 nm silver NPs did not. The former smaller NPs also exacerbated symptoms in a murine model of atopic dermatitis [16]. A similar pro-allergic effect has also been with 20 nm silver NPs, which have been described to degranulate RBL-2H3 and BMMCs by activating PI3K and PLCγ and increasing intracellular calcium levels in a FcεRI-independent manner, apparently involving SR-B1, a receptor for high-density lipoprotein (HDL) [17].

While an exhaustive review about the interactions between mast cells and basophils and NPs is beyond the scope of this introduction, it is clear that this recent emerging field has become a topical arena of research. In particular, engineered nanomaterials are receiving attention from several labs worldwide. In such materials, the engineered nanoparticles (ENPs) can be imparted with new biophysical and immunological properties through a careful design of their corona. ENPs have been investigated in basophil activation tests [18, 19] or immunotherapy [20–22]. Using suitably derivatized ENPs allows more refined control over stoichiometry and the spatial orientation of the ligands, hence a more systematic way of studying the interaction between receptors and their ligands on the cell surface. The extent of receptor aggregation and receptor-ligand binding influences the duration and intensity of intracellular signaling and other downstream events, such as those observed with transmembrane receptor tyrosine kinases [23]. Multivalent ligands, i.e., ligands with multiple receptor binding sites, can cross-link the membrane receptors effectively to control signaling processes [23]. Nanoparticles coated with biological ligands (proteins or/and antibodies) could potentially function as a multivalent ligand with a strongly enhanced ability to cross-link surface receptors.

Moreover, studies involving the biological interaction of ligand-coated nanomaterials have added clinical significance, because the results obtained can inform the design of targeted therapy and diagnosis platforms, as the designed nanoparticles can trigger specific signaling responses. In the current work, we describe the design of a polystyrene nanoparticle surface modified with monoclonal antibodies or ligands and monoclonal antibodies to examine their ability to engage FcεRI receptors for RBL-2H3 cells by measuring their degranulation.

2 Materials

All reagents and supplies should be sterile. Follow local waste disposal regulations.

2.1 Reagents

1. RBL medium: Minimum Essential Medium (MEM), 10% v/v heat-inactivated fetal bovine serum, 100 U/mL penicillin, 100 μg/mL streptomycin, 2 mM/L-glutamine (if not already present in MEM). Store at 4 °C.
2. DPBS without Ca^{2+} and Mg^{2+}.
3. Trypsin–EDTA.
4. Hank's Balanced Salt Solution (HBSS) with sodium bicarbonate, Ca^{2+}, Mg^{2+}, without phenol red.
5. 2,4-Dinitrophenylated bovine serum albumin (DNP-BSA).
6. A23187 (ionophore).
7. 4-*p*-Nitrophenyl-*N*-acetyl β-D-glucosaminidase.

8. Sodium citrate.
9. Glycine.
10. Lysis buffer 1% v/v.
11. 1% (v/v) Triton X-100 in DPBS. Store at 4 °C.
12. Polyethylene glycol (PEG) (MW = 2000 Da).
13. Monoclonal mouse anti-dinitrophenyl (DNP)-IgE (clone SPE-7) antibody.
14. Purified mouse 5-dimethylaminonapthalene-1-sulfonyl (dansyl) (clone 27–74) IgE antibody.
15. 2.5% w/v polybead carboxylated microsphere (50 nm).

2.2 Supplies

1. Graduated pipettes and pipette tips.
2. 25 cm^2/75 cm^2 cell culture flasks.
3. 50 mL centrifuge tubes.
4. Clear 96-well plates.
5. White 96-well plates (for degranulation assay).
6. Sterile empty glass vials.
7. Cryovial.

2.3 Equipment

1. Pipettors for graduated pipettes/tips.
2. Class II biosafety cabinet.
3. Cell culture incubator: humidified incubator set to 37 °C and 5% CO_2.
4. Hemocytometer (e.g., Neubauer modified).
5. Centrifuge capable of holding 50 mL tubes.
6. Plate reader for measuring enzymatic activity (β-hexosaminidase).
7. Balance.

3 Methods

3.1 General Cell Culture

Work in a class II biosafety cabinet using sterile techniques.

1. Using frozen stocks, thaw a vial of cells by placing it in the incubator at 37 °C for 5 min, and then add 1 mL pre-heated RBL medium.
2. Transfer RBL cells to a 25 cm^2 flask containing 8 mL of RBL medium, and keep it in the incubator overnight at 37 °C.
3. When the cells become confluent in the 25 cm^2 flask, detach and transfer RBL cells to a 75 cm^2 flask using trypsin–EDTA (*see* **steps 1–9** of Subheading 3.2).

4. Keep cells in 10 mL RBL medium.
5. When cells become confluent, passage cells using trypsin–EDTA (*see* **steps 1–10** below). Adding 20% of RBL cells to a new flask will lead to confluence after 3 days. Adding 10% of RBL cells will result in confluence after 4 days (add a new medium after 2 or 3 days). Add 10 mL of the medium to the flask.
6. Prepare stocks of cells from different flasks and freeze them. Centrifuge cells at 300 × *g* for 5 min, and resuspend in RBL medium with 10% v/v DMSO to achieve a cell concentration of 5 × 10^6 cells/mL. Do this step quickly, due to the toxicity of DMSO to cells at room temperature. Add 1 mL cells/cryovial, then transfer cryovials to a Mr. Frosty™ (Nalgene) container filled with isopropanol, and store it at −80 °C for several hours or overnight. For long-term storage, transfer the cryovials to a liquid nitrogen Dewar flask. To check the viability of the batch, thaw a vial.

3.2 β-Hexosaminidase Assay

3.2.1 Sensitization

1. Prewarm RBL medium to 37 °C in a water bath.
2. Discard the medium from confluent cells on the flask.
3. Add 10 mL to 75 cm^2 flask from Ca^{2+}/Mg^{2+}-free DPBS (*see* **Note 1**).
4. Rinse the cells by tilting the flask.
5. Remove the Ca^{2+}/Mg^{2+}-free DPBS from the flask.
6. Add 2 mL of trypsin–EDTA to the flask.
7. Swirl the flask gently to ensure the 2 mL trypsin–EDTA solution covered the cells.
8. Transfer the flask into the incubator for 5–15 min, until cell–cluster detachment is observed.
9. After the incubation period, tap the flask carefully until all cells have detached.
10. Add 8 mL of the prewarmed RBL medium to the flask and resuspend the cells.
11. Transfer the cell suspension into a sterile centrifuge (50 mL) tube. Some cells can be left behind in the flask for passaging (*see* Subheading 3.1).
12. Count cells using a hemocytometer and calculate the total cell number.
13. Centrifuge cell suspension in 50 mL tube at 400 × *g* for 5 min and pellet the cells.
14. Resuspend the cells to a concentration of 1 × 10^6 cells/mL using the RBL medium.

15. Sensitize the cells overnight for 16–20 h, using 100 ng/mL of the monoclonal mouse anti-DNP-IgE antibody or with monoclonal mouse dansyl-IgE antibody (*see* **Note 2**).
16. Pipette 150 μL in a 96-well plate at a seeding density of 1.5×10^5 cells/well.
17. Incubate the plate overnight at 37 °C in 5% CO_2 in a humidified incubator.

3.2.2 Stimulation

The procedure described here is based on the method optimized by Chatterjee et al. [24] and modified by Zwierz (*see* **Note 2**).

1. Check the 96-well plate under an optical microscope to see the health state of the cells.
2. Prepare 100 ng/mL DNP-BSA as a positive control and other stimuli (*see* Subheading 3.3) (*see* **Note 3**).
3. Remove the medium from the cells in the 96-well plate.
4. Wash the cell twice with HBSS.
5. Add 150 μL stimuli to appropriate wells of the plate. This step should be done quickly to avoid cell dryness when the medium is removed.
6. Incubate the plate in the incubator at 37 °C for 1 h.

3.2.3 Measurement of Activation

1. Use Triton X-100 (1%) and A23187 ionophore as 100% and positive controls, respectively, and use unsensitized cells without stimuli as a negative control.
2. After incubation time, transfer 30 μL of the cell supernatant to a new 96-well plate.
3. Add 50 μL of β-hexosaminidase substrate (*p*-nitrophenyl *N*-acetyl β-D-glucosaminide in 0.04 mM sodium citrate, pH 4.5) to transferred supernatant.
4. Incubate the plate for 1 h at 37 °C.
5. Add 100 μL of 0.2 M glycine (pH 10) to stop the reaction.
6. Detect β-hexosaminidase activity in the supernatant spectrophotometrically using a plate reader at a wavelength of 405 nm.

3.3 Polystyrene Nanoparticle Preparation

3.3.1 Polystyrene Surface Modified with DNP-BSA

1. Dilute 100 μL of 2.5% w/v stock suspension of polystyrene nanoparticles using dH_2O (1:10), achieving 0.25% nanoparticle suspension.
2. Prepare 1% w/v PEG solution (200 mg; MW 2000 Da) by dissolving in 20 mL dH_2O.
3. Dilute 1% w/v PEG solution using HBSS (1:10), achieving 0.1% w/v PEG solution.
4. Dilute 10 mg/mL DNP-BSA solution using dH_2O to a final concentration of 0.5 mg/mL (*see* **Note 4**).

5. Mix 0.25% PS nanoparticle suspension with 0.5 mg of DNP-BSA solution.
6. Incubate nanoparticle suspension with DNP-BSA for 2 h at room temperature.
7. After the incubation period, add 0.1% w/v PEG solution to suspensions of nanoparticles, further diluting the suspension by 1:10 (the final ratio of 1:100).

3.3.2 Polystyrene Surface Modified with DNP-BSA and IgE

1. Follow the same steps as **1–5** of Subheading 3.3, and then add 50 μg of monoclonal mouse anti-DNP-IgE to nanoparticle suspension with DNP-BSA (*see* **Note 5**).
2. Incubate the nanoparticle suspension with DNP-BSA and anti-DNP-IgE suspension for 2 h at room temperature. Continue the final **step** 7 as described in Subheading 3.3.1.

4 Notes

1. Bovine serum contains large amounts of protease inhibitors (mainly α1-antitrypsin) which would interfere with trypsinization of the cell line; therefore it needs to be removed by washes before the addition of Trypsin–EDTA. Attachment is Ca^{2+}/Mg^{2+} dependent, hence the use of Ca^{2+}/Mg^{2+}-free buffer and EDTA, as the removal of these ions will promote cellular detachment. It is however essential to replace Ca^{2+} in later steps, as this ion will be needed for RBL activation, as described in Subheading 1.
2. Activation of RBL-2H3 cells is determined by measuring the release of mediators, notably the release of β-hexosaminidase, which is used as the standard readout. Different β-hexosaminidase assay procedures have been described. However, it should be noted that the β-hexosaminidase assay described here has been optimized to achieve the maximum possible degranulation taking into consideration nanoparticle colloidal stability. Several sets of parameters were examined such as calcium concentration, preincubation with fibronectin, addition of PEG, and ligand and antibody concentration in order to find an optimal condition that suits both the degranulation assay and nanoparticle characteristics.
3. We usually perform the sensitization overnight (roughly 16 h), but a minimum time of 8 h is advisable [25].
4. The surface area of PS nanoparticles (50 nm) in the sample (100 μL of nanoparticle stock suspension containing 2.5% w/v solid) was calculated to estimate the required amount of BSA that needed to reach the adsorption plateau, based on the approximate-projected surface area of BSA, which in turn was estimated from its molecular dimensions (140 × 40 × 40 Å) [26].

5. The surface area of PS nanoparticles (50 nm diameter) in the sample (100 μL of nanoparticle stock suspension containing 2.5% w/v solid) was calculated to estimate the required amount of IgE that is needed to reach the adsorption plateau. The adsorption plateau has been extensively studied for IgG, taking into consideration the molecular surface during adsorption. To the best of our knowledge, the molecular area of IgE has not been defined. The needed amount of IgE was determined here based on the IgG adsorption capacity calculation, although IgE has a larger size (MW 180 kDa) than IgG (MW 150 kDa).

   The amount of IgG added to PS NPs was determined based on the theoretical number of IgG molecules that can be adsorbed on the surface of one particle. This was firstly calculated by defining the number of NPs in 1 mL of the stock solution using the equation supplied by the manufacturer (Polysciences, Inc.):

$$NP_n = \frac{6.03 \times 10^{10} \times S}{3.3 \times d^3} = 5.746 \times 10^{14}$$

where NP_n is the nanoparticle number, S is the % solid in the nanoparticle stock solution, and d is the diameter (μm).The surface area of one NP was determined using the following equation:

$$SA = 4\pi r^2 \; [nm^2]$$

where SA is the surface area and r is the NP radius.

Based on previous studies [27], the minimal molecular area of T-shaped IgG in the adsorbed closely packed monolayer has been determined using X-ray diffraction data as roughly 7100 $Å^2$ when the IgG molecules attached to the surface by the "leg" (F_c region) or by the "cap" (F_{ab} end of adsorption orientation), roughly 4200 $Å^2$ at the "side on" adsorption, and roughly 12,100 $Å^2$ at the "flat" adsorption. As a result, a hypothetical estimated range of the number of IgG molecules adsorbed per one nanoparticle can be estimated by the following method:

$$\text{AC between} \frac{SA\,[nm^2]}{4200\,[A^2]} \quad \text{or} \quad \frac{SA\,[nm^2]}{12,100\,[A^2]}$$

where AC is the adsorption capacity regarding the number of IgG molecules per 1 NP.

The number of IgG molecules adsorbed on the NP surface calculated in 1 mL of NP suspension (undiluted stock 2.5% w/v) was calculated from AC × NP_n, and then amounting by Avogadro's constant and finally to mg. In our experiment, the final value of mg was divided by 100 as the particle suspension was diluted 1:100.

The selected amounts of IgG based on their adsorbed orientation was between 60 μg and 170 μg (depending on the type of adsorption assumed), and here 50 μg for IgE antibodies were used when mixed with nanoparticles.

Acknowledgments

This work was supported by the Higher Ministry of Education in Saudi Arabia.

References

1. Falcone FH, Haas H, Gibbs BF (2000) The human basophil: a new appreciation of its role in immune responses. Blood 96:4028–4038
2. Turner H, Kinet JP (1999) Signalling through the high-affinity IgE receptor Fc epsilon RI. Nature 402:B24–B30
3. Sánchez-Mejorada G, Rosales C (1998) Signal transduction by immunoglobulin Fc receptors. J Leukoc Biol 63:521–533
4. Kinet JP (1999) The high-affinity IgE receptor (Fc epsilon RI): from physiology to pathology. Annu Rev Immunol 17:931–972
5. Hamawy MM, Mergenhagen SE, Siraganian RP (1995) Protein tyrosine phosphorylation as a mechanism of signalling in mast cells and basophils. Cell Signal 7:535–544
6. Siraganian RP (2003) Mast cell signal transduction from the high-affinity IgE receptor. Curr Opin Immunol 15:639–646
7. Knol EF (2006) Requirements for effective IgE cross-linking on mast cells and basophils. Mol Nutr Food Res 50:620–624
8. Hajkova Z, Bugajev V, Draberova E et al (2010) STIM1-directed reorganization of microtubules in activated mast cells. J Immunol 186:913–923
9. Jabril-Cuenod B, Zhang C, Scharenberg AM et al (2002) Syk-dependent phosphorylation of Shc. J Biol Chem 271:16268–16272
10. Song JS, Gomez J, Stancato LF, Rivera J (1996) Association of a p95 Vav-containing signaling complex with the FcεRI γ chain in the RBL-2H3 mast cell line. Evidence for a constitutive in vivo association of Vav with Grb2, Raf-1, and ERK2 in an active complex. J Biol Chem 271:26962–26970
11. Barsumian EL, Isersky C, Petrino MG, Siraganian RP (1981) IgE-induced histamine release from rat basophilic leukemia cell lines: isolation of releasing and nonreleasing clones. Eur J Immunol 11:317–323
12. Schwartz LB, Austen KF, Wasserman SI (1979) Immunologic release of beta-hexosaminidase and beta-glucuronidase from purified rat serosal mast cells. J Immunol 123:1445–1450
13. Passante E, Frankish N (2009) The RBL-2H3 cell line: its provenance and suitability as a model for the mast cell. Inflamm Res 58:737–745
14. Feltis BN, Elbaz A, Wright PFA et al (2015) Characterizing the inhibitory action of zinc oxide nanoparticles on allergic-type mast cell activation. Mol Immunol 66:139–146
15. Ryan JJ, Bateman HR, Stover A et al (2014) Fullerene nanomaterials inhibit the allergic response. J Immunol 179:665–672
16. Kang HG, Kim S, Lee KH et al (2017) 5 nm silver nanoparticles amplify clinical features of atopic dermatitis in mice by activating mast cells. Small 13:1602363
17. Alsaleh NB, Persaud I, Brown JM (2016) Silver nanoparticle-directed mast cell degranulation is mediated through calcium and PI3K signaling independent of the high affinity IgE receptor. PLoS One 11:e0167366
18. Radauer-Preiml I, Andosch A, Hawranek T et al (2016) Nanoparticle-allergen interactions mediate human allergic responses: protein corona characterization and cellular responses. Part Fibre Toxicol 13:3
19. Molina N, Martin-Serrano A, Fernandez TD et al (2018) Dendrimeric antigens for drug allergy diagnosis: a new approach for basophil activation tests. Molecules 23
20. Gamazo C, Gastaminza G, Ferrer M et al (2014) Nanoparticle based-immunotherapy against allergy. Immunotherapy 6:885–897
21. Balenga NAB, Zahedifard F, Weiss R et al (2006) Protective efficiency of dendrosomes as novel nano-sized adjuvants for DNA

vaccination against birch pollen allergy. J Biotechnol 124:602–614

22. Gómez S, Gamazo C, San Roman B et al (2009) A novel nanoparticulate adjuvant for immunotherapy with Lolium perenne. J Immunol Methods 348:1–8
23. Jiang W, Kim BYS, Rutka JT, Chan WCW (2008) Nanoparticle-mediated cellular response is size-dependent. Nat Nanotechnol 3:145–150
24. Chatterjee S, Velicer LF, Sweeley CC (1975) Glycosphingolipid glycosyl hydrolases and glycosidases of synchronized human KB cells. J Biol Chem 250:4972–4979
25. Wang X, Cato P, Lin H-C et al (2013) Optimisation and use of humanised RBL NF-AT-GFP and NF-AT-DsRed reporter cell lines suitable for high-throughput scale detection of allergic sensitisation in array format and identification of the ECM-integrin interaction as critical factor. Mol Biotechnol 56:136–146
26. Sofia SJ, Premnath V, Merrill EW (1998) Poly (ethylene oxide) grafted to silicon surfaces: grafting density and protein adsorption. Macromolecules 31:5059–5070
27. Kamyshny A, Toledano O, Magdassi S (1999) Adsorption of hydrophobized IgG and gelatin onto phosphatidyl choline-coated silica. Colloids Surf B Biointerfaces 13:187–194

Part IV

Diagnostic Applications

Chapter 15

Flow Cytometric Allergy Diagnosis: Basophil Activation Techniques

Chris H. Bridts, Vito Sabato, Christel Mertens, Margo M. Hagendorens, Luc S. De Clerck, and Didier G. Ebo

Abstract

The basis of flow cytometric allergy diagnosis is the quantification of changes in the expression of basophilic surface membrane markers (Ebo et al., Clin Exp Allergy 34: 332–339, 2004). Upon encountering specific allergens recognized by surface receptor FcεRI-bound IgE, basophils not only secrete and generate quantifiable bioactive mediators but also upregulate the expression of different markers (e.g., CD63, CD203c) which can be detected by multicolor flow cytometry using specific monoclonal antibodies (Ebo et al., Cytometry B Clin Cytom 74: 201–210, 2008). Here, we describe two flow cytometry-based protocols which allow the detection of surface marker activation (Method 1) and changes in intragranular histamine (Method 2), both reflecting different facets of basophil activation.

Key words Basophil activation tests, CD203c, CD63, HistaFlow

1 Introduction

Basophil activation assays have predominantly focused on histamine and leukotriene release assays. However, time-consuming procedures were needed to isolate cells, to incubate with activators, and to quantify (sometimes labile) mediators. Two decades ago, the discovery of the marker CD63 (aka gp53) and subsequently CD203c (aka ENPP-3) facilitated the development of a novel technique to analyze and quantify the allergen-specific in vitro activation of basophils: the so-called basophil activation test (BAT) [1, 2]. Briefly, the BAT relies on a flow cytometric identification and quantification of various specific activation markers on the surface membrane or activated proteins inside the cells. These changes can be detected and quantified on a single-cell level using specific monoclonal antibodies coupled to fluorochromes. Practically, basophils are identified by specific markers such as CCR3/CD3 [3], CD123/HLA-DR [4], or IgE/CD203c [5]. Of these markers, only CD203c is lineage specific. Subsequently, after

Bernhard F. Gibbs and Franco H. Falcone (eds.), *Basophils and Mast Cells: Methods and Protocols*, Methods in Molecular Biology, vol. 2163, https://doi.org/10.1007/978-1-0716-0696-4_15, © Springer Science+Business Media, LLC, part of Springer Nature 2020

activation with an allergen, the upregulation or appearance of specific surface activation markers, such as CD63 and/or CD203c, can be measured. Activation of basophils can also be quantified by means of intracellular molecules, such as phosphorylation of MAPK (p38MAPK) [6, 7] or STAT5 (pSTAT5) [8]. At present, BAT has proven to be a rapid and reliable diagnostic method that allows the diagnosis of allergy to various allergens like food (including pollen-related food allergies), natural rubber latex, hymenoptera venom, and certain drugs. In addition, the technique has proven to be useful in non-IgE-mediated reactions, such as in drug hypersensitivity as well as in the detection of IgG anti-FcεRI or anti-IgE autoantibodies in chronic autoimmune urticaria.

Finally, it was recently shown that intracellular histamine and its release by in vitro activated basophils can successfully be analyzed by multicolor flow cytometry (HistaFlow) [9] applying an enzyme-affinity fluorochrome method, based on the affinity of the histaminase diamine oxidase for its substrate [10]. We developed a flowcytometric technique (HistaFlow) which enables the measurement of histamine release by individual basophils in combination with cell membrane markers. Conjugating diamine oxidase to fluorochromes together with phenotyping activated basophils, histamine content and release can be quantified in whole blood with standardized flow cytometry techniques.

2 Materials

All solutions must be prepared with sterile ultrapure water. We use a Millipore Milli-Q System and water purification is performed with the use of a reversed osmosis system followed by ultra-purification to attain 18 MΩ at room temperature.

2.1 Blood Sample

Peripheral heparinized blood (10 mL) is obtained and immediately brought to the laboratory. Incubation and analysis must be completed within 8 h (*see* **Note 1**).

2.2 Reagents

1. Interleukin-3 (IL-3).

   IL-3 is sometimes used to preactivate ("prime") basophils. It must be noted that IL-3 can upregulate CD203c.

   (a) Stock solution: 10 μg IL-3 in 10 mL water, divide per 100 μL in sterile tubes. Store at −20 °C.

   (b) Storage solution: dilute 100 μL stock solution in 5 mL BAT buffer and divide per 500 μL. Store at −20 °C.

   (c) Working solution: dilute storage solution with an equal volume of buffer; prepare daily before use. The final concentration is 10 ng/mL.

2. BAT buffer.
   - (a) A BAT buffer is prepared under sterile conditions just before use.
   - (b) Dilute 1.9 mL 10× concentrated Hanks' balanced salt solution (HBSS with Ca^{2+} and Mg^{2+}) with 17 mL H_2O.
   - (c) Add 0.6 mL 7.5% $NaHCO_3$.
   - (d) Add 0.4 mL 1 M HEPES.
   - (e) Adjust pH to 7.4.
3. Lysis solution.

   Lysis solution (BD Biosciences) is diluted 1:10 with water and stored at room temperature.
4. Phosphate buffered saline (PBS 1×).
   - (a) PBS (10× concentrated without Ca^{2+} or Mg^{2}) is diluted with water to obtain an isotonic PBS (1×) buffer. Adjust pH to 7.4.
5. PBS-EDTA.

   Dissolve 10 mmol/L Na_2EDTA in PBS (1×). Adjust pH to 7.4.
6. PBS-NaN_3.

   Dissolve 0.5 g/L NaN_3 in PBS (1×). Adjust pH to 7.4. Store at room temperature.
7. PBS-0.1% BSA.

   Mix 50 mL PBS with 0.66 mL BSA (Albumin Bovine Fraction V 7.5%); pH should be 7.4.
8. Anti-human IgE Alexa Fluor© 488 (aIgE).

   Since anti-IgE (Sigma-Aldrich, Clone GE-1) conjugated with an Alexa Fluor© 488 dye is currently not available, this antibody is labeled in house using the procedure by the manufacturer Life Technologies described Subheading 3.
9. CD63-PE or CD63-FITC.

   Anti-CD63-PE or anti-CD63-FITC (BD Biosciences) must be stored at 2–8 °C in the dark. This antibody is ready to use (*see* **Note 2**).
10. Anti-CD123-PE/anti-HLAdr-PERCP/anti-CD63-FITC.

    All antibodies are commercially available and ready to use (BD Biosciences) (*see* **Note 2**).
11. Mouse anti-human IgE (mHu-antiIgE).
    - (a) This antibody (BD Biosciences, clone G7-18) is used as a positive control to stimulate basophils. Stock solution: 0.5 mg/mL in BAT buffer.
    - (b) Working solution: between 1 μg/mL and 10 μg/mL in BAT buffer.

12. CD203c-APC.

    Anti-human CD203c (clone NP4D6) labeled with APC (BD Biosciences or BioLegend) is ready to use (*see* **Note 2**).

13. Allergens.

    Allergens can be protein extracts, recombinant proteins, or drugs. They must be water soluble and used in concentrations which are nontoxic for basophils (*see* **Note 3**).

14. DAO-PE or DAO-V500 (BD Biosciences).
    (a) Diamine oxidase or histaminase (Sigma-Aldrich) has been conjugated with phycoerythrin or Horizon V500 at the laboratories of BD Biosciences.
    (b) Prepare a working solution 1:1 (DAO-PE) or 1:20 (DAO-V500) in PBS/EDTA. Every new batch has to be titrated and measured with a flow cytometer in order to find the optimal staining concentration for intracellular histamine.

15. Lysis/fixing solution (BD Biosciences).
    (a) BD Phosflow Lyse/Fix Buffer (5× concentrate: BD Biosciences) is used to lyse and fix whole blood for use with intracellular staining techniques. Store at room temperature.
    (b) Dilute 1:5 just before use in H_2O.

16. PBS-TX100.

    Working solution is 0.1% Triton X-100 in PBS: add 0.1 mL pure Triton X-100 (Sigma-Aldrich) to 100 mL PBS (1×). Store at room temperature.

17. Calibration particles.

    Rainbow Calibration Particles—six peaks or eight peaks (BD Biosciences). Store undiluted at 4 °C in the dark. Mix well before use. Dilute three to five drops in 1 mL buffer. Use immediately.

3 Methods

3.1 Flow Cytometric Basophil Activation Technique: Membrane Markers (Method 1)

All procedures are carried out at room temperature, in a laminar air flow cabinet when necessary, unless otherwise specified.

3.1.1 Alexa Fluor© 405 Labeling of Mouse Monoclonal Anti-human IgE

1. Labeling procedure: mouse monoclonal anti-human IgE (clone GE-1) is diluted in $NaHCO_3$ (0.1 mol/L in H_2O, pH 8.3) at a concentration of 5 mg/mL.
2. Alexa Fluor 488 carboxylic acid, succinimidyl ester, and triethylammonium salt (Life Technologies, Molecular Probes) can be

stored at −20 °C for a maximum of 3 months. When dissolved in DMSO, the solution becomes unstable. A working solution is made fresh by dissolving 1 mg in 0.1 mL DMSO.

3. Mix 2 mL anti-IgE in $NaHCO_3$ slowly with 0.1 mL Alexa Fluor© 488 working solution for 1 h at room temperature in the dark. Mix regularly.
4. Dialyze overnight with PBS at 4 °C.
5. Add NaN_3 until a final concentration of 0.05% is reached. Store at 2–8 °C.
6. Working solution: 1:50 in PBS-NaN_3. Store at 2–8 °C. The working solution has to be titrated every time a new batch is made (*see* also **Note 2**).

3.1.2 Preparation of Positive and Negative Controls (Exclusion of Nonresponders)

1. Incubate 100 μL blood with 100 μL negative (BAT buffer) or positive control (anti-IgE) for 20 min at 37 °C.
2. Using the procedure described below, stain with aIgE-AF488 and CD203c-APC (*see* **Note 4**) or CD123/HLA-Dr and CD203c-APC (*see* **Note 4**).
3. When basophils do not react, one can decide to stimulate with two concentrations of the allergen to exclude nonreactivity to anti-IgE (*see* **Note 5**).

3.1.3 Allergen Procedure

1. All blood samples, buffers, and allergens are prewarmed without mixing for 15 min at 37 °C in warm water bath (WWB) (*see* **Note 6**).
2. Prepare incubation tubes and add 100 μL allergen solution to the allergen tubes. Dilutions depend on the allergen used (*see* **Note 3**). Use BAT buffer as a negative control or for FMO evaluation (*see* **Note 2**).
3. Add anti-IgE as positive control to a positive control tube. Add 100 μL BAT buffer to a negative control tube.
4. Add to all tubes 100 μL prewarmed whole blood.
5. Incubate for 20 min at 37 °C in WWB.
6. Stop reaction by adding 1 mL PBS-EDTA.
7. Spin at 200 × *g* for 10 min at 4 °C without setting on the brake of the centrifuge. Remove supernatant.
8. Add 20 μL anti-IgE-AF488 and 10 μL anti-CD63-PE (or in case add 10 μL anti-CD123/HLADr/CD63). Add 5 μL anti-CD203c-APC.
9. Incubate on ice for 20 min in the dark.
10. Add 2 mL lysis solution and incubate at room temperature for 20 min.
11. Spin for 10 min at 250 × *g* on room temperature. Remove supernatant.

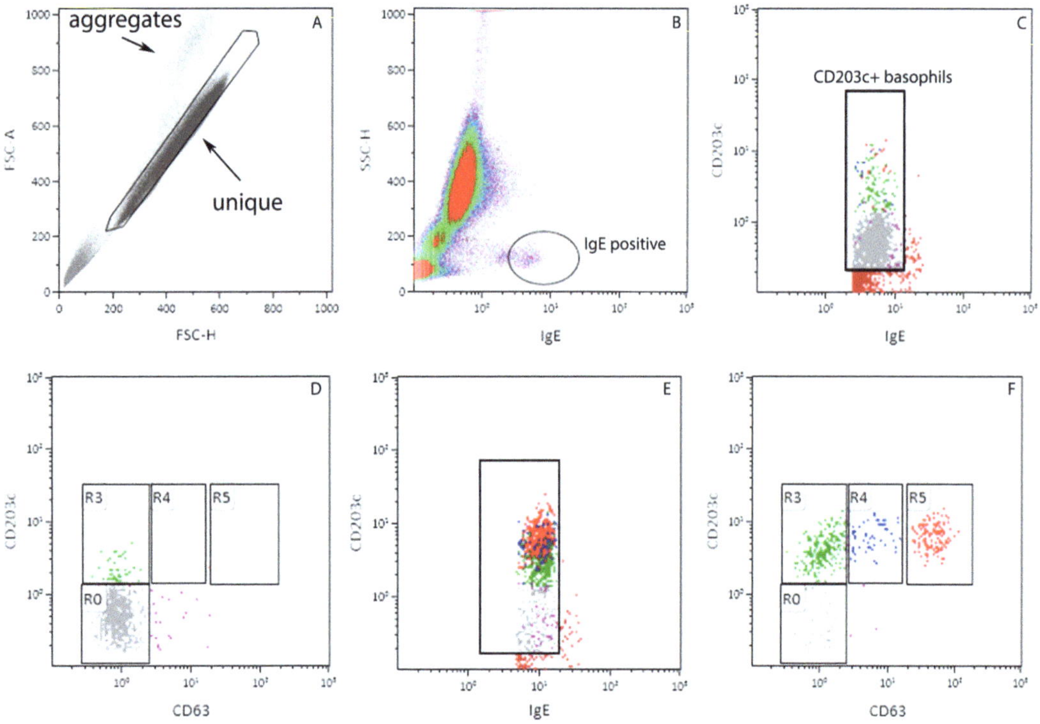

Fig. 1 Basophil selection and activation. Panel **a**: aggregates are avoided by selecting unique cells based on FCS (forward scatter) area and FCS height plot. Plot **b**: IgE high positive cells are gated out. Panel **c**: gated cells are IgE+ and CD203+ (without stimulation). Panel **d**: activated basophils in control setting—R0 are CD203c +CD63−, R3 are CD203chi+CD63−, R4 are CD203chi+CD63low+, and R5 are CD203chi+CD63hi+. Panels **e** and **f** are equal to **c** and **d** but after activation with anti-IgE as a positive control

12. Add 2 mL PBS and spin again for 10 min 250 × *g* at room temperature. Remove supernatant.
13. Add 300 μL PBS-NaN_3. The tubes are now ready to measure with a flow cytometer.
14. Count at least 500 basophils which are defined as aIgE+ and CD203c+.
15. Store data in an FCS formatted file, which can be analyzed with flow cytometric software.
16. Define $CD203c^{hi+}CD63^{-}$ basophils and $CD203c^{hi+}CD63^{hi+}$ as activate basophil populations (*see* Fig. 1).
17. Express as %$CD203c^{hi+}CD63^{-}$ or %$CD203c^{hi+}CD63^{hi+}$ basophils by subtracting % positive control or allergen with % negative control.

3.2 HistaFlow: Basophil Activation Technique with Intracellular Staining of Histamine (Method 2)

All procedures are carried out at room temperature, in a laminar air flow cabinet when necessary, or unless otherwise specified. The procedure is identical as in subheading 3.1 (except volumes) until **step 9**.

3.2.1 Preparation of Positive and Negative Controls (Exclusion of Nonresponders)

1. Incubate 200 μL of blood with 200 μL negative (BAT buffer) or positive control (anti-IgE) for 20 min at 37 °C.
2. Using the procedure described below, stain with aIgE-AF488 and CD203c-APC (*see* **Note 4**).
3. When basophils do not respond, one can decide to stimulate with two concentrations of the allergen to exclude nonreactivity to anti-IgE (*see* **Note 5**).

3.2.2 Allergen Procedure

1. All blood samples, buffers, and allergens are prewarmed without mixing for 15 min at 37 °C in WWB (*see* **Note 6**).
2. Prepare incubation tubes and add 200 μL allergen solution to the allergen tubes. Dilutions depend on the allergen used (*see* **Note 3**). Use BAT buffer as a negative control or for FMO evaluation (*see* **Note 2**).
3. Add anti-IgE as positive control to a positive control tube. Add 100 μL BAT buffer to a negative control tube.
4. Add to all tubes 100 μL prewarmed whole blood.
5. Incubate for 20 min at 37 °C in WWB.
6. Stop reaction on ice and add 1 mL PBS-EDTA.
7. Spin at 500 × *g* for 10 min at 4 °C without setting on the brake of the centrifuge. Remove supernatant.
8. Add 20 μL anti-IgE-AF405, 10 μL anti-CD63-FITC, and 5 μL anti-CD203c-APC.
9. Incubate at 4 °C for 20 min in the dark.
10. Add 2 mL Phosflow Lyse/Fix work solution and incubate at 37 °C in WWB for 20 min.
11. Spin for 10 min at 500 × *g* at room temperature. Remove supernatant.
12. Add 2 mL 0.1% PBS-TX100 and spin again for 10 min 500 × *g* at room temperature. Remove supernatant.
13. Add 100 μL 0.1% PBS-TX100 and 10–40 μL (depending on titration) DAO-V500.
14. Stain for 45 min at 37 °C in WWB. Mix several times during stain procedure.
15. Add 2 mL PBS and spin at 500 × *g* for 10 min at room temperature. Remove supernatant.

16. Add 300 μL mL PBS-NaN_3.
17. The tubes are now ready to be measured with a flow cytometer.
18. Count at least 1000 basophils, defined as $aIgE^+$ and $CD203c^+$.
19. Store data in an FCS formatted file, which can be analyzed with flow cytometric software.
20. Define $CD203c^{hi+}CD63^-$ basophils and $CD203c^{hi+}CD63^{hi+}$ as activated basophil populations. Define histamine content using the compensation setting described below. This setting must be unique for every sample.

3.3 Compensation Procedure

A functional assay like the HistaFlow technique needs an optimal flow cytometric compensation setting which is not always necessary when measuring populations, as long as the populations can be differentiated. In the case of histamine release, we must be aware that not all basophils release the same amount of histamine; hence different populations can (and must be) observed.

One could also explain this by comparing it with taking a picture with a digital camera. You need a correct exposure to see all the details. To do so, one needs a correct exposure time or shutter speed (PMT setting in flow cytometry) and a correct aperture (compensation setting in flow cytometry).

This is in fact what we do by combining the data of the four tubes in the compensation setting of the HistaFlow technique (*see* Table 1).

The first tube is without staining for DAO/CD63. These are the histamine-negative basophils. The second tube is DAO staining alone. These are the DAO (or histamine)-positive basophils.

The third tube is special: DAO as well as CD63 staining after fixation and without stimulation. This will give us the optimal staining for all populations.

The fourth tube is DAO and CD63 after activation (just like a positive control): this tube gives us the possibility to optimize the compensation setting (under an optimal PMT setting). By fine-tuning one will easily observe the different populations releasing basophil.

After staining each tube separately, we can mix them to find out the optimal setting or one can measure every tube individually, and, using software, add all the data in one data file. Afterward we measure all tubes of the same patient with this setting.

Ensure that you repeat this for every donor due to potential differences in histamine content.

Table 1
Compensation method

Tube	1	2	3	4
Type	CD63−DAO−	CD63−DAO+	CD63++DAO+	CD63+ +DAO−
Explanation	Negative cells	DAO+ cells without activation	DAO+ and *intracellular* CD63+ staining *without* stimulation	DAO and CD63 staining *with* stimulation
Incubation/activation procedure (see method 2)				
Whole blood	200 μL	200 μL	200 μL	200 μL
BAT buffer	200 μL	200 μL	200 μL	–
mHu-antiIgE (pos. control)	–	–	–	200 μL
Incubation (see method 2)				
Membrane staining procedure				
CD63	None	None	None	None
IgE/CD203c	20 μL/10 μL	20 μL/10 μL	20 μL/10 μL	20 μL/10 μL
Lyse/Fix (see method 2) *Wash with PBS_0.1% Triton X*				
Intracellular staining procedure (*see* method 2)				
PBS-TX100	100 μL	100 μL	100 μL	100 μL
CD63	–	–	10 μL	10 μL
DAO	–	10 μL	10 μL	10 μL
Population by correct compensation (*see* Fig. 2)	*Grey cell in gate C* (no staining for DAO and CD63)	*Green cells in gate D* (DAO+ staining alone)	*Blue cells in gate E* (all basophils are positive for DAO and CD63; note that there is no CD203c upregulation)	*Red cells in gate F* (histamine-releasing cells)

3.4 Calibration Procedure

1. Measure working solution of diluted beads with the identical flow cytometric settings as for the HistaFlow analysis (*see* Fig. 3).
2. Make a calibration graph (lin-log) with the use of the channel number and the median fluorescence intensity (MFI) as described by the manufacturer (*see* Fig. 2).
3. Use this calibration curve to calculate the MFI of the histamine-containing cell population (Fig. 4).

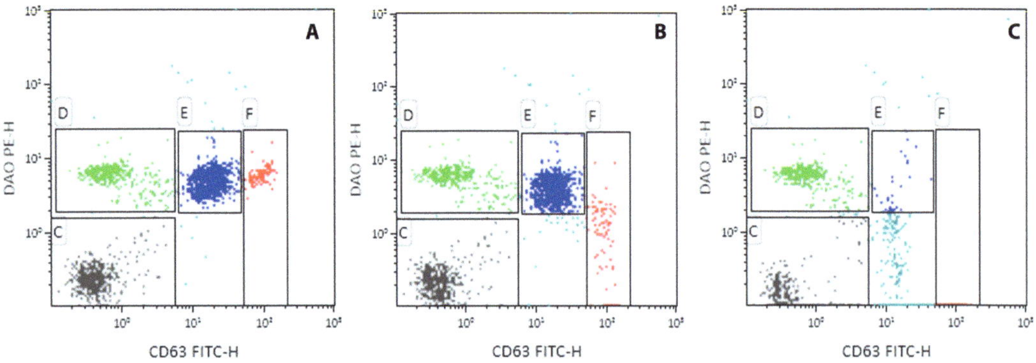

Fig. 2 Correct compensation setting. Basophils were selected as CD203$^+$antiIgE$^+$. Panel **a** shows under-compensation, panel **b** is correctly compensated, and panel **c** is overcompensated

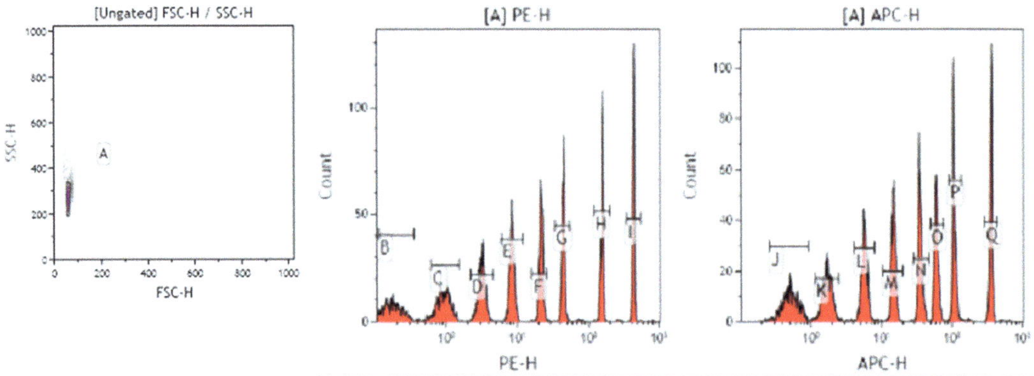

Gate	Number	%Total	X-GMean
All	7 834	78.34	3.39
B	730	7.30	0.16
C	872	8.72	0.91
D	944	9.44	3.12
E	993	9.93	8.41
F	970	9.70	21.64
G	960	9.60	44.34
H	1 012	10.12	155.17
I	1 067	10.67	433.62

Gate	Number	%Gated	X-GMean
All	7 834	100.00	10.89
J	886	11.31	0.50
K	911	11.63	1.70
L	932	11.90	5.70
M	986	12.59	14.75
N	975	12.45	35.46
O	963	12.29	60.61
P	1 022	13.05	108.55
Q	1 065	13.59	354.95

Fig. 3 Selection of standardized beads (left) and calculation of geometric mean of eight different intensities (middle, PE; right, APC). Results are shown in column X_Gmean

4 Notes

1. Although different BAT techniques are described using EDTA or even citrate-treated blood, the best results are obtained with heparinized blood, as basophil activation is very calcium dependent and using chelators or calcium binding products can have deleterious effect on test results. Antihistamines, which target the histamine receptors, have no influence on BATs in contrast to corticosteroids and immunosuppressive drugs.

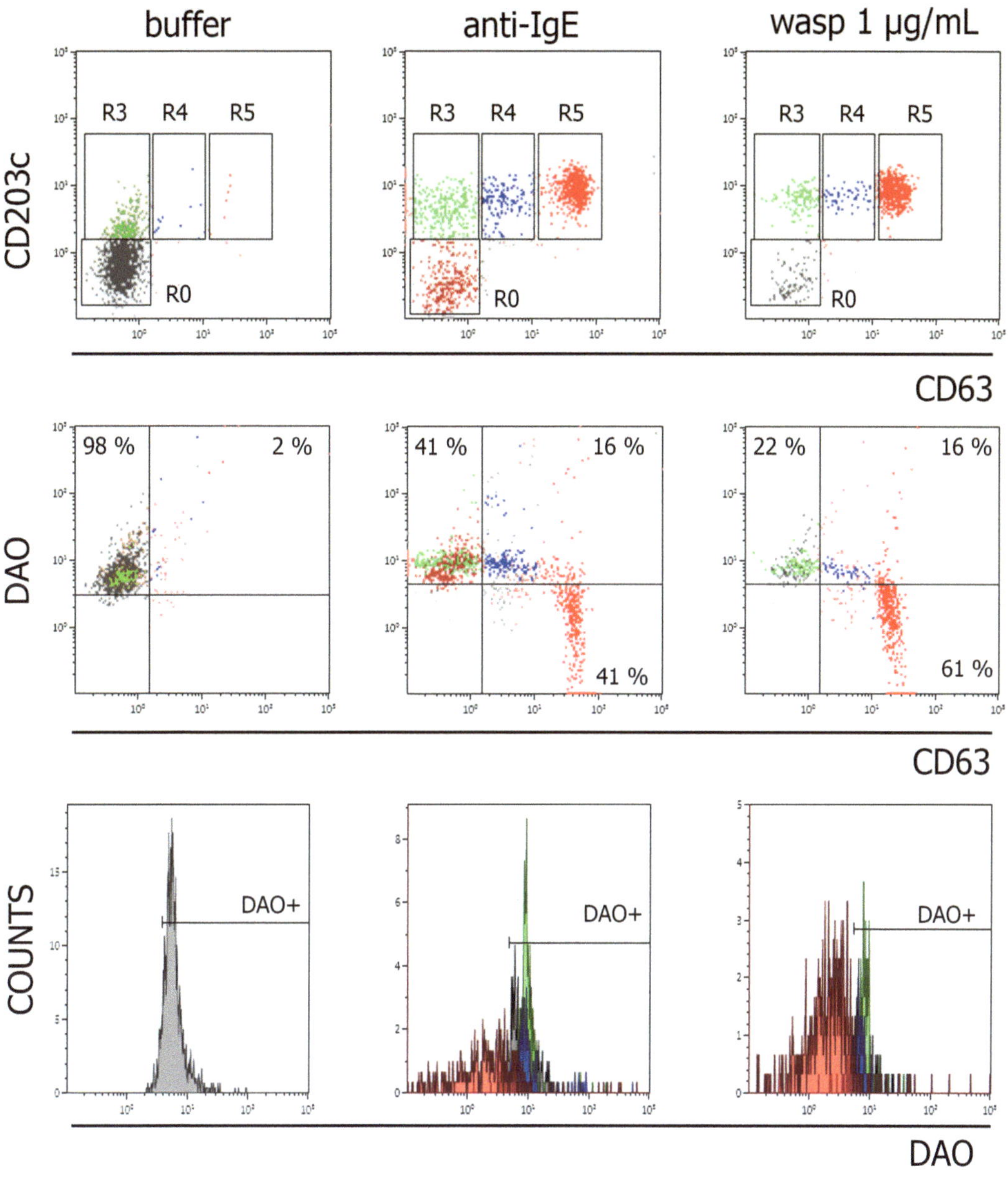

Fig. 4 HistaFlow analysis: basophils are selected as described in Fig. 1. *See* also legend of Fig. 1 for R0–R5 definition. The left column displays results of buffer (negative control), the second column shows results of positive control, and the right column displays results of activation with wasp antigen in a wasp-allergic patient. DAO results (histamine content) show histamine release in 61% of the basophils after activation with wasp antigen

2. All antibodies can be used labeled with different fluorochromes. Be aware that correct titration is necessary. This can be done by using different dilutions of the antibody and using the correct flow cytometric settings, looking for the optimal resolution in comparison with a Fluorescence Minus One (FMO) measurement as a control.

3. Allergens can be protein extracts, recombinant proteins, or drugs. They must be water soluble and used in concentrations which are not toxic for basophils. To test toxicity, incubate basophils with different concentrations of the allergen or mouse anti-human-IgE (positive control) for 20 min at 37 °C, and measure the viability using propidium iodide or the LIVE/DEAD Viability/Cytotoxicity Kit for mammalian cells (e.g., Life Technologies). Do not use preparations for skin testing. These solutions contain preservatives or sometimes modified proteins which can inhibit basophil activation in vitro. Be aware that food allergens are very difficult to standardize. The optimal concentration of an allergen to stimulate basophils must be determined in a dose-response curve experiment including about ten well-defined patients positive to the antigen and ten controls negative to the allergen. Because such a response can result in a bell-shaped curve, in routine practice at least two concentrations have to be used to define a positive result.
4. Basophils can be differentiated with different markers. Basophils are IL-3 receptor (CD123) positive and HLA-DR negative (monocytes and dendritic cells are HLA-DR positive and also CD123 positive). Because CD203c is the only lineage marker known until now for basophils (mast cells are also positive, but they do not circulate in the blood), we prefer to define basophils as $CD203c^{+}/IgE^{+}$. Nevertheless the CD203c density in resting basophils can be too low to differentiate cells.
5. In about 5–10% of the patients, basophils do not respond to a positive control, be it anti-IgE, fMLP, or anti-FcεRI. In such a case, one can decide to stimulate with the allergen. Nevertheless if the allergen also gives a negative response, the procedure has to be stopped because it could be a false-negative response (so-called nonresponder). Results will be inconclusive.
6. Basophil activation results in an immediate response (1 min) that lasts about 20–30 min at maximum response. To obtain highly reproducible results, all solutions are best prewarmed to 37 °C.
7. Wear gloves, lab coat, and safety glasses while handling all human blood products. Dispose of all pipettes into bagged waste collection bins. Wipe work surfaces with disinfectant before and after running tests.

References

1. Ebo DG, Hagendorens MM, Bridts CH, Schuerwegh AJ, De Clerck LS, Stevens WJ (2004) In vitro allergy diagnosis: should we follow the flow? Clin Exp Allergy 34:332–339
2. Ebo DG, Bridts CH, Hagendorens MM, Aerts NE, De Clerck LS, Stevens WJ (2008) Basophil activation test by flow cytometry: present and future applications in allergology. Cytometry B Clin Cytom 74:201–210
3. Hausmann OV, Gentinetta T, Bridts CH, Ebo DG (2009) The basophil activation test in immediate-type drug allergy. Immunol Allergy Clin N Am 29:555–566
4. Ebo DG, Bridts CH, Hagendorens MM, Mertens CH, De Clerck LS, Stevens WJ (2006) Flow-assisted diagnostic management of anaphylaxis from rocuronium bromide. Allergy 61:935–939
5. Sabato V, Verweij MM, Bridts CH, Levi-Schaffer F, Gibbs BF, De Clerck LS et al (2012) CD300a is expressed on human basophils and seems to inhibit IgE/FcepsilonRI-dependent anaphylactic degranulation. Cytometry B Clin Cytom 82:132–138
6. Ebo DG, Dombrecht EJ, Bridts CH, Aerts NE, De Clerck LS, Stevens WJ (2007) Combined analysis of intracellular signalling and immunophenotype of human peripheral blood basophils by flow cytometry: a proof of concept. Clin Exp Allergy 37:1668–1675
7. Verweij MM, De Knop KJ, Bridts CH, De Clerck LS, Stevens WJ, Ebo DG (2010) P38 mitogen-activated protein kinase signal transduction in the diagnosis and follow up of immunotherapy of wasp venom allergy. Cytometry B Clin Cytom 78:302–307
8. Verweij MM, Sabato V, Nullens S, Bridts CH, De Clerck LS, Stevens WJ et al (2012) STAT5 in human basophils: IL-3 is required for its FcepsilonRI-mediated phosphorylation. Cytometry B Clin Cytom 82:101–106
9. Ebo DG, Bridts CH, Mertens CH, Hagendorens MM, Stevens WJ, De Clerck LS (2012) Analyzing histamine release by flow cytometry (HistaFlow): a novel instrument to study the degranulation patterns of basophils. J Immunol Methods 375:30–38
10. Nullens S, Sabato V, Faber M, Leysen J, Bridts CH, De Clerck LS et al (2013) Basophilic histamine content and release during venom immunotherapy: insights by flow cytometry. Cytometry B Clin Cytom 84:173–178

Chapter 16

Basophil Activation Experiments in Immediate Drug Hypersensitivity: More Than a Diagnostic Aid

Didier G. Ebo, Jessy Elst, Athina Van Gasse, Leander De Puysseleyr, Margaretha A. Faber, Margo M. Hagendorens, Lina Mayorga, Christel Mertens, Chris H. Bridts, Luc S. De Clerck, and Vito Sabato

Abstract

Background: Correct diagnosis of immediate drug hypersensitivity reactions (IDHRs) can pose a significant challenge, mainly because of the absence of reliable in vitro tests, uncertainties associated with skin testing, and incomplete understanding of the underlying mechanisms.

Aim: To summarize and hypothesize on the potential of basophil activation test (BAT) as a safe aid to explore the mechanistic endotypes of IDHR, to identify antibody recognition sites, and to monitor drug desensitization.

Methods: A literature search was conducted using the keywords "allergy, basophil activation, CD63, CD203c, diagnosis, drugs, hypersensitivity, flow cytometry"; this was complemented by the authors' own expertise.

Results: At present BAT has mainly been employed as a diagnostic aid. However, evidence is emerging that the technique might also deepen our insights in immune (allergic) and nonimmune (nonallergic) mechanistic processes of IDHR. It is anticipated that BAT might also benefit the identification of antibody recognition sites and benefit our understandings of desensitization strategies.

Conclusion: Although the nondiagnostic application of BAT in IDHR is still in its infancy, with increasing employment, we can expect the technique to become a valuable asset to study many domains of IDHR that remain poorly understood.

Key words Immediate drug hypersensitivity reactions (IDHRs), Basophils, Basophil activation test (BAT), CD63, CD203c

1 Introduction

Correct diagnosis of immediate drug hypersensitivity reactions (IDHRs) can be challenging and can pose significant difficulties, mainly because of the incomplete understanding of underlying mechanistic processes, absence of reliable in vitro tests, uncertainties associated with skin testing, and ethical limitations associated with the gold standard, i.e., full-dose drug challenge tests (DCTs).

Bernhard F. Gibbs and Franco H. Falcone (eds.), *Basophils and Mast Cells: Methods and Protocols*, Methods in Molecular Biology, vol. 2163, https://doi.org/10.1007/978-1-0716-0696-4_16, © Springer Science+Business Media, LLC, part of Springer Nature 2020

For almost two decades, we have witnessed basophil activation tests (BATs), that is, in vitro activation of patients' basophils to benefit diagnosis in patients for whom a safe diagnostic is unavailable or in difficult cases who yield negative or equivocal results in traditional diagnostics. Moreover, evidence has accumulated that BAT has more to offer than a simple complementary diagnostic. For example, several studies have shown that the technique might benefit clarification and (re)classification of IDHR endotypes, help to unveil cross-reactivity between structurally related substances, and monitor desensitization protocols for IDHR. First, we will briefly summarize the technical principles and main diagnostic utilization of BAT. Subsequently, starting from our own priorities and experience, we will focus on the potential of BAT as a complementary instrument to explore and classify specific (allergic) and nonspecific (nonallergic) IDHR endotypes, to establish the accuracy and reliability of traditional diagnostics, and to identify cross-reactivity and safe alternatives and finally its value as a tool to monitor desensitization.

2 Technical Principles

Basophils develop from $CD34^+$ pluripotent progenitor stem cells and represent less than 1% of peripheral blood leukocytes. Like tissue-resident mast cells (MCs), basophils can be activated in distinct ways, which are differentiated whether or not an adaptive immune response, that is, cross-linking of the membrane-bound high-affinity receptor for IgE (FcεRI), is involved [1] (*see* Fig. 1, top). Specific IgE/FcεRI-dependent activation and degranulation mainly occur via (glyco)protein allergens, chemical substances such as drugs and autoantibodies (mostly IgG) directed against the FcεRI receptor or membrane-bound IgE. Nonspecific activation, independent of the cross-linking of IgE/FcεRI complexes, mainly occurs via the engagement of receptors for endogenous (e.g., cytokines, anaphylatoxins, chemokines, neuropeptides) or exogenous (e.g., pathogen-associated molecular patterns) substances. However, activation/degranulation of basophils/MC might also result from alternative means. For example, cutaneous MCs can respond upon the occupation of their Mas-related G-protein-coupled receptor X2 (MRGPRX2) (*see* Fig. 1, bottom and *see* Chapter 18) [2]. Upon activation/degranulation, basophils and MCs will release and synthesize a variety of mediators that are responsible for early and late manifestations of the allergic reaction. Moreover, Gaudenzio et al. [3] elegantly demonstrated that MC degranulation via the cross-linking of IgE/FcεRI complexes and ligation of MRGPRX2 are two different phenomena displaying distinct dynamics and characteristics but are clinically virtually indistinguishable.

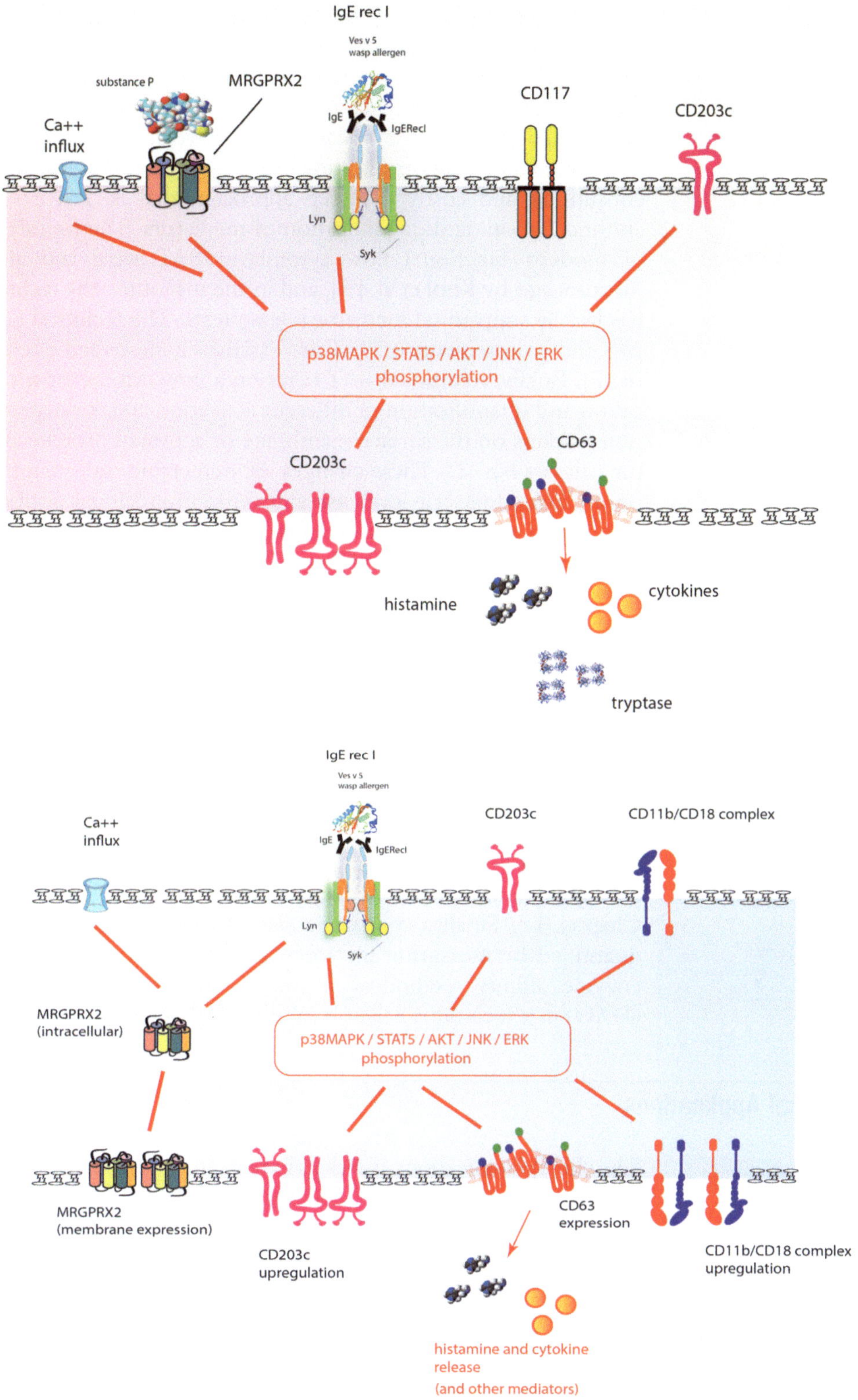

Fig. 1 IgE-dependent and IgE-independent activation of MCs (top) and basophils (bottom). Because basophils, unlike cutaneous MCs, barely express MRGPRX2, these cells will not respond in steady-state conditions of classical BAT. (Ves v 5, wasp allergen as a representative)

For over half a century, in vitro basophil activation studies have focused on histamine and leukotriene release in the supernatant of degranulated cells [4]. Although for many, these mediator release assays remain the "reference tests" for effector cell activation and degranulation, mediator release assays have never succeeded in entering mainstream application, mostly because of their time-consuming and costly two-step approach, that is, cell isolation and incubation, and quantification of mediators. The foundations of modern functional flow cytometric BAT were laid almost 30 years ago by Knol et al. [5], and in the meantime the technique has largely supplanted mediator release tests. The technical principles and requirements of BAT are extensively described elsewhere [6, 7]. Briefly, traditional BAT relies on a flow cytometric identification and quantification of different activation and/or degranulation markers on the surface membrane or activated proteins inside the cell (*see* Fig. 2). These changes are detectable and quantifiable on an individual cell level using specific monoclonal antibodies conjugated to fluorochromes. In most studies, basophils are identified by specific surface markers such as IgE/CD203c, CCR3/CD3, or CD123/HLA-DR. Of these markers, only CD203c is lineage specific. Subsequently, after activation, the appearance or upregulation of specific surface activation and/or degranulation markers, such as CD63 and/or CD203c, is measured. Alternative methods to quantify in vitro basophil activation and degranulation imply the quantification of the expression of inhibitory receptors such as CD300a (IRp60) [8], phosphorylation of intracellular signaling molecules such as p38 mitogen-activated protein kinase (MAPK) [9, 10] signal transducer and activator of transcription (STAT) 5 [11], and labeling of negatively charged exteriorized basophil granule matrix by using avidin probes [12, 13], also *see* Chapter 17. Finally, in vitro activation of basophils can also be quantified by measuring intracellular histamine content using an enzyme affinity method with the histaminase diamine oxidase (DAO) in a technique called HistaFlow [14, 15] (*see* Fig. 2).

3 Clinical Applications

The clinical applications as well as limitations of BAT in IDHR have been reviewed elsewhere [7, 16–20]. From these reviews, it appears that the technique, mainly used in IDHR to neuromuscular blocking agents (NMBA), β-lactam antibiotics, and fluoroquinolones, displays a good positive predictive value and can benefit correct diagnosis and management of IDHR. However, thorough analysis of the data indicates different particularities need to be kept in mind when using the technique to allow its further entrance in mainstream diagnosis. Firstly, it is evident that BAT only adds to diagnosis in IDHR that involves the degranulation of basophils. BAT

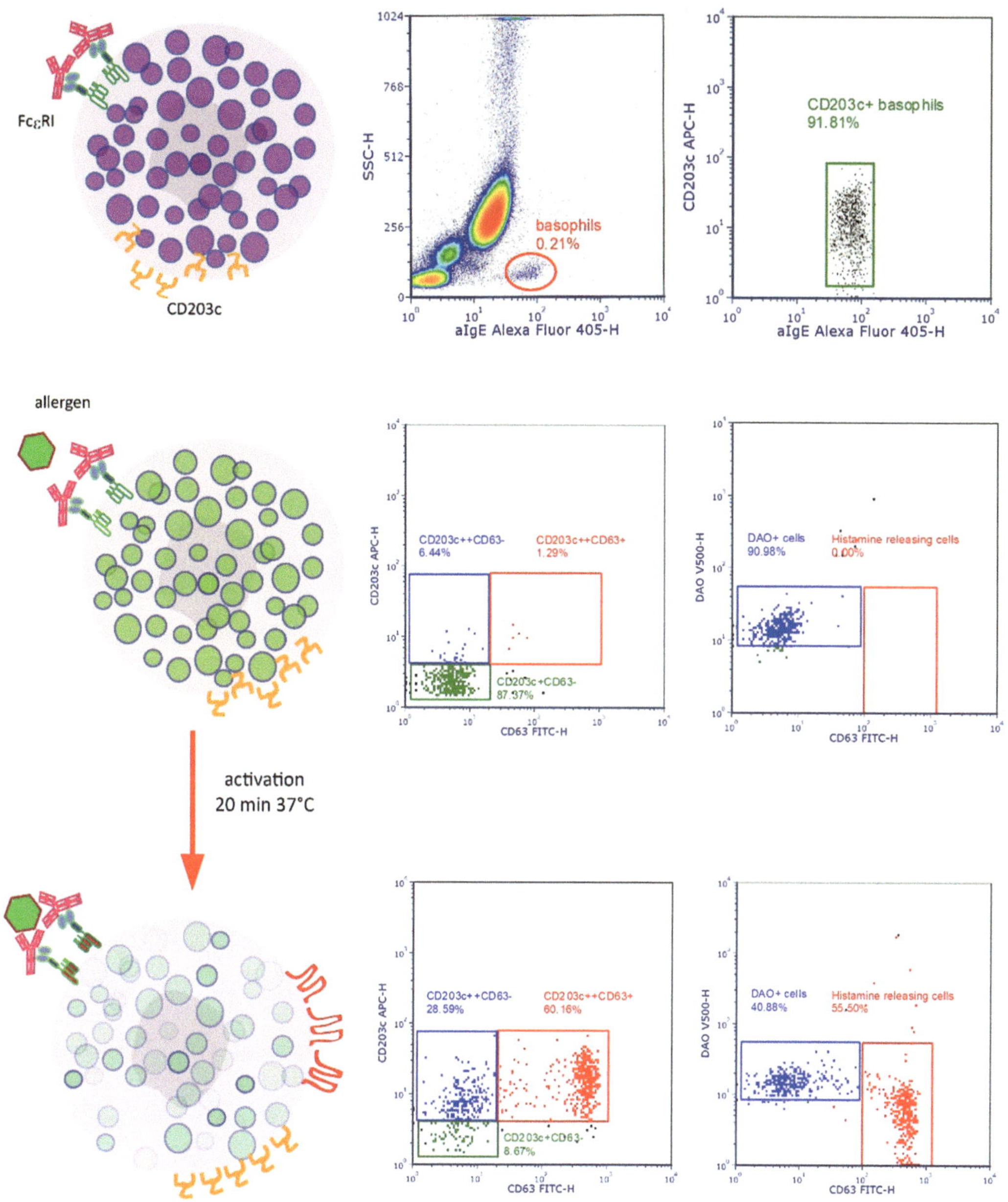

Fig. 2 Basophils are identified by specific surface markers such as IgE/CD203c. After activation, the appearance or upregulation of specific surface activation and/or degranulation markers, such as CD63 and/or CD203c, is measured. Activation/degranulation of basophils can also be quantified by measuring intracellular histamine content using an enzyme affinity method with the histaminase DAO in a technique called HistaFlow

cannot document IDHR resulting from alternative mechanisms such as enzymatic inhibition by nonselective cyclo-oxygenase (COX) inhibitors as it happens with hypersensitivity to nonsteroidal anti-inflammatory drugs (NSAIDs) and angiotensin-converting enzyme (ACE) inhibitors [1, 21]. Alternatively, the fact that positive basophil responses do not discriminate the cross-linking of adjacent allergen-specific IgE (sIgE)/FcεRI complexes and nonspecific mediator release should not be considered as a drawback. On the contrary, BAT might benefit the disentangling complexity of IDHR and unveil nonimmune mechanistic processes that currently do not meet the eye [22]. Secondly, optimal stimulation conditions and test outcomes are not universal. Findings for one substance might not necessarily apply to other compounds [23]. Moreover, validation of BAT cannot be considered appropriate when it failed to identify and to investigate drug-specific physicochemical properties and cytotoxic effects that could have profoundly affected the outcome of the test. For example, false-negative results for quinolones might occur as a result of photodegradation [24] or when basophil degranulation is inhibited via the occupation of other receptors (e.g., nicotinic and muscarinic) [25]. Critical appraisal of BAT validation studies have revealed limitations such as small study magnitude, heterogeneous inclusions, incorrect pooling of data, and application of obsolete arbitrarily chosen decision thresholds, all hampering the correct assessment and appreciation of BAT. Therefore, as with any study of this nature, it is mandatory to enroll a sufficient number of well-identified patients and (exposed) control individuals, to perform drug-specific experiments [22, 26–28] and to calculate correct cutoffs by receiver operating characteristics [29]. Thirdly, it is clear that the performance of BAT varies and that its place in the diagnostic work-up is not uniform and depends on the availability and performance of other diagnostics. In the diagnostic approach of rocuronium and atracurium hypersensitivity, BAT merits the status of secondary diagnostic test downstream skin testing but upstream quantification of sIgE [18, 30]. In contrast, for β-lactam antibiotics, BAT can be applied when skin testing and quantification yield equivocal or negative results [31], especially when DPT is contraindicated [30, 32]. Evidently, BAT can be helpful when other in vitro assays are lacking [22, 33, 34] or when skin testing is unreliable because of nonspecific histamine release [35, 36]. Finally, BAT has been shown to benefit the identification of an excipient in cases where the latter rather than the active compound is responsible for the IDHR [37].

4 Exploring Mechanistic Endotypes

As already exemplified, like MCs, basophils can be triggered by specific IgE-dependent and various nonspecific IgE-independent ways. However, unlike tissue-resident MCs, circulating basophils are easily accessible in sufficient numbers offering the potential to investigate pathways of drug-induced effector cell activation and to classify and harmonize IDHR. For example, an IgE/FcεRI-dependent pathogenesis can be proven by measuring a reduced/abolished basophil activation by the application of phosphoinositide 3-kinase inhibitors such as wortmannin [38] as shown in the evaluation for different drug hypersensitivities [39–41] or pretreatment of the cells by omalizumab [40]. Furthermore, comparative studies among basophils, cultured MCs, and skin testing might help to discriminate between IgE/FcεRI-dependent IDHR and—clinically indistinguishable—reactions resulting from alternative IgE/FcεRI-independent effector cell activation, such as through the off-target occupation of MRGPRX2. Because basophils, unlike cutaneous MCs, barely express MRGPRX2, these cells will not respond in steady-state conditions of classical BAT. Therefore, unlike IgE/FcεRI-dependent reactions, IDHR from MRGPRX2 occupancy will probably yield opposite negative BAT and positive skin test results. Skin test responses are potentially reflecting nonspecific mediator release, as can be seen in healthy control individuals [22, 36]. In other words, congruent skin test and basophil activation responses are likely to reflect IgE/FcεRI-dependent reaction, especially when drug-specific IgE antibodies are detectable [27, 30, 42]. Moreover, given the contradictory observations for rocuronium by McNeil et al. [43] and Lansu et al. [44], it seems premature to reclassify anaphylaxis to NMBA as nonimmunological adverse drug reactions [45]. Clearly, as already addressed previously [30, 46], additional studies are required before shifting the IgE/MRGPRX2 paradigm in IDHR. Even more, because of the adaptive evolution of the human MRGPRX2 gene [47], it is anticipated that conditioned basophils, that is, basophils expressing MRGPRX2 [48] and human MC models [49, 50], might be mandatory to verify data mainly gained from mouse models.

5 Detection and Identification of Antibody Recognition Structures

Traditionally, antibody recognition studies and exploration of cross-reactivity between drugs and related substances start with direct IgE-binding and quantitative (hapten) IgE inhibition studies using solid-phase assays. However, solid-phase assays exhibit different limitations. Firstly, not all drugs might contain suitable functional groups to allow successful coupling to the solid-phase support [51]. Secondly, coupling of a drug to the solid phase might hide relevant IgE combining sites. For example, a comparison of the IgE binding of different drug conjugates disclosed IgE antibodies to quinolones only to react with the conjugate linking the piperazine substituent at position 7 to bis-oxirane-activated Sepharose. Conjugates prepared by coupling the carboxyl group at position 3 to the support via carbodiimide/human serum albumin were unreactive, indicating hiding of the main allergenic

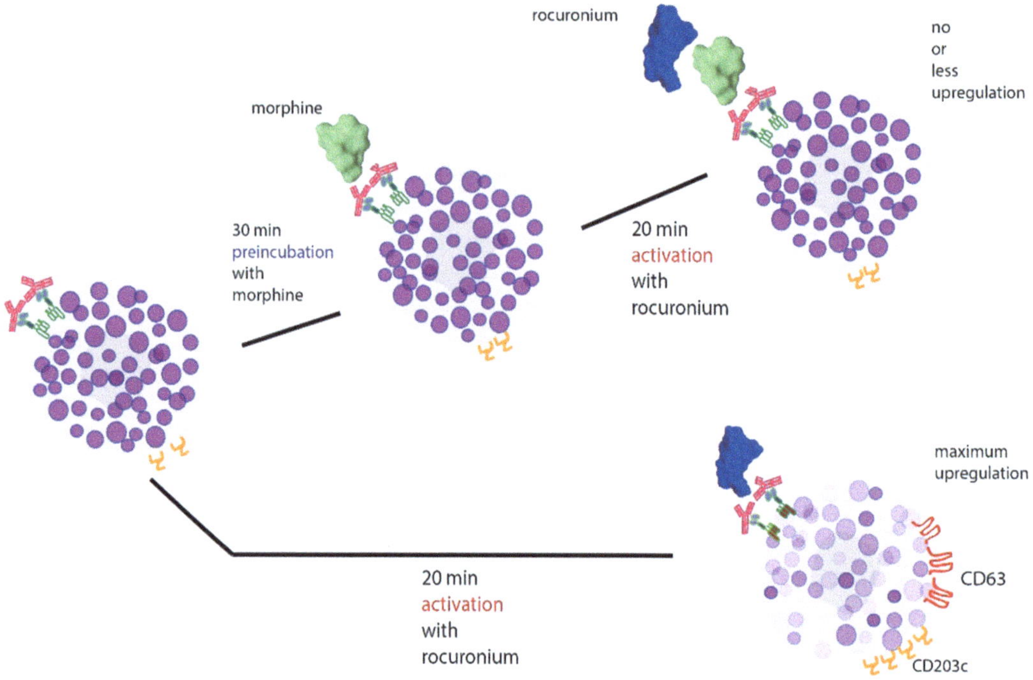

Fig. 3 Principle of combining activation and quantitative (hapten) inhibition experiments in a single run. In this example rocuronium is the activator, whereas morphine is the inhibitor

structures composed of all, or part of, the side of the molecule containing the 3-carboxy group [52]. Thirdly, IgE binding/inhibition data might not correctly predict skin test responses nor the outcome of an administration. For example, we demonstrated cisatracurium to yield negative skin test responses and to be safe in rocuronium-allergic patients who had displayed a negative cisatracurium skin test but had their sIgE to rocuronium strongly inhibited by benzylisoquinoline [53]. Unlike traditional solid-phase assays, BAT is a liquid-phase technique without the need for coupling, and BAT more closely mirrors the in vivo situation, as the cells are only activated upon the cross-linking of adjacent allergen-specific IgE antibodies. Therefore, BAT might capture data that are inaccessible for traditional nonfunctional solid-phase techniques of which results represent IgE-binding and not activation and/or degranulation of effector cells. Actually, as shown in Fig. 3, BAT enables safely combining multiple activation and quantitative (hapten) inhibition experiments in a single run. Figure 4 displays a representative plot of basophil inhibition by morphine and cisatracurium in a rocuronium-allergic patient. Morphine contains a tertiary substituted ammonium structure that is readily recognized by IgE in sera from patients allergic to rocuronium and suxamethonium but not atracurium and cisatracurium [53, 54]. Whether BAT could benefit the evaluation of the near-

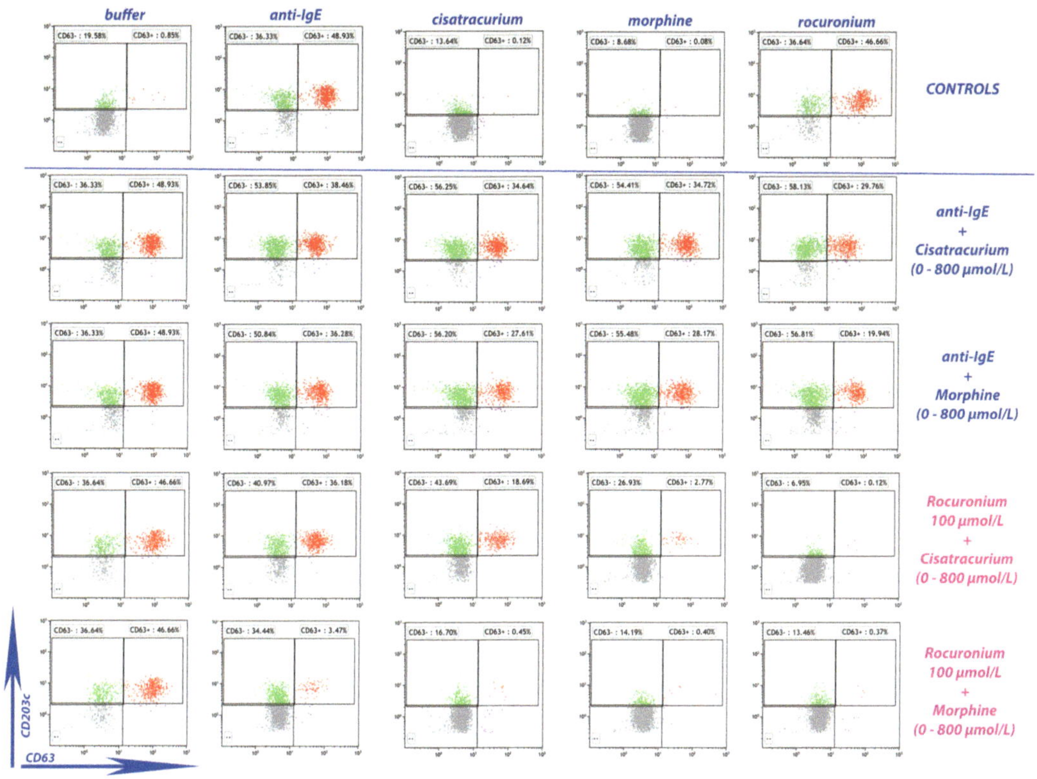

Fig. 4 Representative plot of anti-IgE and rocuronium stimulated basophils inhibited by morphine and cisatracurium in a rocuronium-allergic patient. Morphine, like NMBA, contains a tertiary substituted ammonium structure that is readily recognized by IgE in sera from patients allergic to rocuronium and suxamethonium but not atracurium and cisatracurium

neighbor effect of structures to promote or weaken antigenic recognition remains elusive, but it is likely the technique enabling a functional evaluation of the epitope-paratope binding, to add rigor to the observations from former inhibition experiments [55]. Moreover, the inclusion of drug metabolites, especially those with the capacity to react to proteins, has shown to increase BAT sensitivity in β-lactams and pyrazolone IDHR [41, 56]. Needless to stress, BAT, being an in vitro test, allows safe investigation of compounds that await approval for human use.

6 Drug Challenge Tests (DCTs): From Bed to Bench?

DCT is often considered as the gold standard for drug hypersensitivity diagnosis [57]. However, full-dose DCTs are hampered by the risk of severe, life-threatening reactions and might be contraindicated (i.e., intake of β-blockers or ACE inhibitors) or not possible for obvious ethical or practical reasons (i.e., hypersensitivity to NMBA). Besides, even DCTs do not show absolute predictive values [58]. In these cases, BAT might constitute an attractive complementary diagnostic, especially when traditional diagnostics remain negative. In a series of 72 rocuronium-allergic patients, 8 (11.1%) had their diagnosis documented by unequivocal BAT and did not require an additional and difficult-to-perform DCT [30]. In β-lactam antibiotics, BAT has shown to be useful to identify the culprit drug in patients' IDHR after amoxicillin-clavulanic acid (AX-CLV) combination intake. This is important since for diagnosing patients who selectively react to CLV, DCT must be performed with both AX alone and AX-CLV combination, ruling out the response to AX. In a study performed in 58 CLV selective patients, BAT was positive in 87% of patients suffering anaphylactic shock and with negative skin test. Moreover, the combined results of skin test and BAT diagnosed around 90% of CLV-allergic patients indicate that BAT could be particularly useful for patients with negative skin test and with more severe reactions in whom the performance of DPT is not allowed [59]. The usefulness of the combination of BAT and ST results for increasing diagnosis sensitivity has been shown for several drugs [30, 60].

7 Excipients

It is increasingly appreciated that excipients can trigger IDHR, including life-threatening anaphylaxis. Nevertheless, excipient testing is frequently overlooked entailing a risk for dramatic re-exposure or unnecessary avoidance of an innocent active substance. To date, literature regarding the evaluation and validation of BAT in IDHR has principally investigated commercially available drug formulations with only poor interest for excipients. Nevertheless, evidence has accumulated that the technique might help to incriminate various kinds of (hidden) excipients rather than the active compound. Moreover, in some of these patients, the reliability of BAT was confirmed with subsequent uneventful challenges with drug formulations without the offending excipients such as Cremophor [37], carboxymethylcellulose [61], succinate [62], and maltitol [63].

8 Desensitization

Generally correct management of drug hypersensitivity implies strict avoidance of future administration of the culprit and potentially cross-reactive compounds. However, for particular cases, substitution by safe alternatives might not always be the optimal approach, mainly because of the lower efficacy of the substitute. In such cases, desensitization, that is, induction of a temporary state of tolerance to the offender, might be considered to allow optimal treatment and improve clinical outcome [64, 65]. Despite the rapid and clinical efficiency of various desensitization regimens, the underlying mechanisms contributing to the transient state of clinical tolerance remain poorly understood but probably involve temporary "refractoriness" of MCs and basophils similar to that seen in the early phase of venom-specific immunotherapy [66, 67]. Alternatively, there is evidence that BAT constitutes a safe and valuable instrument to monitor efficacy and predict adverse events of desensitization with drugs such as platinum compounds [68] and biologicals such as brentuximab [69] and pertuzumab [70].

9 Conclusion

Since almost two decades, we have witnessed the introduction of BAT as a safe and valuable diagnostic in IDHR, especially in difficult cases or situations where no reliable alternative diagnostic is available. As a matter of fact, BAT has been shown to benefit diagnosis and to allow studying of cross-reactivity and identification of safe alternatives for the future. Furthermore, BAT enables deepening of our insights in mechanistic processes and exploring of the antibody recognition sites of drugs and their metabolites. Whether BAT might substitute DCT and monitor desensitization requires further investigation.

References

1. Ebo DG, Clarke RC, Mertes PM, Platt PR, Sabato V, Sadleir PH (2019) Molecular mechanisms and pathophysiology of perioperative hypersensitivity and anaphylaxis: a narrative review. Br J Anaesth 123(1):38
2. Ali H (2017) Emerging roles for MAS-related G protein-coupled receptor-X2 in host defense peptide, opioid, and neuropeptide-mediated inflammatory reactions. Adv Immunol 136:123–162. https://doi.org/10.1016/bs.ai.2017.06.002
3. Gaudenzio N, Sibilano R, Marichal T, Starkl P, Reber LL, Cenac N et al (2016) Different activation signals induce distinct mast cell degranulation strategies. J Clin Invest 126 (10):3981–3998. https://doi.org/10.1172/jci85538
4. Demoly P, Lebel B, Arnoux B (2003) Allergen-induced mediator release tests. Allergy 58 (7):553–558
5. Knol EF, Mul FP, Jansen H, Calafat J, Roos D (1991) Monitoring human basophil activation via CD63 monoclonal antibody 435. J Allergy Clin Immunol 88(3 Pt 1):328–338
6. Bridts CH, Sabato V, Mertens C, Hagendorens MM, De Clerck LS, Ebo DG (2014) Flow cytometric allergy diagnosis: basophil activation techniques. Methods Mol Biol 1192:147–159. https://doi.org/10.1007/978-1-4939-1173-8_11
7. Ebo DG, Sainte-Laudy J, Bridts CH, Mertens CH, Hagendorens MM, Schuerwegh AJ et al (2006) Flow-assisted allergy diagnosis: current applications and future perspectives. Allergy 61 (9):1028–1039. https://doi.org/10.1111/j.1398-9995.2006.01039.x
8. Sabato V, Verweij MM, Bridts CH, Levi-Schaffer F, Gibbs BF, De Clerck LS et al (2012) CD300a is expressed on human basophils and seems to inhibit IgE/FcepsilonRI-dependent anaphylactic degranulation. Cytometry B Clin Cytom 82(3):132–138. https://doi.org/10.1002/cyto.b.21003
9. Ebo DG, Dombrecht EJ, Bridts CH, Aerts NE, de Clerck LS, Stevens WJ (2007)

Combined analysis of intracellular signalling and immunophenotype of human peripheral blood basophils by flow cytometry: a proof of concept. Clin Exp Allergy 37(11):1668–1675. https://doi.org/10.1111/j.1365-2222.2007.02819.x

10. Aerts NE, Dombrecht EJ, Bridts CH, Hagendorens MM, de Clerck LS, Stevens WJ et al (2009) Simultaneous flow cytometric detection of basophil activation marker CD63 and intracellular phosphorylated p38 mitogen-activated protein kinase in birch pollen allergy. Cytometry B Clin Cytom 76(1):8–17. https://doi.org/10.1002/cyto.b.20437
11. Verweij MM, Sabato V, Nullens S, Bridts CH, De Clerck LS, Stevens WJ et al (2012) STAT5 in human basophils: IL-3 is required for its FcepsilonRI-mediated phosphorylation. Cytometry B Clin Cytom 82(2):101–106. https://doi.org/10.1002/cyto.b.20629
12. Mukai K, Chinthrajah RS, Nadeau KC, Tsai M, Gaudenzio N, Galli SJ (2017) A new fluorescent-avidin-based method for quantifying basophil activation in whole blood. J Allergy Clin Immunol 140(4):1202–1206.e3. https://doi.org/10.1016/j.jaci.2017.03.052
13. Joulia R, Mailhol C, Valitutti S, Didier A, Espinosa E (2017) Direct monitoring of basophil degranulation by using avidin-based probes. J Allergy Clin Immunol 140(4):1159–1162.e6. https://doi.org/10.1016/j.jaci.2017.03.030
14. Ebo DG, Bridts CH, Mertens CH, Hagendorens MM, Stevens WJ, De Clerck LS (2012) Analyzing histamine release by flow cytometry (HistaFlow): a novel instrument to study the degranulation patterns of basophils. J Immunol Methods 375(1–2):30–38. https://doi.org/10.1016/j.jim.2011.09.003
15. Cop N, Uyttebroek AP, Sabato V, Bridts CH, De Clerck LS, Ebo DG (2016) Flow cytometric analysis of drug-Induced basophil histamine release. Cytometry B Clin Cytom 90(3):285–288. https://doi.org/10.1002/cyto.b.21226
16. Mayorga C, Celik G, Rouzaire P, Whitaker P, Bonadonna P, Cernadas JR et al (2016) In vitro tests for drug hypersensitivity reactions. An ENDA/EAACI drug allergy interest group position paper. Allergy 71(8):1103–1134. https://doi.org/10.1111/all.12886
17. Ariza A, Montanez MI, Fernandez TD, Perkins JR, Mayorga C (2016) Cellular tests for the evaluation of drug hypersensitivity. Curr Pharm Des 22(45):6773–6783. https://doi.org/10.2174/1381612822666161107165917
18. Ebo DG, Faber M, Elst J, Van Gasse AL, Bridts CH, Mertens C et al (2018) In vitro diagnosis of immediate drug hypersensitivity during anesthesia: a review of the literature. J Allergy Clin Immunol Pract 6(4):1176–1184. https://doi.org/10.1016/j.jaip.2018.01.004
19. Hemmings O, Kwok M, McKendry R, Santos AF (2018) Basophil activation test: old and new applications in allergy. Curr Allergy Asthma Rep 18(12):77. https://doi.org/10.1007/s11882-018-0831-5
20. Mayorga C, Ebo DG, Lang DM, Pichler WJ, Sabato V, Park MA et al (2019) Controversies in drug allergy: in vitro testing. J Allergy Clin Immunol 143(1):56–65. https://doi.org/10.1016/j.jaci.2018.09.022
21. Ebo DG, Leysen J, Mayorga C, Rozieres A, Knol EF, Terreehorst I (2011) The in vitro diagnosis of drug allergy: status and perspectives. Allergy 66(10):1275–1286. https://doi.org/10.1111/j.1398-9995.2011.02661.x
22. Van Gasse AL, Sabato V, Uyttebroek AP, Elst J, Faber MA, Hagendorens MM et al (2017) Immediate moxifloxacin hypersensitivity: is there more than currently meets the eye? Allergy 72(12):2039–2043. https://doi.org/10.1111/all.13236
23. Kleine-Tebbe J, Erdmann S, Knol EF, MacGlashan DW Jr, Poulsen LK, Gibbs BF (2006) Diagnostic tests based on human basophils: potentials, pitfalls and perspectives. Int Arch Allergy Immunol 141(1):79–90. https://doi.org/10.1159/000094495
24. Mayorga C, Andreu I, Aranda A, Dona I, Montanez MI, Blanca-Lopez N et al (2013) Fluoroquinolone photodegradation influences specific basophil activation. Int Arch Allergy Immunol 160(4):377–382. https://doi.org/10.1159/000343023
25. Watson BM, Oliveria JP, Nusca GM, Smith SG, Beaudin S, Dua B et al (2014) Inhibition of allergen-induced basophil activation by ASM-024, a nicotinic receptor ligand. Int Arch Allergy Immunol 165(4):255–264. https://doi.org/10.1159/000370068
26. Uyttebroek AP, Sabato V, Cop N, Decuyper II, Faber M, Bridts C et al (2016) Diagnosing cefazolin hypersensitivity: lessons from dual labelling flow cytometry. J Allergy Clin Immunol Pract 4(6):1243–1245
27. Uyttebroek AP, Sabato V, Leysen J, Bridts CH, De Clerck LS, Ebo DG (2014) Flowcytometric diagnosis of atracurium-induced anaphylaxis. Allergy 69(10):1324–1332. https://doi.org/10.1111/all.12468
28. Leysen J, Bridts CH, De Clerck LS, Vercauteren M, Lambert J, Weyler JJ et al (2011) Allergy to rocuronium: from clinical suspicion to correct diagnosis. Allergy 66

(8):1014–1019. https://doi.org/10.1111/j.1398-9995.2011.02569.x

29. Gardner IA, Greiner M (2006) Receiver-operating characteristic curves and likelihood ratios: improvements over traditional methods for the evaluation and application of veterinary clinical pathology tests. Vet Clin Pathol 35 (1):8–17

30. Van Gasse AL, Elst J, Bridts CH, Mertens C, Faber M, Hagendorens MM et al (2019) Rocuronium hypersensitivity: does off-target occupation of the MRGPRX2 receptor play a role? J Allergy Clin Immunol Pract 7 (3):998–1003. https://doi.org/10.1016/j.jaip.2018.09.034

31. Torres MJ, Blanca M, Fernandez J, Romano A, Weck A, Aberer W et al (2003) Diagnosis of immediate allergic reactions to beta-lactam antibiotics. Allergy 58(10):961–972

32. Ebo DG, Stevens WJ, Bridts CH, Matthieu L (1998) Contact allergic dermatitis and life-threatening anaphylaxis to chlorhexidine. J Allergy Clin Immunol 101(1 Pt 1):128–129. https://doi.org/10.1016/s0091-6749(98)70205-2

33. Ebo DG, Mertens C, Braes M, Mennes I, Bridts CH, Sabato V (2019) Clindamycin anaphylaxis confirmed by in vivo and in vitro testing. J Allergy Clin Immunol Pract 7 (1):331–333. https://doi.org/10.1016/j.jaip.2018.06.008

34. Decuyper II, Ebo DG, Uyttebroek AP, Hagendorens MM, Faber MA, Bridts CH et al (2016) Quantification of specific IgE antibodies in immediate drug hypersensitivity: more shortcomings than potentials? Clin Chim Acta 460:184–189. https://doi.org/10.1016/j.cca.2016.06.043

35. Leysen J, De Witte L, Sabato V, Faber M, Hagendorens M, Bridts C et al (2013) IgE-mediated allergy to pholcodine and cross-reactivity to neuromuscular blocking agents: lessons from flow cytometry. Cytometry B Clin Cytom 84(2):65–70. https://doi.org/10.1002/cyto.b.21074

36. Uyttebroek AP, Sabato V, Bridts CH, De Clerck LS, Ebo DG (2015) Moxifloxacin hypersensitivity: uselessness of skin testing. J Allergy Clin Immunol Pract 3(3):443–445. https://doi.org/10.1016/j.jaip.2014.12.012

37. Ebo DG, Piel GC, Conraads V, Stevens WJ (2001) IgE-mediated anaphylaxis after first intravenous infusion of cyclosporine. Ann Allergy Asthma Immunol 87(3):243–245. https://doi.org/10.1016/s1081-1206(10)62234-x

38. Knol EF, Koenderman L, Mul E, Verhoeven AJ, Roos D (1990) Differential mechanisms in the stimulus-secretion coupling in human basophils: evidence for a protein-kinase-C-dependent and a protein-kinase-C-independent route. Agents Actions 30(1–2):49–52

39. Torres MJ, Ariza A, Mayorga C, Dona I, Blanca-Lopez N, Rondon C et al (2010) Clavulanic acid can be the component in amoxicillin-clavulanic acid responsible for immediate hypersensitivity reactions. J Allergy Clin Immunol 125(2):502–505.e2. https://doi.org/10.1016/j.jaci.2009.11.032

40. Iwamoto T, Hirai H, Yamaguchi N, Kobayashi N, Sugimoto H, Tabata T et al (2014) Carboplatin-induced severe hypersensitivity reaction: role of IgE-dependent basophil activation and FcepsilonRI. Cancer Sci 105 (11):1472–1479. https://doi.org/10.1111/cas.12538

41. Ariza A, Garcia-Martin E, Salas M, Montanez MI, Mayorga C, Blanca-Lopez N et al (2016) Pyrazolones metabolites are relevant for identifying selective anaphylaxis to metamizole. Sci Rep 6:23845. https://doi.org/10.1038/srep23845

42. Uyttebroek AP, Sabato V, Bridts CH, De Clerck LS, Ebo DG (2015) Immunoglobulin E antibodies to atracurium: a new diagnostic tool? Clin Exp Allergy 45(2):485–487. https://doi.org/10.1111/cea.12448

43. McNeil BD, Pundir P, Meeker S, Han L, Undem BJ, Kulka M et al (2015) Identification of a mast-cell-specific receptor crucial for pseudo-allergic drug reactions. Nature 519 (7542):237–241. https://doi.org/10.1038/nature14022

44. Lansu K, Karpiak J, Liu J, Huang XP, McCorvy JD, Kroeze WK et al (2017) In silico design of novel probes for the atypical opioid receptor MRGPRX2. Nat Chem Biol 13(5):529–536. https://doi.org/10.1038/nchembio.2334

45. Spoerl D, Nigolian H, Czarnetzki C, Harr T (2017) Reclassifying anaphylaxis to neuromuscular blocking agents based on the presumed patho-mechanism: IgE-mediated, pharmacological adverse reaction or "innate hypersensitivity"? Int J Mol Sci 18(6). https://doi.org/10.3390/ijms18061223

46. Van Gasse AL, Sabato V, Faber MA, Hagendorens MM, Ebo DG (2017) An alternative explanation for immediate hypersensitivity reactions to opioids. J Allergy Clin Immunol Pract 5(6):1806. https://doi.org/10.1016/j.jaip.2017.08.016

47. Yang S, Liu Y, Lin AA, Cavalli-Sforza LL, Zhao Z, Su B (2005) Adaptive evolution of MRGX2, a human sensory neuron specific

gene involved in nociception. Gene 352:30–35. https://doi.org/10.1016/j.gene.2005.03.001
48. Sabato V, Van Gasse A, Cop N, Claesen K, Decuyper I, Faber M et al (2017) The Mas-related G protein-coupled receptor MRGPRX2 is expressed on human basophils and up-regulated upon activation. J Allergy Clin Immunol 139:AB168
49. Cop N, Decuyper II, Faber MA, Sabato V, Bridts CH, Hagendorens MM et al (2017) Phenotypic and functional characterization of in vitro cultured human mast cells. Cytometry B Clin Cytom 92(5):348–354. https://doi.org/10.1002/cyto.b.21399
50. Cop N, Ebo DG, Bridts CH, Elst J, Hagendorens MM, Mertens C et al (2018) Influence of IL-6, IL-33, and TNF-alpha on human mast cell activation: lessons from single cell analysis by flow cytometry. Cytometry B Clin Cytom 94(3):405–411. https://doi.org/10.1002/cyto.b.21547
51. Harle DG, Baldo BA, Fisher MM (1985) Assays for, and cross-reactivities of, IgE antibodies to the muscle relaxants gallamine, decamethonium and succinylcholine (suxamethonium). J Immunol Methods 78(2):293–305
52. Baldo BA (2014) IgE and drug allergy: antibody recognition of "small" molecules of widely varying structures and activities. Antibodies 3:56–91
53. Ebo DG, Venemalm L, Bridts CH, Degerbeck F, Hagberg H, De Clerck LS et al (2007) Immunoglobulin E antibodies to rocuronium: a new diagnostic tool. Anesthesiology 107(2):253–259. https://doi.org/10.1097/01.anes.0000270735.40872.f2
54. Baldo BA, Fisher MM, Pham NH (2009) On the origin and specificity of antibodies to neuromuscular blocking (muscle relaxant) drugs: an immunochemical perspective. Clin Exp Allergy 39(3):325–344. https://doi.org/10.1111/j.1365-2222.2008.03171.x
55. Pham NH, Baldo BA, Puy RM (2001) Studies on the mechanism of multiple drug allergies. Structural basis of drug recognition. J Immunoass Immunochem 22(1):47–73. https://doi.org/10.1081/ias-100102897
56. Barbero N, Fernandez-Santamaria R, Mayorga C, Martin-Serrano A, Salas M, Bogas G et al (2019) Identification of an antigenic determinant of clavulanic acid responsible for IgE-mediated reactions. Allergy 74:1490–1501. https://doi.org/10.1111/all.13761
57. Garvey LH, Ebo DG, Kroigaard M, Savic S, Clarke R, Cooke P et al (2019) The use of drug provocation testing in the investigation of suspected immediate perioperative allergic reactions: current status. Br J Anaesth 123(1):e126–ee34. https://doi.org/10.1016/j.bja.2019.03.018
58. Demoly P, Romano A, Botelho C, Bousquet-Rouanet L, Gaeta F, Silva R et al (2010) Determining the negative predictive value of provocation tests with beta-lactams. Allergy 65(3):327–332. https://doi.org/10.1111/j.1398-9995.2009.02228.x
59. Salas M, Fernandez-Santamaria R, Mayorga C, Barrionuevo E, Ariza A, Posadas T et al (2017) Use of the basophil activation test may reduce the need for drug provocation in amoxicillin-clavulanic allergy. J Allergy Clin Immunol Pract 6(3):1010–1018.e. https://doi.org/10.1016/j.jaip.2017.08.009
60. Mayorga C, Dona I, Perez-Inestrosa E, Fernandez TD, Torres MJ (2017) The value of in vitro tests to DiminishDrug challenges. Int J Mol Sci 18(6). https://doi.org/10.3390/ijms18061222
61. Dumond P, Franck P, Morisset M, Sainte Laudy J, Kanny G, Moneret-Vautrin DA (2009) Pre-lethal anaphylaxis to carboxymethylcellulose confirmed by identification of specific IgE—review of the literature. Eur Ann Allergy Clin Immunol 41(6):171–176
62. Walker AI, Rawer HC, Sieber W, Przybilla B (2011) Immediate-type hypersensitivity to succinylated corticosteroids. Int Arch Allergy Immunol 155(1):86–92. https://doi.org/10.1159/000318678
63. Rodriguez Trabado A, Camara Hijon C, Garcia-Trujillo JA, Magriz Trascon I, Fernandez Pereira LM (2017) A case of immediate hypersensitivity reaction to maltitol. Case Rep Med 2017:2127167. https://doi.org/10.1155/2017/2127167
64. Cernadas JR, Brockow K, Romano A, Aberer W, Torres MJ, Bircher A et al (2010) General considerations on rapid desensitization for drug hypersensitivity—a consensus statement. Allergy 65(11):1357–1366. https://doi.org/10.1111/j.1398-9995.2010.02441.x
65. Castells MC (2017) Drug hypersensitivity and desensitizations. Immunol Allergy Clin N Am 37(4):xvii–xviii. https://doi.org/10.1016/j.iac.2017.08.002
66. Ebo DG, Hagendorens MM, Schuerwegh AJ, Beirens LM, Bridts CH, De Clerck LS et al (2007) Flow-assisted quantification of in vitro activated basophils in the diagnosis of wasp

venom allergy and follow-up of wasp venom immunotherapy. Cytometry B Clin Cytom 72 (3):196–203. https://doi.org/10.1002/cyto.b.20142

67. Nullens S, Sabato V, Faber M, Leysen J, Bridts CH, De Clerck LS et al (2013) Basophilic histamine content and release during venom immunotherapy: insights by flow cytometry. Cytometry B Clin Cytom 84(3):173–178. https://doi.org/10.1002/cyto.b.21084
68. Giavina-Bianchi P, Galvao VR, Picard M, Caiado J, Castells MC (2017) Basophil activation test is a relevant biomarker of the outcome of rapid desensitization in platinum compounds-allergy. J Allergy Clin Immunol Pract 5(3):728–736. https://doi.org/10.1016/j.jaip.2016.11.006
69. de la Varga MR, Gutierrez Fernandez D, Anez GA, Foncubierta Fernandez A, Andres Garcia JA, Medina VF (2017) Use of the basophil activation test in monitoring clinical tolerance after desensitization to brentuximab vedotin. Ann Allergy Asthma Immunol 118 (6):745–747. https://doi.org/10.1016/j.anai.2017.04.015
70. Gonzalez-de-Olano D, Morgado JM, Juarez-Guerrero R, Sanchez-Munoz L, Letellez-Fernandez J, Malon-Gimenez D et al (2016) Positive basophil activation test following anaphylaxis to pertuzumab and successful treatment with rapid desensitization. J Allergy Clin Immunol Pract 4(2):338–340. https://doi.org/10.1016/j.jaip.2015.10.007

Chapter 17

Basophil Activation Techniques: Staining of Exteriorized Granule Matrix

Jessy Elst, Vito Sabato, Margo M. Hagendorens, Michel van Houdt, Margaretha A. Faber, Chris H. Bridts, Didier G. Ebo, and Christel Mertens

Abstract

The basis of traditional flow cytometry allergy diagnosis is measurement of the expression of basophilic surface activation and/or degranulation markers. Basophils, upon encounter with a specific allergen that cross-links surface FcRI-bound IgE antibodies, not only secrete and release quantifiable bioactive mediators but also upregulate the expression of different markers (e.g., CD63, CD203c) which can be detected by multicolor flow cytometry using specific monoclonal antibodies. Here, we describe a novel technique that relies upon the staining of exteriorized anionic proteoglycans from a basophil granule matrix by cationic fluorescent avidin probes.

Key words Basophils, Basophil activation test (BAT), Proteoglycans, Granule matrix, CD203c

1 Introduction

Almost three decades ago, the discovery of the marker CD63 (gp53) and subsequently CD203c (ENPP-3) paved the way for the development of a novel technique to analyze and quantify the allergen-specific in vitro activation of basophils: the so-called basophil activation test (BAT) [1]. Briefly, the BAT relies on a flow cytometric measurement of various specific activation and/or degranulation markers on the surface membrane or activated proteins inside the cells. These changes are detected and quantified on an individual cell level through specific monoclonal antibodies coupled to fluorochromes. Practically, basophils are identified by specific markers such as CCR3/CD3, CD123/HLA-DR, or IgE/CD203c. Of these markers, only CD203c is lineage specific. Subsequently, after activation with an allergen, the upregulation or appearance of specific surface activation markers, such as CD63 and/or CD203c, is measured. Activation of basophils can also be quantified by means of intracellular molecules, such as the

Bernhard F. Gibbs and Franco H. Falcone (eds.), *Basophils and Mast Cells: Methods and Protocols*, Methods in Molecular Biology, vol. 2163, https://doi.org/10.1007/978-1-0716-0696-4_17,

phosphorylation of MAPK (p38MAPK) or STAT5 (pSTAT5) as well as quantification of the intracellular histamine content in a technique called HistaFlow®. For a description of these, *see* Chapter 15. Although these techniques have added insights into the mechanisms culminating in degranulation and have benefited correct allergy diagnosis, it has been anticipated that staining of exteriorized anionic proteoglycans from basophil granule matrix could be of added value [2, 3].

2 Materials

All solutions should be prepared with sterile ultrapure water. We use a Millipore Milli-Q System and the water is purified with a reversed osmosis system followed by ultra-purification to attain 18 MΩ at room temperature.

2.1 Blood Sample

Peripheral heparinized blood (10 mL) is obtained and immediately transported to the laboratory. Incubation and analysis are done within 8 h (*see* **Note 1**).

2.2 Reagents

1. BAT buffer: BAT buffer is prepared sterile just before use. Dilute 1.9 mL 10× Hanks' balanced salt solution (HBSS, with Ca^{2+} and Mg^{2+}) with 17.1 mL H_2O. Add 0.6 mL $NaHCO_3$ 7.5% and 0.4 mL 1 M HEPES. Adjust pH to 7.4.
2. Lysing solution: BD FACS™ lysing solution (BD Biosciences, cat.nr. 349202) is diluted 1/10 with water and stored at room temperature.
3. Phosphate buffered saline (PBS 1×):10× concentrated (without Ca^{2+} or Mg^{2+}) is diluted with water to reach an isotonic PBS (1×) buffer. Adjust pH to 7.4.
4. PBS-EDTA: dissolve 10 mM Na_2EDTA in 1×PBS. Adjust pH to 7.4.
5. PBS-NaN_3: dissolve 0.5 g/L NaN_3 in 1×PBS. Adjust pH to 7.4. Store at room temperature.
6. Anti-human IgE Alexa Fluor© 405 (aIgE): since aIgE (Sigma-Aldrich, Clone GE-1) conjugated with an Alexa Fluor© 405 dye is not available, this antibody is labeled in house using the procedure described by the manufacturer Life Technologies.
7. CD63-PE: CD63-PE (BD Biosciences, catalogue #557305) must be stored at 2–8 °C. This antibody is ready to use (*see* **Note 2**).

8. Mouse anti-human IgE (mHu-anti-IgE). This antibody (BD Biosciences, clone G7-18: catalogue #555894) is used as a positive control to stimulate basophils. Stock solution: 0.5 mg/mL in BAT buffer. Working solution: between 1 μg/mL and 10 μg/mL in BAT buffer.
9. CD203c-APC: CD203c (clone NP4D6) labeled with APC (BD Biosciences catalogue #562973 or BioLegend catalogue #324610). Ready to use (*see* **Note 2**).
10. Allergens. (*see* **Note 3**).
11. Avidin Alexa Fluor© 488. Prepare avidin Alexa Fluor© 488 (Thermo Fisher Scientific) solution of 2 mg/mL by dissolving the powder in the appropriate amount of PBS. The reconstituted solution is stable for approximately 6 months with the addition of sodium azide to a final concentration of 5 mM or thimerosal to 0.2 mM. For longer storage, divide the solution into aliquots and freeze at −20 °C. Protect from light, and avoid repeated freezing and thawing of solutions. Working solution: 1/20 in PBS-NaN_3 (*see* **Note 2**).

3 Methods

All procedures are carried out at room temperature, in a laminar airflow cabinet when necessary, unless otherwise specified (*see* **Note** 7).

3.1 Labeling Procedure

1. Dilute mouse monoclonal anti-human IgE (clone GE-1, Sigma-Aldrich catalogue #I6510) in 0.1 M $NaHCO_3$ (pH 8.3) to a concentration of 5 mg/mL.
2. Alexa Fluor 405 carboxylic acid, succinimidyl ester, and triethylammonium salt (Life Technologies, Molecular Probes, A30000) can be stored at −20 °C for a maximum of 3 months. When dissolved in DMSO, the solution becomes unstable. A working solution is made fresh by dissolving 1 mg in 0.1 mL DMSO.
3. Mix 2 mL anti-IgE in $NaHCO_3$ slowly with 0.1 mL Alexa Fluor© 405 working solution during 1 h at room temperature in the dark. Mix regularly.
4. Dialyze overnight with PBS at 4 °C.
5. Add NaN_3 until a final concentration of 0.05% is reached. Store at 2–8 °C.
6. Working solution: 1/50 in PBS-NaN_3, store at 2–8 °C. Working solution has to be titrated every time a new batch is made (*see* **Note 2**).

3.2 Control of Available Basophils and Their Stimulation (Exclusion of Nonresponders)

1. Incubate 100 μL heparinized blood with 100 μL negative (BAT buffer) or positive control (anti-IgE) for 20 min at 37 °C.
2. Using the procedure described in Subheading 3.3, stain with aIgE-AF405 and CD203c-APC (*see* **Note 4**) or CD123/HLA-DR and CD203c-APC (*see* **Note 4**).
3. If basophils do not respond to the positive control, one can still decide to stimulate with two concentrations of the allergen to exclude nonreactivity to anti-IgE (*see* **Note 5**).

3.3 Allergen Procedure

1. All blood samples, buffers, and allergens are prewarmed without mixing for 15 min to 37 °C in a warm water bath (WWB) (*see* **Note 6**).
2. Prepare incubation tubes and add 100 μL allergen solution to the allergen tubes. Dilutions depend on the allergen used (*see* **Note 3**). Use BAT buffer as a negative control or for FMO evaluation (*see* **Note 2**).
3. Add anti-IgE as positive control to a positive control tube. Add 100 μL BAT buffer to a negative control tube.
4. Add 100 μL prewarmed whole blood to each tube.
5. Incubate for 20 min at 37 °C in WWB.
6. Stop reaction by adding 1 mL PBS-EDTA.
7. Spin at 200 × *g* for 10 min at 4 °C with no brake.
8. Remove supernatant.
9. Add 20 μL aIgE-AF405, 20 μL avidin-AF488, and 10 μL aCD203c-APC.
10. Incubate on ice for 20 min in the dark.
11. Add 2 mL lysis solution and incubate at room temperature for 20 min.
12. Spin for 10 min at 250 × *g* at room temperature.
13. Remove supernatant.
14. Add 2 mL PBS and spin again for 10 min 250 × *g* at room temperature.
15. Remove supernatant.
16. Add 300 μL PBS-NaN_3. The tubes are now ready to measure with a flow cytometer.
17. Count at least 500 basophils that are defined as $aIgE^+$ and $CD203c^+$.
18. Store data in an FCS-formatted file, which can be analyzed with flow cytometric software.
19. Define $CD203c^{hi+}$/$avidin^-$ basophils and $CD203c^{hi+}$/$avidin^{hi+}$ as non-degranulating and degranulating subpopulations, respectively (*see* Fig. 1).

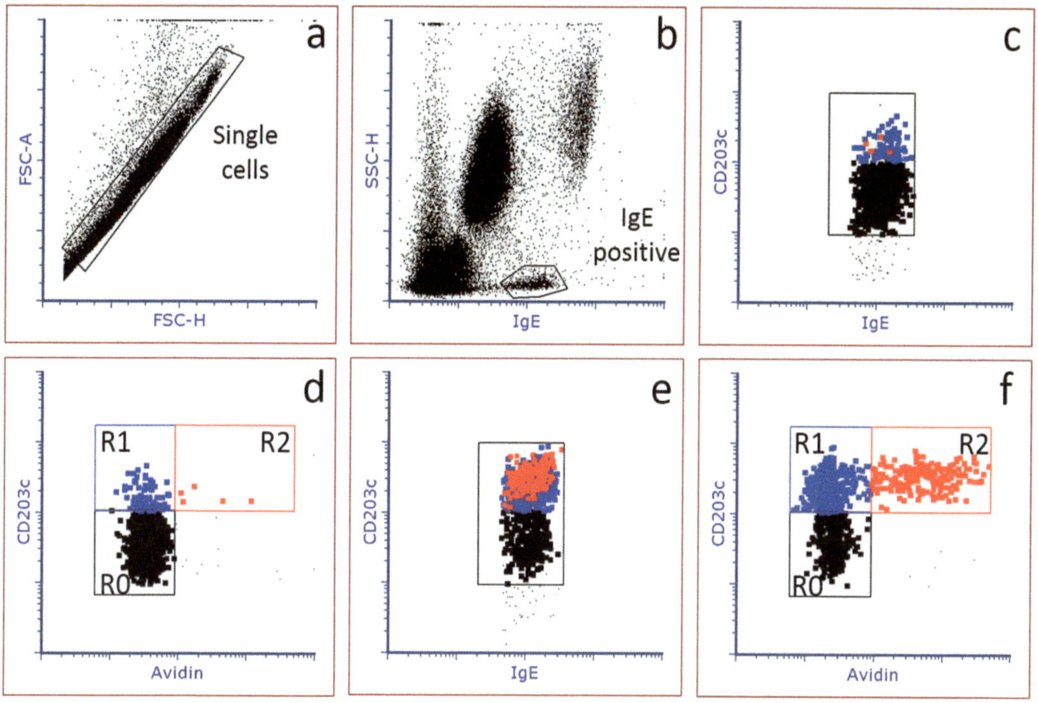

Fig. 1 Basophil selection and activation. Panel **a**: aggregates are avoided by selecting single cells based on FCS (forward scatter) area and FCS height plot. Plot **b**: IgE high positive cells are gated out. Panel **c**: gated cells are IgE^{+} and $CD203c^{+}$ (without stimulation). Panel **d**: activated basophils in control setting: R0 are $CD203c^{+}$ $avidin^{-}$, R1 are $CD203c^{hi+}$ $avidin^{-}$, and R2 are $CD203c^{hi+}$ $avidin^{+}$. Panels **e** and **f** are equal to **c** and **d** but activated with anti-IgE as a positive control

20. Percentage of net upregulation is obtained by subtracting the % obtained with negative control (BAT buffer) from the % obtained with positive control or allergen.

4 Notes

1. Although different BAT techniques use EDTA- or even citrate-treated blood, the best results are obtained with heparinized blood, as basophil activation is very calcium dependent. The use of chelators or calcium-binding products can have deleterious effect on test results. Antihistamines have no deleterious effect on BATs.
2. All antibodies can be used labeled with different fluorochromes. Be aware that correct titration is mandatory. This can be achieved via different dilutions of the antibody and by using the correct flow cytometric settings, looking for the optimal resolution in comparison with a Fluorescence Minus One (FMO) measurement as a control.

3. Allergens can be protein extracts, recombinant proteins, or drugs. They must be water soluble and used in concentrations that are not toxic for basophils. To test toxicity, incubate basophils with different concentrations of the allergen or mHu-antiIgE (positive control) for 20 min at 37 °C, and measure the viability using propidium iodide or the LIVE/DEAD Viability/Cytotoxicity Kit for mammalian cells (Life Technologies). Do not use preparations for skin testing. These solutions can contain preservatives or sometimes modified proteins that can inhibit basophil activation. Be aware that food allergens are very difficult to standardize. The optimal concentration of an allergen to stimulate basophils must be determined in dose-response experiments including ten well-defined patients positive to the antigen and ten controls negative to the allergen. Because such a response can result in a bell-shaped curve, in routine practice, at least two concentrations might be needed to define a positive result.
4. Basophils can be identified using different markers. Basophils are IL-3 receptor (CD123) positive and HLA-DR negative (monocytes and dendritic cells are both HLA-DR positive and also CD123 positive). Because CD203c is the only lineage marker known until now for basophils (mast cells are also CD203c positive, but they do not circulate in the blood), we prefer to define basophils as CD203c+/IgE+. Nevertheless, the CD203c density can be too low to differentiate cells.
5. In about 5–10% of the patients, basophils do not respond to a positive control, be it anti-IgE or anti-FcεRI. In such a case, one can decide to stimulate with the allergen. However, if the allergen also gives a negative response, the test should be considered as false negative (so-called nonresponder). Results will be inconclusive.
6. Basophil activation results in an immediate response (1 min) and lasts about 20–30 min at maximum response. To obtain reproducible results, all solutions are best prewarmed to 37 °C.
7. Wear gloves, lab coat, and safety glasses while handling all human blood products. Dispose of all pipettes into bagged waste collection bins. Wipe work surfaces with disinfectant before and after running tests.

References

1. Knol EF, Mul FPJ, Jansen H, Calafat J, Roos D (1991) Monitoring human basophil activation via CD63 monoclonal antobody-435. J Allergy Clin Immunol 88:328–338
2. Joulia R, Mailhol C, Valitutti S, Didier A, Espinosa E (2017) Direct monitoring of basophil degranulation by using avidin-based probes. J Allergy Clin Immunol 140:1159–1162
3. Mukai K, Chinthrajah RS, Nadeau KC, Tsai M, Gaudenzio N, Galli SJ (2017) A new fluorescent-avidin-based method for quantifying basophil activation in whole blood. J Allergy Clin Immunol 140:1202–1206

Chapter 18

Measurement and Functional Analysis of the Mas-Related G Protein-Coupled Receptor MRGPRX2 on Human Mast Cells and Basophils

Jessy Elst, Vito Sabato, Margo M. Hagendorens, Michel van Houdt, Margaretha A. Faber, Chris H. Bridts, Didier G. Ebo, and Christel Mertens

Abstract

Basophils and mast cells (MCs) are important effector cells in the immune system. For a long time, it has been known that these cells can be activated though the cross-linking of IgE antibodies bound to their high-affinity receptor (FcεRI). However, evidence has accumulated suggesting that these cells can also be activated by various IgE-independent mechanisms. Occupation of MAS Related GPR Family Member X2 (MRGPRX2), a G protein-coupled receptor, is described as an alternative IgE-independent activation mechanism. Here we describe a flow cytometric technique to analyze MRGPRX2 expression and its functionality on cultured human MCs and conditioned basophils, that is, basophils with upregulated surface expression of MRGPRX2.

Key words Flow cytometry, Basophils, Cultured mast cells, MRGPRX2

1 Introduction

Flow-assisted analysis of in vitro activated basophils has entered the field of allergy diagnosis about two decades ago. Today, the application of basophil activation test (BAT) has mainly to be sought in the domain of drug hypersensitivity [1]. For long, most of these studies have focused on basophil degranulation in response to the cross-linking of allergen-specific IgE antibodies (sIgE) coupled to their high-affinity receptor (FcεRI) that is present on the surface of the cell. However, during the last years, evidence has accumulated suggesting that immediate drug hypersensitivity reactions might also result from alternative, IgE-independent degranulation of mast cells (MCs) [2]. Amongst these alternative mechanisms, it seems that off-target occupancy of the MRGPRX2 receptor by drugs might trigger (severe) hypersensitivity reactions that are clinically indistinguishable from genuine IgE-mediated reactions.

Bernhard F. Gibbs and Franco H. Falcone (eds.), *Basophils and Mast Cells: Methods and Protocols*, Methods in Molecular Biology, vol. 2163, https://doi.org/10.1007/978-1-0716-0696-4_18,

However, most of this information has been gathered from genetically modified animals and transfected cell lines, and findings in these surrogate models might not simply be extrapolatable to humans. The receptor is constitutively expressed on human MCs [3] and is inducible on human basophils [4]. Here, we describe how MRGPRX2 expression can be measured on human MCs that are cultured out of small volumes of peripheral blood and the so-called "conditioned" basophils, that is, cells that have been preactivated and that, unlike resting basophils, do express MRGPRX2.

2 Materials

2.1 Mast Cells

Human MCs are cultured as described elsewhere [5]. Briefly, $CD34^+$ progenitors were isolated using the EasySep Human CD34 Selection Kit (Stemcell Technologies, Vancouver, Canada) out of Histopaque-isolated peripheral blood mononuclear cells. Isolated $CD34^+$ progenitors are cultured in a serum-free methylcellulose-based medium (MethoCult SF H4236, Stemcell Technologies) supplemented with penicillin (100 units/mL), streptomycin (100 μg/mL), low-density lipoprotein (LDL, 10 μg/mL: Stemcell Technologies), 2-mercaptoethanol (55 μmol/L), stem cell factor (SCF, 100 ng/mL, Miltenyi Biotec, Bergisch Gladbach, Germany), and interleukin-3 (IL-3, 100 ng/mL, PeproTech, Rocky Hill, USA) during 4–5 weeks. Mature MCs are defined as $CD117^+$ and $CD203c^+$.

2.2 Blood for Basophil Activation Test

Peripheral heparinized blood (10 mL) is obtained and immediately transported to the laboratory. Incubation and analysis should be performed within 8 h for MRGPRX2 on basophils.

2.3 Reagents

All solutions should be prepared with sterile ultrapure water. We use a Millipore Milli-Q System and the water is purified with a reversed osmosis system followed by ultra-purification to attain 18 MΩ at room temperature.

1. Tyrode buffer: dissolve 9.6 g Tyrode's salts in 980 mL H_2O. Add 20 mL 1 M HEPES. The solution must be filtered to make it sterile. Mix 9.7 mL Tyrode buffer with 300 μL 7.5% $NaHCO_3$ right before use; pH should be 7.4.
2. Lysis/Fixing solution: BD Phosflow Lyse/Fix solution (BD Biosciences, cat.nr. 558049) is diluted 1/5 with water just before use.
3. Phosphate buffered saline: 10× PBS (without Ca^{2+} or Mg^{2+}) is diluted with H_2O to obtain an isotonic PBS buffer. Adjust pH to 7.4.
4. PBS-EDTA: dissolve 10 mmol/L Na_2EDTA in PBS (1×). Adjust pH to 7.4.

5. PBS-NaN_3: dissolve 0.5 g/L NaN_3 in PBS. Adjust pH to 7.4. Store at room temperature.
6. PBS-0.1% BSA: Mix 7.4 mL PBS with 100 μL BSA (bovine albumin fraction V 7.5%). Adjust pH to 7.4.
7. Substance P: stock solution is 742 μM in 0.05 M acetic acid. Working solution is 148.4 μM in Tyrode's buffer (*see* **Note 2**).
8. CD117-APC: monoclonal anti-human CD117 (clone: 104D2) labeled with APC (BD Biosciences) is ready to use (*see* **Note 1**).
9. CD203c-PeCy7: monoclonal anti-human CD203c (Clone: NP4D6) labeled with PE-Cy7 (eBioscience) is ready to use (*see* **Note 1**).
10. CD63-FITC: anti-CD63-FITC (Clone: H5CB6, BD Bioscience) stored at 2–8 °C in the dark. This antibody is ready to use (*see* **Note 1**).
11. Anti-MRGPRX2-PE: anti-MRGPRX2-PE (Clone: K125H4, BioLegend), stored at 2–8 °C in the dark. This antibody is ready to use (*see* **Note 1**).
12. Activation buffer for BAT: Activation buffer is prepared STERILE just before use. Dilute 1.9 mL 10× concentrated Hanks' balanced salt solution (HBSS with Ca^{2+} and Mg^{2+}) with 17.1 mL H_2O. Add 0.6 mL 7.5% $NaHCO_3$. Add 0.4 mL 1 M HEPES. Adjust pH to 7.4.
13. Monoclonal mouse anti-human-IgE (mHu-anti-IgE): This antibody (BD Biosciences, clone G7-18: catalog #555894) is used as a positive control to stimulate basophils. Stock solution: 0.5 mg/mL in activation buffer. Working solution: between 1 μg/mL and 10 μg/mL in activation buffer (*see* **Note 2**).
14. CD203c-APC: CD203c (clone NP4D6) labeled with APC (BD Biosciences catalogue #562973 or BioLegend catalogue #324610). Ready to use (*see* **Note 1**).
15. Anti-human IgE Alexa Fluor 405 (aIgE): since aIgE (Sigma-Aldrich, Clone GE-1) conjugated with an Alexa Fluor 405 dye is not available, this antibody is labeled in house using the procedure described by the manufacturer Life Technologies (*see* chapter 17).

3 Methods

All procedures are carried out at room temperature, in a laminar airflow cabinet when necessary, unless otherwise specified.

3.1 Expression of MRGPRX2 on Cultured Mast Cells (PBCMCs)

1. Prepare three tubes with 100 μL of MCs in PBS-NaN_3 at a density of 0.5×10^6 cells/mL.

2. Stain:

   Tube 1: (FMO MRGPRX2): 10 μL CD117-APC, 10 μL CD203c-PE-Cy7, and 10 μL CD63-FITC.

   Tube 2: (FMO CD203c): 10 μL CD117-APC, 10 μL CD63-FITC, and 5 μL MRGPRX2-PE.

   Tube 3: 10 μL CD117-APC, 10 μL CD203c-PE-Cy7, 10 μL CD63-FITC, and 5 μL MRGPRX2-PE.

3. Incubate 20 min on ice in the dark (*see* **Note 3**).
4. Wash cells with 2 mL PBS-NaN_3 by spinning at 500 × *g* for 10 min and removing the supernatant.
5. Resolve the cell pellet in 300 μL PBS-NaN_3 and acquire data on a flow cytometer.
6. Count at least 500 MCs, which are defined as $CD117^+$ and $CD203c^+$. MCs have a high FCS and express both CD117 and CD203c (*see* Fig. 1).

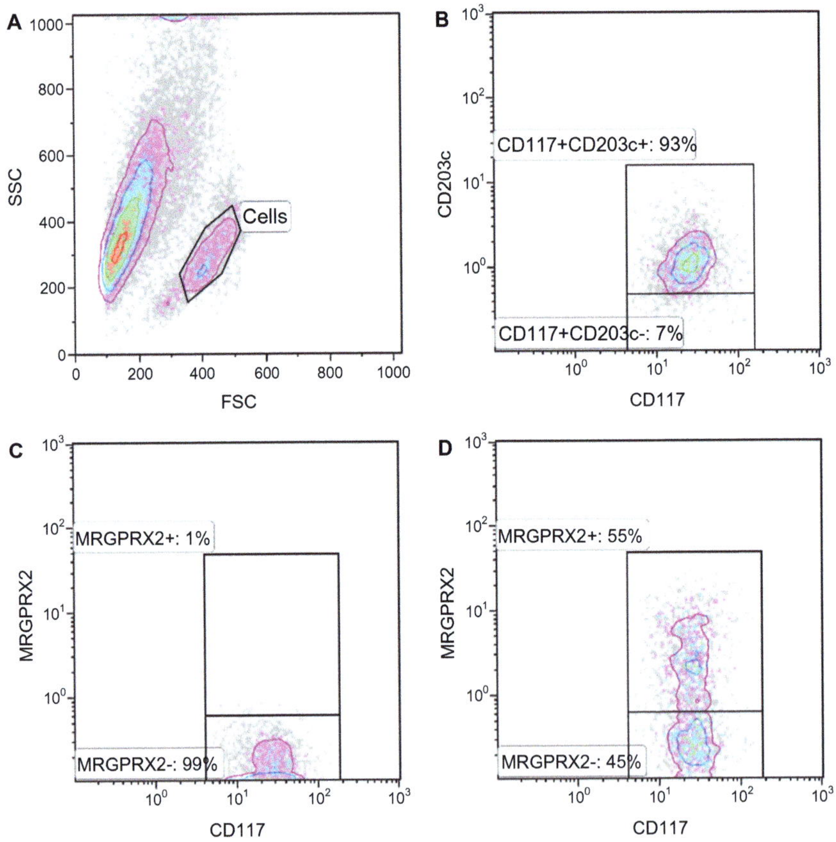

Fig. 1 Mast cell and MRGPRX2. (**a**) Cells are gated based on forward scatter (FSC) and side scatter (SSC). (**b**) Gated cells are $CD117^+$ and $CD203^+$. (**c**) FMO sample for MRGPRX2. (**d**) MRGPRX2-positive mast cells. FMO, Fluorescence Minus One

7. Store data in an FCS-formatted file, which can be analyzed with flow cytometric software.
8. Define MRGPRX2-positive MCs (*see* Fig. 1).

3.2 Mast Cell Activation via MRGPRX2

1. Prepare incubation tubes with 100 μL of MC in Tyrode's buffer at a density of 0.5×10^6 cells/mL.
2. Prewarm the cells, buffer, and substance P for 15 min at 37 °C in a warm water bath (WWB).
3. Add 100 μL of the natural MRGPRX2 ligand substance P as positive control or Tyrode's buffer as negative control.
4. Incubate 3 or 20 min at 37 °C in WWB.
5. Stop the reaction by cooling the cells on ice for 5 min and removing the supernatant (centrifugation 5 min at $500 \times g$ at 4 °C).
6. Add 100 μL PBS-BSA and stain with 10 μL CD117-APC, 10 μL CD203c-PE-Cy7, 10 μL CD63-FITC, and MRGPRX2-PE (for the FMO sample without MRGPRX2).
7. Incubate 20 min on ice in the dark.
8. Fixate the cells with 1 mL Lyse/Fix buffer, vortex, and incubate for 20 min in the dark at room temperature.
9. Spin the samples at $500 \times g$ for 10 min. Remove the supernatant and wash the cell pellet with 2 mL PBS-NaN_3. Spin again at $500 \times g$ for 10 min and remove the supernatant.
10. Resolve the cell pellet in 300 μL PBS-NaN_3 and acquire data on a flow cytometer.
11. Count at least 500 MCs that are defined as $CD117^+$ and $CD203c^+$.
12. Store data in an FCS-formatted file, which can be analyzed with flow cytometric software.
13. Define the activation of the MC (*see* Fig. 2).

3.3 MRGPRX2 on Basophils After the So-Called Conditioning

As already exemplified in the introductory paragraph, resting human basophils barely express the MRGPRX2 receptor. Therefore, to demonstrate the effect of MRGPRX2 ligation, basophils need to be conditioned, viz., their MRGPRX2 expression has to be enhanced. Despite numerous efforts, for the time being, this conditioning has only been successful via FcεRI cross-linking. Here, we describe MRGPRX2 appearance after incubation of the cells with anti-IgE.

1. All blood samples, buffers, and activators are prewarmed without mixing for 15 min at 37 °C in a WWB.

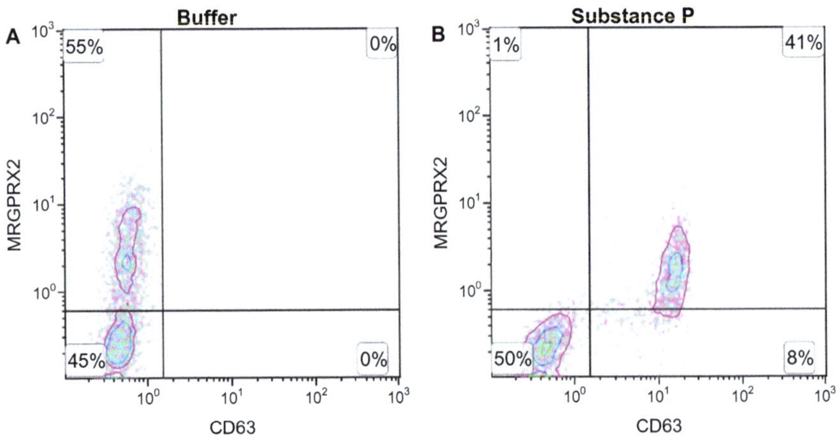

Fig. 2 Activation of mast cells. (**a**) Mast cells incubated with Tyrode's buffer. (**b**) Mast cells stimulated with positive control substance P for 20 min. The cells that are expressing the MRGPRX2 receptor show appearance of the lysosomal degranulation marker CD63

2. Prepare incubation tubes and substance P. Use activation buffer as a negative control or aIgE for FMO evaluation (*see* **Note 1**).
3. Add 200 μL separately or simultaneously mHu-anti-IgE and substance P; add 200 μL BAT buffer to a negative control tube.
4. Incubate between 3 min and 20 min at 37 °C in WWB.
5. Stop reaction by cooling on ice.
6. Spin at 500 × *g* for 10 min at 4 °C without brake. Remove supernatant.
7. Add 20 μL aIgE-AF405, 10 μL anti-CD63-FITC, 5 μL aCD203c-APC, and 5 μL MRGPRX2-PE (for FMO without MRGPRX2).
8. Incubate at 4 °C for 20 min in the dark.
9. Add 2 mL Phosflow Lyse/Fix work solution and incubate at 37 °C in WWB for 20 min.
10. Spin for 10 min at 500 × *g* at room temperature. Remove supernatant.
11. Add 2 mL PBS and spin at 500 × *g* for 10 min at room temperature. Remove supernatant.
12. Add 0.5 mL PBS-NaN_3.
13. The tubes are now ready to be measured on a flow cytometer.
14. Count at least 1000 basophils which are defined as $aIgE^+$ and $CD203c^+$.
15. Store data in an FCS-formatted file, which can be analyzed with flow cytometric software (*see* Fig. 3).

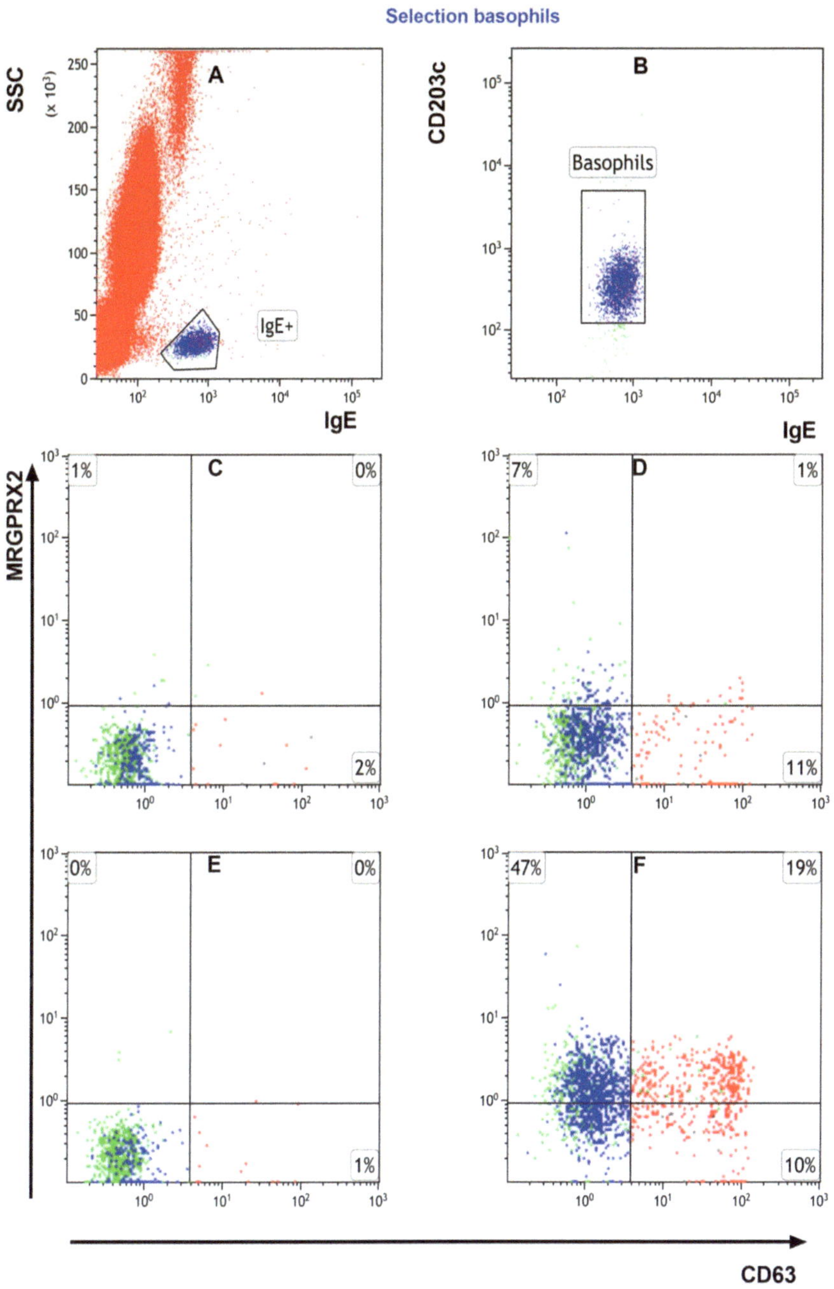

Fig. 3 Basophil selection and MRGPRX2 expression. (**a**) IgE brightly positive cells are gated out. (**b**) Gated basophils are IgE$^+$ and CD203$^+$. (**c**) Basophils stimulated with buffer. (**d**) Basophils stimulated with mHu-anti-IgE after 3 min. (**e**) Basophils stimulated with substance P after 3 min. (**f**) Basophils stimulated with mHu-anti-IgE and substance P after 3 min. Note the dissociation between the rather uniform appearance of MRGPRX2 and more restricted bimodal upregulation of CD63

4 Notes

1. All antibodies labeled with different fluorochromes can be used. Be aware that correct titration is necessary. This can be done by using different dilutions of the antibody and using the correct flow cytometric settings, looking for the optimal resolution in comparison with a Fluorescence Minus One (FMO) sample as a control.
2. The concentrations of substance P and mHu-anti-IgE are chosen after performing a dose-response curve. Be aware that for each new activator, a new dose-response curve is required.
3. Fixation is not performed before the phenotyping of the cells, because fixation has an effect on the expression and density of some phenotype markers.

References

1. Decuyper II, Mangodt EA, Van Gasse AL, Claesen K, Uyttebroek A, Faber M et al (2017) In vitro diagnosis of immediate drug hypersensitivity anno 2017: potentials and limitations. Drugs R D 17(2):265–278. https://doi.org/10.1007/s40268-017-0176-x
2. Gaudenzio N, Sibilano R, Marichal T, Starkl P, Reber LL, Cenac N et al (2016) Different activation signals induce distinct mast cell degranulation strategies. J Clin Invest 126(10):3981–3998. https://doi.org/10.1172/JCI85538
3. McNeil BD, Pundir P, Meeker S, Han L, Undem BJ, Kulka M et al (2015) Identification of a mast-cell-specific receptor crucial for pseudo-allergic drug reactions. Nature 519:237–241. https://doi.org/10.1038/nature14022
4. Sabato V, Van Gasse A, Cop N et al (2017) The Mas-related G protein-coupled receptor MRGPRX2 is expressed on human basophils and up-regulated upon activation. J Allergy Clin Immun 139(2):Ab168-Ab. https://doi.org/10.1016/j.jaci.2016.12.550
5. Cop N, Decuyper II, Faber MA, Sabato V, Bridts CH, Hagendorens MM et al (2017) Phenotypic and functional characterization of in vitro cultured human mast cells. Cytometry B Clin Cytom 92(5):348–354. https://doi.org/10.1002/cyto.b.21399

Chapter 19

Mast Cell Activation Test (MAT)

Rajia Bahri and Silvia Bulfone-Paus

Abstract

The mast cell (MC) activation assay is a robust in vitro tool for exploring MC reactivity in allergy. Here we describe the use of the mast cell activation test (MAT) that makes use of human primary MCs generated from peripheral blood progenitors, sensitized overnight with patients' sera and activated with allergens. Flow cytometry is used to assess the changes in expression of the activation marker CD63, and the percentage of cell degranulation is defined as the percentage of $CD63^{+}$-positive MCs.

Key words Mast cell, Allergic disease, Allergic patient sera, Mast cell activation, Peripheral blood progenitor, Flow cytometry

1 Introduction

Mast cells (MCs) are granulated long-lived immune resident tissue cells and central effector cells in IgE-mediated allergic-type I hypersensitivity reactions [1]. Allergen-specific IgE binds FcεRI, a high-affinity immunoglobulin IgE receptor displayed at the mature MC membrane. Upon allergen cross-linking follows the activation of the IgE-prebound FcεRI and the immediate release of the granule content [2]. MCs secrete a wide range of preformed and de novo synthesized mediators that include histamine, proteases, leukotrienes, prostaglandins, chemokines (e.g., CCL-5, CCL-2), cytokines (e.g., IL-8, TNF-α), and growth factors that are the triggers of allergic responses such as anaphylaxis, asthma, angioedema, urticaria, nausea, and hay fever.

Several cell lines expressing the high-affinity IgE receptor are available to study MC activities in vitro: the humanized rat basophilic leukemia cells, RBL-2H3 [3], the human leukemic mast cells, LAD2 [4], LUVA cells [5], and wild-type ROSA cells [6]. Due to their tumor origin, maturation stage, in vitro cell growth, and low or variable capacity to induce FcεRI-mediated degranulation and to produce mediators, the use of primary human MCs for an allergen-

Bernhard F. Gibbs and Franco H. Falcone (eds.), *Basophils and Mast Cells: Methods and Protocols*, Methods in Molecular Biology, vol. 2163, https://doi.org/10.1007/978-1-0716-0696-4_19,

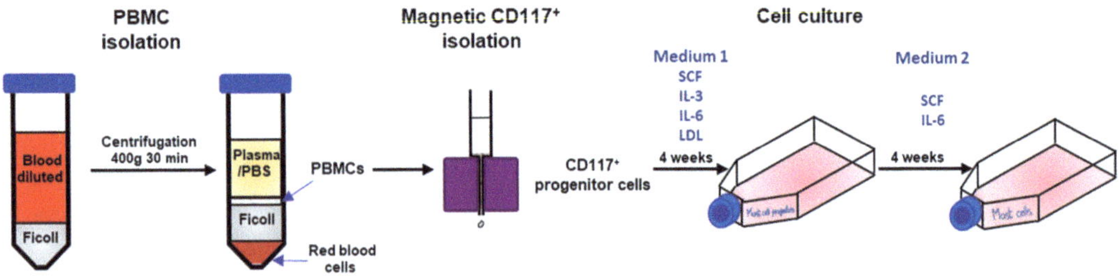

Fig. 1 Human primary MC generation. PBMCs are isolated from human peripheral blood by Ficoll–Paque gradient centrifugation. CD117$^+$ progenitor cells, magnetically separated from PMBCs, are cultured for 4 weeks in medium 1 supplemented with LDL (low-density lipoproteins), IL3, IL-6, and SCF, and after week 4 medium 1 is replaced by medium 2 containing IL-6 and SCF. Mature MCs are obtained at 8–10 weeks of culture

induced mast cell activation assay appears superior to cell lines [7–12].

In adults, MCs originate from embryonic or bone marrow hematopoietic stem cell progenitors [13]. Using peripheral blood hematopoietic progenitors, we differentiate MCs with some adjustments as previously described [14–16]. We developed a serum-free MC culture protocol starting with CD117$^+$ progenitor cells isolated from human peripheral blood (Fig. 1). After 8–10 weeks of culture, cells differentiate to highly pure and mature granulated MCs. More than 90% of the cells express CD117, tryptase, and chymase and exhibit a functional FcεRI [12]. Allogenic blood-derived MCs are sensitized with allergic patients' sera and then stimulated with the allergen of interest (e.g., peanut, grass pollen, wasp venom) (Fig. 2). MC degranulation relies on the measurement by flow cytometry of CD63, an activation marker which is increased on the MC surface after granule release. The mast cell activation test (MAT) shows high sensitivity and dose-dependent responses to allergens (Fig. 3b) [12]. This approach is a robust alternative to the use of primary MCs isolated from human tissues which are limited in number and quantity.

2 Materials

2.1 Equipment

1. Class II biosafety cabinet.
2. Cell culture CO_2 incubator, 37 °C with 5% CO_2.
3. Centrifuge with swinging bucket rotor for 96-well plate and 50—15 mL tube.
4. Flow cytometer.
5. Magnetic field for Miltenyi Biotec LS Columns.
6. 10, 100, and 1000 μL micropipettes.
7. Sterile pipet tips, 10, 20, 200, and 1000 μL.

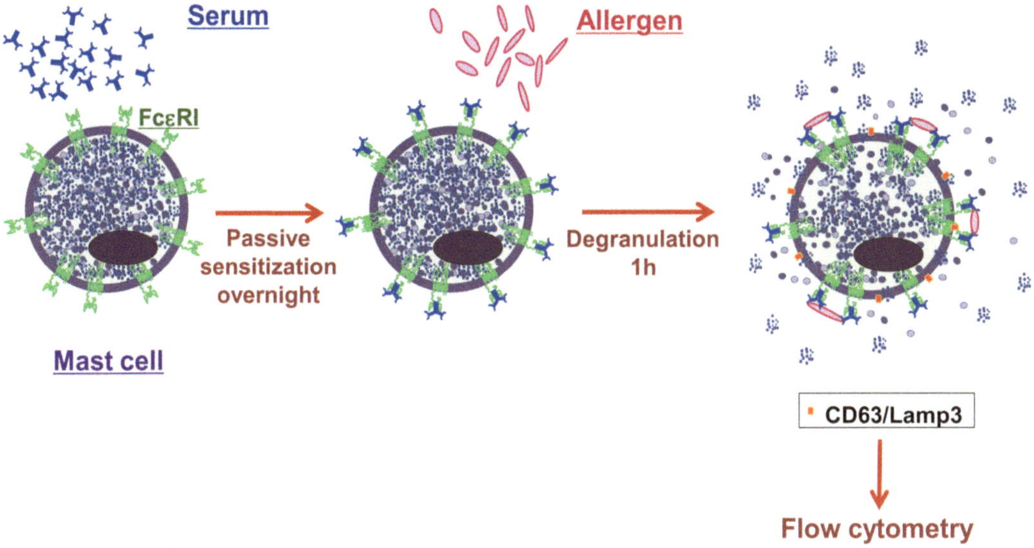

Fig. 2 Mast cell activation test (MAT). MCs are sensitized overnight with patient's serum and then activated for 1 h with the allergen. To quantify MC degranulation, surface upregulation of the activation marker CD63 is measured by flow cytometry

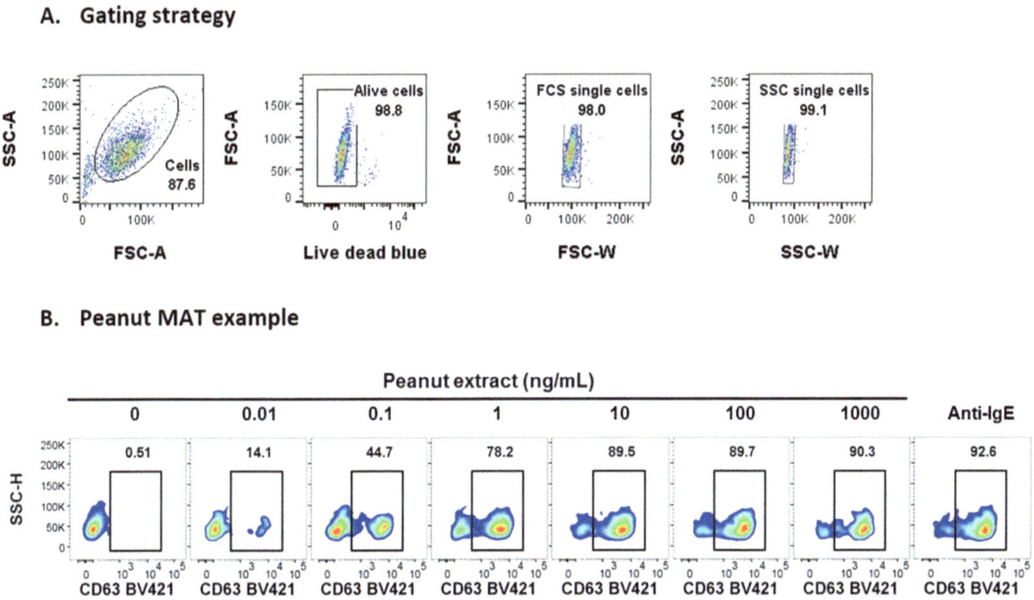

Fig. 3 Flow plots. MCs are isolated from debris and aggregates using the forward scatter (FSC) versus the side scatter (SSC). Then after gating on alive MC population, live/dead negative doublets are discriminated based on forward scatter area (FSC-A) versus forward scatter width (FSC-W) and on side scatter area (SSC-A) versus side scatter width (SSC-W) (Fig. 3a). $CD63^{+}$ gate is set up to use as negative control unstimulated MCs (Fig. 3b). (**b**) shows an example of peanut MAT using sera from the Manchester Allergy, Respiratory and Thoracic Surgery (ManARTS) Biobank (UK NHS Human Research Authority, reference [15] /NW/040)

8. 50–100 μL multichannel pipette.
9. Pipet boy.
10. 5, 10, and 25 mL disposable sterile serological pipets.
11. Sterile scissors.
12. 2–8 °C refrigerator.
13. 50 mL tubes.
14. 15 mL tubes.
15. LS columns (Miltenyi Biotec, 130-042-401).
16. Cell counting chamber.
17. Cell culture flask 25 cm^2, 75 cm^2, and 150 cm^2.
18. Six-well cell culture plate with a flat bottom with lid tissue culture-treated.
19. 96-well cell culture plate with a round bottom with lid tissue culture-treated.

2.2 Human Biological Material

Laboratory procedures are conducted using the personal protective equipment required for human body samples according to the local laboratory guidelines and practices.

1. Buffy coat or apheresis cones from healthy donors.
   Human blood samples are ordered and handled following the ethical guidelines of the institution.
2. Patient allergic and healthy sera.

All patients should sign a consent form and sera are collected and used in accordance with the national medical ethics.

2.3 Reagents

2.3.1 Human Peripheral Blood-Derived Mast Cell Differentiation

1. Ficoll–Paque Plus.
2. Dulbecco's Phosphate Buffered Saline 1× (D-PBS) without calcium chloride and magnesium chloride.
3. CD117 MicroBead Kit human (Miltenyi Biotec, 130-091-332).
4. Ultrapure 0.5 M ethylenediaminetetraacetic acid solution (EDTA), pH 8.0 (Invitrogen, 15575-038).
5. StemSpan™ SFEM, serum-free medium for culture and expansion of hematopoietic cells (Stemcell Technologies, 09650).
6. Human low-density lipoproteins (LDLs) prescreened for hematopoietic cell growth (Stemcell Technologies, 02698).
7. Penicillin streptomycin with 10,000 units of penicillin and 10 mg/mL streptomycin (100×).
8. Recombinant human IL-3 (PeproTech, 200-03).
9. Recombinant human IL-6 (PeproTech, 200-06).

10. Recombinant human Stem Cell Factor (SCF) (PeproTech, 300-07).
11. 2-Mercaptoethanol.
12. Bovine albumin (BSA) fraction V solution (7.5%).
13. Insulin–transferrin–selenium (ITS) (100×).
14. Iscove's Modified Dulbecco's Medium with GlutaMAX and HEPES (IMDM).
15. 0.4% Trypan blue solution.

2.3.2 Degranulation Assay and Flow Cytometry Staining

1. Peanut extract (from Clare Mills laboratory, MIB University of Manchester).
2. Wasp venom was from Hal Allergy company.
3. Timothy grass pollen allergen extract was kindly provided as well by Euroimmun AG (Luebeck, Germany).
4. Goat anti-human IgE (KPL, 5210-0158).
5. Human IgE melanoma (Merck Millipore, 401152).
6. Fc Receptor Blocking Solution, Human TruStain FcX™ (BioLegend, 422302).
7. LIVE/DEAD® Fixable Blue Dead Cell Stain Kit (Thermo Fisher Scientific, L34962).
8. Human anti-CD63 antibody BV421 conjugated (clone H5C6) (BioLegend, 353012).

2.3.3 Disinfectant

1. 70% v/v ethanol disinfectant in a spray bottle used before procedures.
2. 1% v/v distel disinfectant in a spray bottle used after or during procedures.
3. Rely+On Virkon powder for disinfecting liquid waste (Dupont, VK734).

2.4 Solution and Media

MACS buffer: 0.5% BSA (3.33 mL of BSA 7.5% stock solution), 2 mM EDTA (200 μL of 0.5 M EDTA stock solution), 50 mL D-PBS (*see* **Note 1**).

Medium 1: 500 mL of StemSpan™ SFEM supplemented with 100 U/mL penicillin and 100 μg/mL streptomycin, 100 ng/mL human SCF, 50 ng/mL human IL-6, 10 ng/mL human IL-3, and 10 μg/mL LDL. Medium 1 is used from day 1 to day 29 (*see* **Note 2**).

Medium 2: 500 mL of IMDM supplemented with 100 U/mL penicillin and 100 μg/mL streptomycin (5 mL of stock solution 100×), 100 ng/mL human SCF, 50 ng/mL human IL-6, 50 mMol/L 2-mercaptoethanol, 0.5% BSA (33.33 mL of stock solution 7.5%), and 1% ITS (5 mL of stock solution 100×). Medium 2 is used starting from day 29 (*see* **Note 3**).

Live/dead stock solution: bring at room temperature one vial of each from the LIVE/DEAD® Fixable Blue Dead Cell Stain Kit. Add 50 μL of DMSO (component B) to the fluorescent reactive dye vial (component A). Aliquot and store at −20 °C.

2.5 Software

1. FlowJo.
2. Excel.

3 Methods

3.1 Human Peripheral Blood-Derived Mast Cell Differentiation (8–10 Weeks)

3.1.1 Peripheral Blood Mononuclear Cell (PBMC) Isolation from Peripheral Blood

1. Order buffy coat or apheresis cones.
2. All procedures must be done in a class II biosafety cabinet.
3. Work aseptically and use sterile reagents.
4. Clean the cabinet surface as well as the equipment (e.g., scissors, pipettes, tubes, racks) with 70% ethanol.
5. D-PBS and Ficoll should be at room temperature.
6. Add 15 mL Ficoll per 50 mL tube (the number of tubes to prepare depends on the final volume of diluted blood).
7. Upon receipt, check carefully and record the donor information (*see* **Note 4**).
8. Spray the buffy coat pouch or apheresis cone and scissors with 70% ethanol.
9. Collect blood content.
   *For buffy coats.
   With one hand hold the pouch and cut with sterile scissors one of the below tubing ports above a sterile flask 75 cm^2.
   Cut the upper tubing port.
   Pour all the pouch contents in the sterile flask.
   *For apheresis cones.
   Cut with sterile scissors the below tubing port and put the cone on top of a sterile 50 mL tube.
   Cut the upper tubing port.
   Leave the content drop into the 50 mL tube till the last drop.
   Put a sterile nonfilter 1000 μL tip into the upper tubing port and add in it 10 mL of D-PBS using the pipet boy and 10 mL serological pipets.
   Wait for the last drop (*see* **Note 5**).
10. Determine the blood volume collected and dilute it 1:1 with D-PBS.

11. Carefully overlay by slowly pipetting 25 mL of diluted blood on the top of the 15 mL Ficoll layer in the 50 mL tubes (*see* **Note 6**).
12. Close tubes and gently transport them to the centrifuge without breaking the layers.
13. Centrifuge at 20 °C, 400 × *g*, 30 min acceleration 1 without brake (*see* **Notes 7** and **8**).
14. Carefully remove tubes from the centrifuge and bring them back into the biosafety cabinet.
15. The red blood cell pellet layer is at the bottom of the tube (red), then the Ficoll layer (transparent), and the top layer plasma/PBS (light yellow). The white cloudy PBMC ring is located at the interphase between the Ficoll and the plasma/PBS layer (Fig. 1). Introduce slowly a sterile pipet through the plasma/PBS layer and aspirate PBMCs; collect them in a 50 mL tube. Discard the rest of the layers in a liquid waste containing disinfectant according to the laboratory guidelines.
16. Add cold D-PBS to PBMCs up to 50 mL. Resuspend cells gently.
17. Centrifuge at 4 °C at 300 × *g* for 10 min.
18. Discard the supernatant and resuspend the cell pellet in 30 mL of cold MACS buffer.
19. Count cell number and viability by diluting 20 times in D-PBS (50 μL of cell suspension in 950 μL of D-PBS) and then 2 times in Trypan blue (*see* **Note 9**).

3.1.2 $CD117^+$ Cell Isolation from PBMCs

1. Centrifuge the remaining 30 mL PBMC suspension at 4 °C at 300 × *g* for 10 min.
2. Discard the supernatant and resuspend the cell pellet in 300 μL of cold MACS buffer per 10^8 cells.
3. Add 100 μL per 10^8 cells of FcR blocking reagent and 100 μL per 10^8 cells of CD117 microbeads.
4. Mix gently and incubate at 4 °C in the refrigerator for 20 min. Mix every 5 min.
5. Add 2 mL per 10^8 cells of cold MACS buffer.
6. Centrifuge at 4 °C at 300 × *g* for 10 min.
7. Discard the supernatant and resuspend the cell pellet in 500 μL per 10^8 cells of cold MACS buffer.
8. Place the LS column on the magnetic field and use a 50 mL empty tube to collect the flow-through.
9. Rinse the LS column with 3 mL of cold MACS buffer.
10. After the last drop, add PBMC suspension into the column.

11. After the last drop, wash with 3 mL of cold MACS buffer. Repeat twice for a total of three washes.
12. After the last drop, remove the column from the magnetic field and place it on the top of a sterile 15 mL tube.
13. Add 5 mL of cold MACS buffer into the column and immediately push the plunger into the column.
14. Count cell number and viability (*see* **Notes 10** and **11**).
15. Centrifuge at 4 °C at 300 × *g* for 5 min.
16. Discard the supernatant and resuspend the cell pellet in medium 1 at 0.5 × 10^6 cells/mL. Start the cell culture in a six-well plate.
17. Incubate at 37 °C with 5% CO_2.

3.1.3 Cell Culture

Cells grow very fast the first weeks in medium 1 and then when medium 2 is added at day 29 (4 weeks), cells start to die, and the remaining cells at weeks 6–7 are MCs; they mature till weeks 8–10 of culture.

1. Cells should be checked daily and maintained at 0.5–1 × 10^6 cell/mL by diluting them in medium 1. Culture surface should be adapted to the cell number at 0.2 × 10^6 cells/cm^2 maximum (*see* **Notes 12** and **13**).
2. At day 29 start to add progressively medium 2 to cell culture, do not use medium 1 anymore. Starting week 5, dead cells and debris can be removed by low centrifugation at room temperature at 100 × *g* for 10 min (*see* **Note 14**).
3. After week 7, cells stop to die and the cell number stabilizes. Cells should be given fresh medium 2 at least once per week.
4. Between weeks 8 and 10, check MC phenotype and function. Use flow cytometry to detect membrane CD117 and FcεRIα on mature MCs.

In addition, the FcεRI function is tested by sensitizing MCs with 1 μg/mL human IgE overnight and activating them with 10 μg/mL anti-human IgE. The activation surface marker, CD63, is analyzed by flow cytometry (*see* next paragraph) (*see* **Note 15**).

3.2 Degranulation (48 h) (Fig. 2)

All procedures must be done under a sterile environment in a class II biosafety cabinet.

3.2.1 Day 1: Mast Cell Overnight Sensitization with Sera (1:10) or IgE (1 μg/mL) in a 96-Well Plate

1. Calculate the MC number required for the experiment. Use 50,000 MCs per well in 100 μL final volume and ideally work in triplicates.
2. Count MCs and take the cell number needed for the assay.

3. Centrifuge at room temperature at 300 × *g* for 5 min and resuspend cell pellet at a cell concentration of 1 × 10^6 cells/mL in 37 °C pre-warmed medium 2.
4. In a 96-well round bottom plate, put 50 μL of MC suspension prepared in **step 3**.
5. Add 10 μL of patient's sera (dilution 1:10).

   For the positive control serum independent, add 10 μL of human IgE, 10 μg/mL (prediluted in medium 2 using the stock solution 1 mg/mL), final concentration 1 μg/mL in 100 μL.
6. Add 40 μL of 37 °C pre-warmed medium 2 in each well to have a final volume of 100 μL.
7. Mix well with a multichannel pipette.
8. Incubate overnight at 37 °C with 5% CO_2.

3.2.2 Day 2: Activation with Allergen or Anti-IgE

1. Prepare first stimuli with different concentrations of allergen, 0.01, 0.1, 1, 10, 100, and 1000 ng/mL, diluted in 37 °C pre-warmed medium 2 for each serum tested. As positive control use 10 μg/mL anti-human IgE prepared in 37 °C pre-warmed medium 2, and for the negative control use just medium 2.

   Keep solutions at 37 °C.
2. Take out the 96-well plate containing sensitized MCs from the incubator. Wash wells by adding 200 μL of 37 °C pre-warmed medium 2.
3. Centrifuge at room temperature at 300 × *g* for 5 min.
4. Take out all the supernatants with the multichannel without disturbing cell pellets (*see* **Note 16**).
5. Wash wells once more by adding 200 μL of 37 °C pre-warmed medium 2.
6. Centrifuge at room temperature at 300 × *g* for 5 min.
7. Take out all the supernatants with the multichannel without disturbing cell pellets.
8. Add 100 μL of the stimulus.
9. Mix gently and incubate the plate 1 h at 37 °C with 5% CO_2.

3.2.3 Flow Staining

1. Prepare the Fc block solution by diluting the Human TruStain FcX™ 1:50 in MACS buffer.
2. Prepare the staining solution by diluting the anti-CD63-BV421 antibody 1:50 in MACS buffer.
3. Remove the plate from the incubator and centrifuge at room temperature at 300 × *g* for 5 min.
4. Promptly flick off the supernatant and resuspend the cell pellet in 30 μL of Fc block solution.

5. Incubate for 5 min at 4 °C.
   (*See* **Note 17**).
6. Add 30 μL anti-CD63-BV421 staining solution.
7. Mix and incubate 30 min at 4 °C.
8. Wash by adding 200 μL of D-PBS.
9. Centrifuge at 4 °C at 300 × *g* for 5 min.
10. Promptly flick off the supernatant.
11. Add 200 μL of D-PBS to the cell pellet.
12. Centrifuge at 4 °C at 300 × *g* for 5 min.
13. Promptly flick off the supernatant.
14. Add 50 μL of live/dead blue solution diluted 1:1000 in D-PBS.
15. Mix and incubate 20 min at 4 °C in the dark (*see* **Note 18**).
16. Add 200 μL of MACS buffer.
17. Centrifuge at 4 °C at 300 × *g* for 5 min.
18. Promptly flick off the supernatant.
    Resuspend the cell pellet in 250 μL of MACS buffer (*see* **Note 19**).
19. Transfer samples into appropriate tubes for cytometer.
20. On a flow cytometer, set up voltages and compensations with the unstained and single-staining controls.
21. Analyze cells on a flow cytometer.

3.2.4 Flow Analysis Data Using FlowJo Software (Fig. 3)

1. Exclude debris and aggregates using the size, forward scatter (FSC) versus the granularity, and side scatter (SSC) (Fig. 3a).
2. Gate on the live mast cell population from the dead population which is positive for the live/dead staining (Fig. 3a).
3. Discriminate doublets based on forward scatter area (FSC-A) versus forward scatter width (FSC-W) (Fig. 3a).
4. Discriminate doublets based on side scatter area (SSC-A) versus side scatter width (SSC-W) (Fig. 3a).
5. Set up $CD63^+$ gate to use as negative control unstimulated MCs (Fig. 3b).

4 Notes

1. MACS buffer should be prepared freshly before $CD117^+$ cell separation or flow cytometry staining.
2. Medium 1 can be aliquoted and stored at −20 °C without cytokines and LDL. Ideally, cytokines and LDL should be added just before usage, once cytokines and LDL are added, medium 1 can be stored up to 1 week at 4 °C.

3. Medium 2 can be aliquoted and stored at −20 °C without cytokines. Ideally, cytokines should be added just before usage, once cytokines are added, medium 2 can be stored up to 1 week at 4 °C.
4. If the buffy coat or apheresis cones were stored at 4 °C before, allow them to reach room temperature.
5. If the dropping from the apheresis cone is blocked, carefully move it in a sterile way upside down.
6. Hold the 50 mL tube containing Ficoll in a 45 °C position; flow down very slowly the diluted blood on the tube side without disturbing the Ficoll surface. PBMC separation depends on the quality of the overlay.
7. It is very important to leave the brake off and lower the acceleration to not disrupt the layers during the centrifugation.
8. During that time you can prepare the solutions needed for the next steps, e.g., MACS buffer and medium 1.
9. Freshly isolated PBMC viability should be 95% at least.
10. Usually, 2–7 × 10^6 $CD117^+$ cells can be isolated per donor.
11. After the $CD117^+$ cell isolation, cell purity can be checked by performing a CD117 surface staining using the flow cytometry.
12. If you notice some adherent cells, change the cell culture plate or flask.
13. Cells should not be centrifuged before day 29.
14. Do not centrifuge more than 300 × g; cells are sensitive to high speed.
15. MCs can be kept in culture for up to 12 months.
16. Supernatant should be completely removed; residual volume could interfere with the final concentration of stimuli.
17. For single-staining compensation controls, mix some stimulated and non-stimulated MCs and divide them in three wells: one for the unstained control, one stained only with anti-CD63-BV421, and one only stained with the live/dead.
18. Live/dead blue solution is very light sensitive.
19. For later analysis, MCs can be fixed for 20 min in a 4% formaldehyde solution in D-PBS.

Acknowledgments

This research is supported by the following grants: Medical Research Council (MRC) Confidence-in-Concepts award to the University of Manchester (reference MC-PC-15038).

References

1. Reber LL, Hernandez JD, Galli SJ (2017) The pathophysiology of anaphylaxis. J Allergy Clin Immunol 140(2):335–348
2. Galli SJ, Tsai M (2012) IgE and mast cells in allergic disease. Nat Med 18(5):693–704
3. Ladics GS et al (2008) Assessment of three human FcepsilonRI-transfected RBL cell-lines for identifying IgE induced degranulation utilizing peanut-allergic patient sera and peanut protein extract. Regul Toxicol Pharmacol 51 (3):288–294
4. Santos AF et al (2018) A novel human mast cell activation test for peanut allergy. J Allergy Clin Immunol 142(2):689–691.e9
5. Laidlaw TM et al (2011) Characterization of a novel human mast cell line that responds to stem cell factor and expresses functional FcepsilonRI. J Allergy Clin Immunol 127 (3):815–822. e1–5
6. Saleh R et al (2014) A new human mast cell line expressing a functional IgE receptor converts to tumorigenic growth by KIT D816V transfection. Blood 124(1):111–120
7. Passante E (2014) Mast cell and basophil cell lines: a compendium. Methods Mol Biol 1192:101–113
8. Thomas K et al (2008) Current and future methods for evaluating the allergenic potential of proteins: international workshop report 23-25 October 2007. Food Chem Toxicol 46 (9):3219–3225
9. Kirshenbaum AS et al (2014) A ten-year retrospective analysis of the distribution, use and phenotypic characteristics of the LAD2 human mast cell line. Int Arch Allergy Immunol 164(4):265–270
10. Kulka M et al (2008) Neuropeptides activate human mast cell degranulation and chemokine production. Immunology 123(3):398–410
11. Guhl S et al (2010) Mast cell lines HMC-1 and LAD2 in comparison with mature human skin mast cells - drastically reduced levels of tryptase and chymase in mast cell lines. Exp Dermatol 19(9):845–847
12. Bahri R et al (2018) Mast cell activation test in the diagnosis of allergic disease and anaphylaxis. J Allergy Clin Immunol 142 (2):485–496.e16
13. Gentek R et al (2018) Hemogenic endothelial fate mapping reveals dual developmental origin of mast cells. Immunity 48(6):1160–1171.e5
14. Schmetzer O et al (2014) A novel method to generate and culture human mast cells: peripheral CD34+ stem cell-derived mast cells (PSCMCs). J Immunol Methods 413:62–68
15. Saito H et al (2006) Culture of human mast cells from peripheral blood progenitors. Nat Protoc 1(4):2178–2183
16. Joulia R et al (2015) Mast cells form antibody-dependent degranulatory synapse for dedicated secretion and defence. Nat Commun 6:6174

Part V

Whole Cells: Functional Studies

Chapter 20

Quantity and Quality of Basophil RNA Depend on the RNA Extraction Technique

Victor G. Bechmann, Anette Teo Hansen, Lars K. Poulsen, and Bettina M. Jensen

Abstract

Basophils have been suggested to express low quantities of RNA, challenging the study of gene expression within these cells. However, the purification technique employed might have an impact on the quantity and quality of RNA purified from basophils. This chapter describes a method which gives an optimal RNA output using a TRIzol-based method in contrast to a commercial kit.

Key words Basophil, TRIzol, RNA, PCR

1 Introduction

Basophil granulocytes comprise <1% of blood leukocytes. In addition, they have been suggested to only contain a small amount of RNA resulting in low quantities after RNA purification [1, 2]. These two factors make studies of basophil RNA challenging. However, the method used for RNA purification might affect RNA quantity and quality. To test if basophils do contain very low levels of RNA compared to white blood cells in general, two different RNA purification techniques were investigated for the quantity and quality of the purified RNA. Peripheral blood mononuclear cells (PBMCs) were purified from buffy coat blood stored overnight at 8 °C. Basophils (n = 8 donors) were purified by negative selection. RNA was purified from cells either snap-frozen in liquid N_2 and then treated according to Qiagen RNeasy Mini Kit [3] or dissolved in TRIzol [4] containing glycogen followed by phenol/chloroform extraction. RNA quantity was determined using a Qubit RNA Assay Kit and RNA quality was investigated using a StepOnePlus Real-Time PCR.

Independently of basophil purity (66–97%, median 90%), sufficient RNA was obtained for quantification from all eight donors

Bernhard F. Gibbs and Franco H. Falcone (eds.), *Basophils and Mast Cells: Methods and Protocols*, Methods in Molecular Biology, vol. 2163, https://doi.org/10.1007/978-1-0716-0696-4_20,

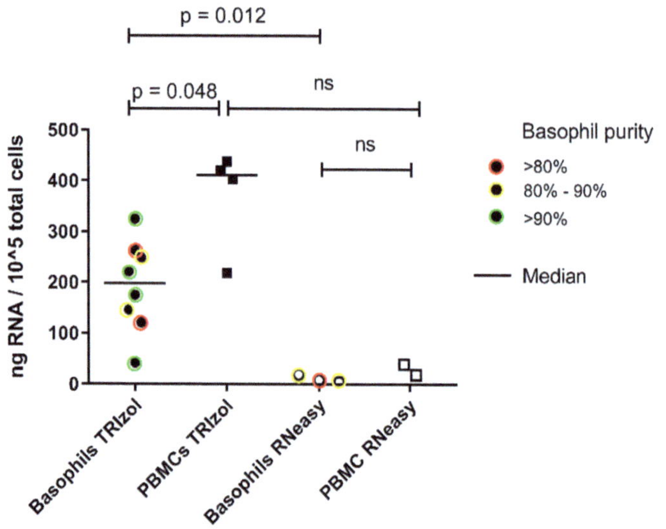

Fig. 1 Extracting RNA from basophils using the TRIzol method yields significantly more RNA than using the RNeasy method. A Mann-Whitney test was used to compare the groups

using the TRIzol method (median ± SD: 198 ± 90 ng RNA/10^5 total cells). However, with RNeasy, only three donors had quantifiable RNA levels, and the concentration was significantly lower (8 ± 7 ng RNA/10^5 total cells, $p = 0.012$) (*see* Fig. 1).

Using PBMCs (four donors), again RNA could be quantified from all four donors using TRIzol purification (412 ± 101 ng RNA/10^5 total cells), but only two out of four with RNeasy, in addition to a highly reduced concentration (29 ± 15 ng RNA/10^5 total cells) (*see* Fig. 1).

If basophils indeed carry a low level of RNA, highly pure samples would be expected to have a low RNA quantity compared to less pure preparations, but no correlation was found between RNA quantity and purity ($r_s = 0.02381$, $p = 0.9768$). A correlation was, however, observed between the total cell number and extracted RNA ($r_s = 0.9207$, $p = 0.0024$). Comparing the RNA quantity obtained from basophils with the quantity from PBMCs (using the TRIzol method) illustrated a twofold difference indicating that basophils do not have an abnormally low RNA level.

Real-time PCR revealed significantly higher C_T values (FcεRIα, β-actin) or even undetectable gene expression (HDC, HuPO) in the RNeasy-purified RNA, illustrating a higher quality of TRIzol-purified RNA (Fig. 2). We therefore believe that basophils in resting condition express RNA at a level comparable to other white blood cells, but the quantity and quality of purified RNA are highly dependent on the purification technique. The TRIzol method is given below (*see* **Notes 1** and **2**).

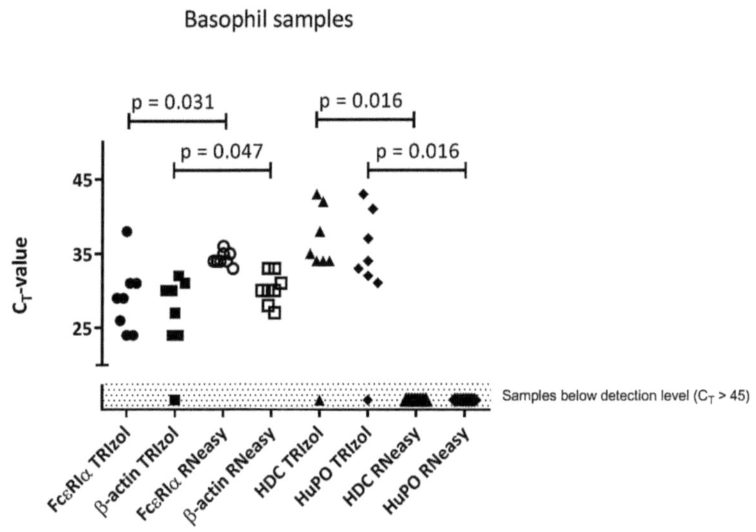

Fig. 2 RNA from the TRIzol method is of higher quality as indicated by significantly smaller C_T values. A Wilcoxon matched-pairs signed rank test was used to compare C_T values

2 Materials

1. UltraPure Glycogen (20 mg/mL) (Life Technologies).
2. TRIzol reagent (Ambion/Invitrogen).
3. RNAseZap (Ambion/Invitrogen).
4. 1-Bromo-3-chloropropane (BCP).
5. Phenol/chloroform/isoamyl alcohol (pH 6.6—*see* **Note 3**) (abbreviated: phenol/chloroform in the text).
6. Isopropanol.
7. 80% ethanol.
8. Diethyl pyrocarbonate (DEPC)-treated water (RNase-free).
9. Centrifuge.
10. Microcentrifuge tubes, 1.5 mL, RNase-free.
11. Gloves.

3 Methods

3.1 Preparation of Samples

1. Add cell suspension containing between 100,000 and 1,000,000 cells to a 1.5 mL RNase-free tube.
2. Centrifugate at 500 × *g* for 10 min at 11 °C.
3. Remove the supernatant by pipetting.
4. Add 800 μL TRIzol (resuspend with a pipette to dissolve the pellet) and 4 μL glycogen (*see* **Note 4**).
5. Store at −80 °C until further use.

3.2 RNA Purification

The RNA purification procedure will take 3–4 h. The samples should be handled immediately after every centrifugation step, as yield will decrease if the two phases have time to mix again (*see* **Note 5**).

1. Clean working area, pipettes, etc. with RNAseZap. Use gloves.
2. Thaw samples if they have been stored in the freezer.
3. Add 160 μL BCP and vortex for 10–15 s followed by a 2–3 min incubation at room temperature.
4. Centrifuge samples at 15,000 × *g* for 15 min at 4 °C.
5. Extract top phase to another 1.5 mL tube (RNase free) without disturbing the bottom phase or the debris (the bottom phase can be stored for later DNA and protein isolation).
6. Add an equivalent amount of phenol/chloroform (remember to go below the buffer phase (at the top) in the phenol/chloroform bottle) to the extracted phase, and vortex for 10–15 s followed by centrifugation at 15,000 × *g* for 15 min at 4 °C, and transfer the top phase to a new vial. This step should be performed at least three times in total—until there is a very distinct and debris-free interface.
7. Remove the top phase from the last phenol/chloroform extraction and transfer it to a 1.5 mL tube. Add a 1:1 volume of BCP, and vortex for 10–15 s, and centrifuge at 15,000 × *g* for 15 min at 4 °C (this is to get rid of any leftover phenol).
8. Remove the top phase to a new 1.5 mL tube and precipitate the RNA with a 1:1 volume of cold (−20 °C) isopropyl alcohol without vortexing, but only inverting the tube 3–4 times followed by centrifugation at 15,000 × *g* for 15 min at 4 °C.
9. Remove the supernatant and wash the cell pellet with 500 μL 80% cold (4 °C) ethanol (without pipetting) and centrifuge at 15,000 × *g* for 5 min at 4 °C.
10. Remove the supernatant again and leave the lid of the tube open in order for the pellet to air-dry in the fume hood. This step can take 5–30 min. Check if the pellet is dry by flicking a finger on the side of the tube and observing whether drops appear.
11. When the pellet is dry, add at least 20 μL DEPC-treated (RNase-free) (*see* **Note 6**) water to dissolve the pellet (carefully resuspend the RNA by pipetting up/down a few times). If a smaller volume is used, the glycogen concentration will exceed 4 mg/mL (*see* **Note 4**). The RNA solution is now ready for quantification and qualification. For storage keep the RNA at −80 °C.

4 Notes

1. The TRIzol method requires experience in order to achieve a high RNA quantity. Therefore, it is recommended to run purification rounds using, for example, a cell line to obtain a repeatable RNA quantity before handling real research samples.
2. Compounds such as the TRIzol reagent, BCP, phenol/chloroform/isoamyl alcohol, and isopropanol require several precautions concerning waste, laboratory safety, and personal equipment as large well-ventilated fume hood space. Please read the safety documents before starting using the products.
3. For DNA isolation a higher pH is needed; therefore the Tris alkaline buffer should be added to the phenol/chloroform bottle. This enables the DNA and RNA to be in the aqueous phase. If the pH is acidic, only RNA will be in the aqueous phase [5].
4. Glycogen is co-precipitated with the RNA, thereby giving a visible pellet even when working with low RNA quantities. If the final glycogen concentration (purified RNA dissolved in water) is kept at or below 4 mg/mL, cDNA synthesis or PCR should not be affected.
5. If more than ten samples are prepared, half of them should be left in the centrifuge—still running—while the top phase is removed on the first set of samples and the reagent is added.
6. The RNA concentration is dependent on the amount of DEPC-treated water—too much water might dilute the RNA too much for further use. In general, 10–50 μL water is used. If glycogen is added, the final concentration should not exceed 4 mg/mL (*see* **Note 4**).

References

1. Reimer JM, Magnusson S, Juremalm M, Nilsson G, Hellman L, Wernersson S (2006) Isolation of transcriptionally active umbilical cord blood-derived basophils expressing FcεRI, HLA-DR and CD203c. Allergy 61:1063–1070
2. Grundström J, Reimer JM, Magnusson SE, Nilsson G, Wernersson S, Hellman L (2012) Human cord blood derived immature basophils show dual characteristics, expressing both basophil and eosinophil associated proteins. PLoS One 7:1–10
3. Qiagen RNeasy mini kit handbook. https://www.qiagen.com/es/resources/resourcedetail?id=14e7cf6e-521a-4cf7-8cbc-bf9f6fa33e24&lang=en. Accessed 21 June 2019
4. TRIzol reagent user guide. https://assets.thermofisher.com/TFS-Assets/LSG/manuals/trizol_reagent.pdf. Accessed 21 June 2019
5. Phenol:Chloroform:IAA, 25:24:1, pH 6.6 (100 mL) datasheet. https://www.thermofisher.com/document-connect/document-connect.html?url=https%3A%2F%2Fassets.thermofisher.com%2FTFS-Assets%2FLSG%2Fmanuals%2Fsp_9730.pdf&title=UGhlbm9sOkNobG9yb2Zvcm06SUFBLCAyNToyNDoxLCBwSCA2LjYgKDEwMCBtbCk. Accessed 05 June 2019

Chapter 21

Measuring Histamine and Cytokine Release from Basophils and Mast Cells

Bettina M. Jensen, Ewa A. Bartko, Katrine Baumann, and Per S. Skov

Abstract

Basophils and mast cells are known for their capability to release both preformed and newly synthesized inflammatory mediators. In this chapter, we describe how to stimulate and detect histamine released from basophils in whole blood, purified basophils, in vitro cultured mast cells, and in situ skin mast cells (the latter by microdialysis), using either a solid phase assay or flow cytometry. We also give an example of an activation protocol for basophil and mast cell cytokine release and discuss approaches for cytokine detection.

Key words Basophils, Mast cells, Histamine, Microdialysis, Diamine oxidase, Flow cytometry, Cytokines

1 Introduction

Basophils and mast cells are granulated cells, the former usually residing in the blood and the latter in other tissues, which play an important role in inflammatory responses. Mast cells have long been known to be one of the 'troublemakers' of allergic reactions (characterized by the presence of allergen-specific IgE) due to the expression of high-affinity IgE receptors (FcεRI) which, when cross-linked, causes mast cells to degranulate. Even though basophils also express FcεRI, they seem to have a more secondary role in the allergic reaction. Nevertheless, both basophils and mast cells can produce a large repertoire of inflammatory mediators when activated via FcεRI. The IgE-mediated activation of basophils and mast cells is, however, not the only way to stimulate these cells to degranulation. Other well-known triggers of activation are complement components, TLR ligands, and neurogenic factors (primarily from mast cells). In addition, formyl-methionyl-leucyl-

The original version of this chapter was revised. The correction to this chapter is available at https://doi.org/10.1007/978-1-0716-0696-4_30

Bernhard F. Gibbs and Franco H. Falcone (eds.), *Basophils and Mast Cells: Methods and Protocols*, Methods in Molecular Biology, vol. 2163, https://doi.org/10.1007/978-1-0716-0696-4_21,

phenylalanine (fMLP) is a well-known basophil activator [1–3], whereas compound 48/80 has been used as a mast cell activator (mediating its action via MRGPRX2) [4–6], also *see* Chapter 18.

In the following, we will describe some of the main mediators released by basophils and mast cells, primarily after cross-linking of FcεRI, and give examples of methods which can be used to detect these mediators.

1.1 Histamine

Histamine is a biogenic amine derived from the decarboxylation of the amino acid histidine, a reaction catalyzed by the enzyme L-histidine decarboxylase. Histamine is found pre-stored in the granules within basophils and mast cells and can therefore quickly be released after stimulation. Activation caused by cross-linking of FcεRI will result in histamine release [7]. However, non-IgE-mediated stimuli such as complement 3a and 5a, substance P, fMLP, or compound 48/80 can also result in histamine release [8]. Activation through TLRs has been reported to cause histamine release in certain phenotypes of mast cells [9–13].

1.2 Cytokines

In contrast to histamine, cytokines belong to the group of newly generated mediators, meaning that basophils and mast cells in general do not contain large amounts of pre-stored cytokines. Nevertheless, TNFα might be found in resting mast cells [14].

As is the case for histamine release, cross-linking of FcεRI will result in cytokine release from both basophils and mast cells, but the range and quantity of cytokines released from the two cell types are not identical. Mast cells appear to release a broader range of cytokines when compared to basophils, and depending on the mast cell phenotype, the following cytokines have been associated with FcεRI stimulation: IL-1β, IL-5, IL-6, IL-8, IL-10, IL-13, IFNγ, GM-CSF, TNFα, and VEGF [7, 15]. The major cytokines released from basophils after FcεRI stimulation are IL-4, IL-13, and VEGF, but new cytokines are still reported such as TNFα [16].

TLRs have been described to activate both basophils and mast cells to release cytokines [8–12].

2 Materials

2.1 Histamine Release and Quantification

1. 96-well glass fiber-coated plates (RLA 210).
2. HR-Test Reagent Kit (RLA700) (RefLab ApS, Copenhagen, Denmark).
3. HISTAREADER (RefLab ApS, Copenhagen, Denmark) or a highly sensitive fluorometer.
4. Piperazine-*N*,*N*′-bis(2-ethanesulfonic acid) (PIPES) buffer: 3.02 g PIPES sodium salt, 19.05 g $CH_3COONa \cdot 3H_2O$ (sodium acetate trihydrate), 0.491 g CH_3COOK (potassium acetate anhydrous), 0.088 g $CaCl_2 \cdot 2H_2O$ (calcium chloride dihydrate),

0.224 g $MgCl_2 \cdot 6H_2O$ (magnesium chloride hexahydrate), distilled water up to 1 L, adjust pH to 7.4 with 1 M Tris.

5. Histamine release buffer (HR-buffer): PIPES containing 0.5% human serum albumin.
6. Distilled water with a conductivity of <20 μS/cm.
7. Reagents for maximal histamine release: phorbol 12-myristate 13-acetate (PMA) (1 μg/mL final concentration) and ionomycin (5 μg/mL final concentration).
8. Humidified 5% CO_2 incubator at 37 °C.
9. Microdialysis probes with a molecular weight cutoff (MWCO) of 2 kDa (e.g., low flux probes 2 kDa from EP Medical, Copenhagen, Denmark).
10. 23 G guide cannulas.
11. Syringes (1 mL) for probe perfusion.
12. Catheter connectors (e.g., Perifix, B. Braun Medical, Melsungen, Germany).
13. Microinfusion syringe pump (e.g., NE-1200, New Era Pump Systems, Farmingdale, NY, USA).
14. Collection vials (e.g., PCR tubes).
15. Surgical tape.
16. V-bottom 96-well microplates.
17. Ruler.
18. Scissors.
19. Syringes for injection of stimulants.

2.2 Detection of Intracellular Histamine by Flow Cytometry

1. Microcentrifuge tubes, 1.5 mL.
2. Porcine kidney diaminoxidase (DAO).
3. LYNX Rapid RPE Antibody Conjugation Kit®.
4. Phosphate-buffered saline (PBS).
5. LAD2 culture medium: StemPro®-34 SFM, StemPro®-Nutrient Supplement, 2 mM L-glutamine, 100 μg/mL penicillin/streptomycin, and 100 ng/mL rhSCF.
6. FACS tubes (5 mL polystyrene tubes).
7. BD Cytofix/Cytoperm™ solution.
8. BD Perm/Wash™ buffer—dilution 1:10 in demineralized water (1× BD Perm/Wash™).
9. BD CellFIX™—dilution 1:10 in demineralized water (1× BD CellFIX™).
10. Flow buffer: PBS containing 0.5% human serum albumin and 2 mM ethylenediaminetetraacetic acid (EDTA).
11. Cluster tubes, 1.2 mL.
12. Flow cytometer.

2.3 Cytokine Release

1. Basophil culture medium: RPMI 1640, 100 U/mL penicillin/100 μg/mL streptomycin, 1 mM L-glutamine, 5% v/v human AB serum.
2. 48-well cell culture plate.
3. Humidified 5% CO_2 incubator at 37 °C.
4. Microdialysis probes with a high MWCO (e.g., high flux probes 3000 kDa from EP Medical, Copenhagen, Denmark).
5. 21 G guide cannulas.
6. Perfusate optimized for cytokine recovery (e.g., Ringer's lactate containing 1% human serum albumin).

3 Methods

3.1 Histamine Release and Quantification

Stimulation of basophils and mast cells to release histamine varies depending on the cell source, and therefore different techniques are described below. However, the method to quantify histamine is the same, and the one described in this chapter is dependent on glass fiber-coated plates and fluorometric detection. Other protocols for histamine quantification, like histamine ELISAs, can also be used according to the manufacturer's guidelines, but they will not be described here.

All the reagents used for cell activation are made up 2× the final concentration (e.g., if cells will be stimulated with 100 ng/mL anti-IgE, make solutions of 200 ng/mL anti-IgE). Avoid concentrations above 1 mg/mL since such solutions might cause unspecific binding to the glass fibers. A typical anti-IgE titration curve could range from 1 ng/mL to 1 μg/mL (*see* **Note 1**). Do not include fetal bovine serum in the buffer since it contains diamine oxidase which degrades histamine.

In order to express the results as % histamine release, one can include a stimulation of the cells using a solution of phorbol 12-myristate 13-acetate (PMA) (1 μg/mL final concentration) and ionomycin (5 μg/mL final concentration) in order to obtain a "maximal" histamine release.

It is also possible to lyse the cells and thereby estimate the total histamine content. This can be done by lysing 100 μL cell suspension with 30 μL 7% $HClO_4$ (vortex after adding the $HClO_4$) for 30 min at room temperature, followed by neutralization with 100 μL PIPES buffer. The lysate is the centrifuge at 16,000 × *g* for 5 min at 20 °C.

3.1.1 Histamine Release in Whole Blood

1. Heparinized blood (use within 24 h after venipuncture; *see* **Note 2**) is centrifuged (600 × *g*, 10 min, 20 °C); mark the level of the total volume (blood cells and plasma) and remove the plasma. Be careful not to disturb the white blood cell layer

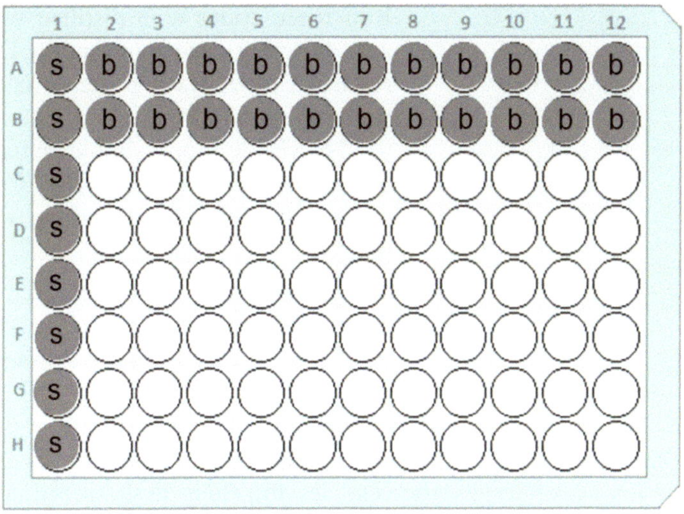

Fig. 1 Design of a glass fiber-coated plate. Gray wells are for buffer (s, histamine standard; b, blanks). White wells are for samples. Always reconstitute the plate by adding 25 μL PIPES buffer to all wells before adding samples

(the buffy coat) on top of the red blood cells (leave 5–10 mm plasma above the cell layer). Reconstitute the plasma volume with PIPES buffer (add buffer up to the marked level), and mix the blood by inverting the tube. This is now termed "washed blood."

2. Use a 96-well glass fiber-coated plate (*see* Fig. 1). Add 50 μL PIPES buffer to the first column (includes a histamine standard) and rows A and B (these are kept for background values) on the plate. Add 50 μL stimulant of interest to the other wells. Add 50 μL washed blood to all wells (also the first column and rows A and B).
3. Immediately after adding the washed blood to the plate, cover it with a lid (do not cover it by tape) and incubate at 37 °C for 1 h. During this time, stimulated basophils will release histamine which will then bind to the glass fibers present in the plate. The histamine release can sometimes be improved by adding IL-3 to the blood (*see* **Note 3**); however, not all blood donors have responding basophils (*see* **Note 4**).
4. After incubation, the blood cells are removed by washing the plate; add 100 μL distilled water (conductivity <20 μS/cm) to the wells, aspirate by inverting plate over a sink, add 200 μL distilled water, and aspirate again by inverting the plate over a sink. Repeat this two more times and finish by tapping the plate semidry on paper.

   An ELISA plate washer can also be used for the washing procedure. You must be able to adjust the nozzles for dispensing and aspiration so they are 2–3 mm above the fiber matrix.

Run three cycles with distilled water (conductivity <20 μS/cm). It is very important that the ELISA plate washer is clean of regular ELISA wash buffers. Salt solutions will disturb the histamine binding.

No visible blood should remain in the plate. If so, repeat one more wash cycle or, if the blood is at the side of the well, remove it with cotton swabs without disturbing the fiber matrix.

5. The plate can now either be stored (upside down and protected from light) for 1 month or used immediately for histamine quantification according to Subheading 3.1.5.

3.1.2 Histamine Release from Purified Basophils

In order to obtain pure basophils, it is preferable to use a negative selection protocol since positive selection may cause activation of the cells, leaving them desensitized and thereby unable to release histamine during the experiment.

1. Centrifuge the purified basophils and resuspend in HR-buffer to a concentration of 100,000 cells/mL.
2. Prepare reagents in HR-buffer and add 50 μL to a v-shaped 96-well plate. Add 50 μL cell suspension to the plate, cover it with a lid (do not cover it by tape), and incubate at 37 °C for 1 h. During this time, stimulated basophils will release histamine.
3. Reconstitute a glass fiber-coated 96-well plate by adding 25 μL PIPES buffer to all wells.
4. Centrifuge the cell plate (10 min, 300 × *g*, 4 °C, acceleration 5, brake 3) and transfer 25 μL supernatant to wells designed for samples in the reconstituted glass fiber-coated plate. Add HR-buffer to the first lane (histamine standard) and to the blanks (A2–A12, B2–B12) (*see* Fig. 1).
5. Cover the glass fiber plate with a lid and incubate for 30 min at 37 °C.
6. Wash the plate with distilled water; add 150 μL distilled water (conductivity <20 μS/cm) to the wells, aspirate by inverting plate over a sink, add 200 μL distilled water, and aspirate again by inverting plate over a sink. Repeat this two more times and finish by tapping the plate semidry on paper. An ELISA plate washer can also be used for the washing procedure as described in Subheading 3.1.1.
7. The plate can now either be stored (upside down and protected from light) for 1 month or used right away for histamine quantification according to Subheading 3.1.5.

3.1.3 Histamine Release from In Vitro Cultured Human Mast Cells

Stem cell-derived human mast cells or cell lines such as LAD2 do not have IgE bound to their high-affinity IgE receptors and therefore need to be sensitized if IgE-dependent histamine release is to be studied. IgE sensitization can be performed by culturing the cells overnight with 100 ng/mL to 1 μg/mL human IgE.

1. Centrifuge the human mast cells and resuspend in HR-buffer to a concentration of 100,000 cells/mL.
2. Prepare reagents in HR-buffer and add 50 μL to a v-shaped 96-well plate. Add 50 μL cell suspension to the plate, cover it with a lid (do not cover it by tape), and incubate at 37 °C for 1 h. During this time, stimulated mast cells will release histamine.
3. Assay histamine release as described earlier (Subheading 3.1.2, **steps 3–7**).

3.1.4 Histamine Release from In Situ Mast Cells

Microdialysis is a minimally invasive sampling technique used for continuous measurement of substances in the extracellular fluid of virtually any tissue. It has traditionally been used to assess neurotransmitters and hormones in neuroscience and to measure exogenous compounds (e.g., pharmaceuticals) to investigate distribution within the body [17–19]. Microdialysis is, however, also a highly relevant technique when applied to the skin for investigation of the various functions of tissue-residing mast cells. While human primary tissue-residing mast cells can be isolated, it is not clear to what degree tissue dispersion and subsequent isolation with digestive enzymes affect mast cell functions [20]. Microdialysis in the skin can be performed in vivo in humans and experimental animals [21–23] but can also be used to investigate mast cell functions in excised skin since these mast cells remain responsive several hours after the skin is removed (own unpublished data). The experimental procedure outlined below describes recovery of histamine from skin-resident mast cells by microdialysis in general. Specific procedures related to sampling in vivo (e.g., on patients) or ex vivo (e.g., in human skin obtained from cosmetic surgeries) can be found in [24].

1. Adjust the length of the microdialysis membrane if necessary (the microdialysis membrane of the probes should be adjusted to the same length, if not already so from the manufacturer). Handle the microdialysis probes with care and avoid touching the membrane.
2. Flush the microdialysis probes with perfusate to ensure proper function. Make sure that the central guide wire is not flushed out of the hollow membrane while doing so.
3. Connect the inlet tubing to the syringes prefilled with perfusate (PIPES) mounted in the microinfusion pump. Be careful not to introduce any air bubbles into the system, as this will have a negative impact on the microdialysis sampling.

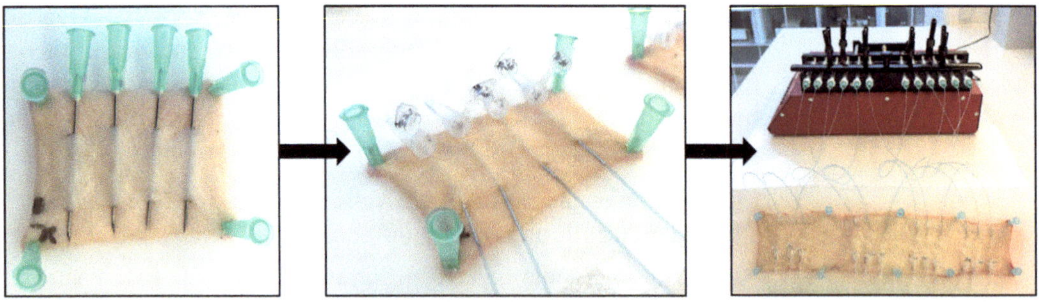

Guide cannulas inserted into the skin spanning 2 cm intradermally.

The probes inserted through the tip of the guide cannulas, which are then withdrawn to leave the probe in the skin. Collection vials are placed at the tip of the membrane.

The inlet tubing is connected to the perfusion syringes mounted in the pump and the sampling can be commenced.

Fig. 2 Microdialysis setup using ex vivo skin. The setup for in vivo skin is illustrated in [24]

4. Introduce the guide cannulas into the upper dermis at an intradermal length of 2 cm, as superficially as possible (*see* Fig. 2). Do not penetrate the skin barrier with the guide cannulas apart from at the skin sites of probe inlet and outlet.
5. Introduce the microdialysis membranes through the tip of the inserted guide cannulas and gently pull out the cannulas to leave the probe inside the skin (*see* Fig. 2). Be careful not to come into contact with the sharp edges of the guide cannula as this may damage the membrane.
6. Fix the inlet tubing to the skin using a piece of surgical tape just above the membrane to ensure stable positioning of the probe during sampling.
7. Perfuse the microdialysis probes in parallel with PIPES using a flow rate of 3 μL/min for 30 min to flush out trauma-induced histamine which may be released in response to probe insertion (discard the dialysate).
8. Perform a basal sampling to measure the background levels of histamine in the skin (e.g., for 10–30 min).
9. Challenge the skin sites just next to the inserted membranes by injecting 50 μL of the stimulant intradermally. Remember to include the appropriate controls, e.g., vehicle (negative control) and a positive control (e.g., anti-IgE to ensure mast cell responsivity).
10. After challenge, sample dialysates in suitable time intervals (e.g., 10–30 min) for the required time (e.g., 1 h).
11. Reconstitute a glass fiber-coated 96-well plate by adding 25 μL PIPES buffer to all wells and transfer 25 μL of the dialysate to wells designed for samples in the reconstituted glass fiber-coated plate. Add PIPES buffer to the first lane (histamine standard) and to the blanks (A2–A12, B2–B12) (Fig. 1).
12. Analyze histamine concentrations in the dialysate samples according to Subheading 3.1.2, **steps 5–7**.

3.1.5 Histamine Quantification

1. Allow the HR-Test Reagent Kit HISTAREADER to reach room temperature (20–25 °C) before use.
2. Prepare coupling reagent by transferring 1 OPA Disc to the bottle containing 10 mL diluent for coupling reaction (sufficient for one plate). Leave the reaction mix for 45 min (avoid light). Invert the bottle gently before use. The disc will *not* dissolve.
3. In order to remove unspecific bound compounds that still might be present, add 150 μL wash solution to the plates. Incubate the plate for 30 min at 37 °C.
4. Aspirate the wash buffer by inverting plate over a sink and then rinse the plate five times; add 200 μL distilled water and aspirate again by inverting the plate over a sink. Repeat this four more times and finish by tapping the plate on highly absorbent paper towel until it is free of any residual water.
5. Add 75 μL coupling reagent to all used wells and leave plate for exactly 10 min; then, stop the reaction by adding 75 μL stop solution into the wells.
6. Read the plate using the HISTAREADER or a highly sensitive fluorometer within the same day. *See* **Note 5** for calculation of % histamine release.

3.2 Detection of Intracellular Histamine in LAD2 Cells Using Flow Cytometry

The method described below is modified from the histamine detection in basophils due to the morphological differences between basophils and mast cells (e.g., cell size and structure of granules). The protocol consists of four steps: DAO-PE preparation, cell activation, intracellular staining, and flow cytometry analysis.

For histamine detection in basophils using DAO, (*see* Chapter 15).

3.2.1 DAO-PE Preparation

The DAO-PE conjugation described below is performed accordingly to the manufacturer instructions [25].

1. Prepare 1 mg/mL DAO solution in PBS.
2. Transfer maximal volume of 600 μL of DAO 1 mg/mL to the new tube and add 1 μL of modifier for each 10 μL of DAO. Mix it gently.
3. Transfer the mixed enzyme-modifier sample straight onto the LYNX lyophilized mix. Resuspend gently.
4. Incubate the mixture in the dark, at RT for 3 h or overnight.
5. Stop the reaction by adding 1 μL of Quencher for each 10 μL of the enzyme and leave the DAO-PE for 30 min, at RT in the dark.
6. Keep the newly conjugated DAO-PE at 4 °C (*see* **Note 6**).

3.2.2 Cell Activation

To distinguish histamine-positive and histamine-negative cells, the following protocol consists of subsequent types of samples:

- Negative control—unstimulated, non-permeabilized cells.
- Positive control—unstimulated, permeabilized cells.
- Samples—stimulated, permeabilized cells.

1. Harvest LAD2 cells to a falcon tube, count cells, and determine viability. Obtain the amount of 2×10^5 cells/condition. Centrifuge at $300 \times g$ for 10 min at RT.
2. Remove supernatant and resuspend cell pellet in LAD2 medium to obtain 2×10^5 cells/50 μL.
3. Prepare stimulants in LAD2 medium.
4. Add 50 μL of LAD2 medium (negative and positive controls) or 50 μL of stimulus (samples) to 5 mL FACS tubes. If surface staining is required, prepare antibody mix and add it directly to the tubes.
5. Transfer 50 μL of cell suspension to all tubes and incubate cells for 30 min at 37 °C.
6. After incubation, wash cells by adding 100 μL of flow buffer. Centrifuge at $500 \times g$, for 5 min at 4 °C. Remove supernatant and proceed with intracellular staining.

3.2.3 Intracellular Staining

1. Add 250 μL of BD Cytofix/Cytoperm™ solution to the positive control and samples and 150 μL of 1× BD CellFIX™ to the negative control. Vortex each tube for 1 s and incubate for 20 min at 4 °C.
2. Wash cells by adding 1 mL of 1× BD Perm/Wash™ buffer to the positive control and samples and 1 mL of flow buffer to the negative control. Centrifuge for 5 min, at $500 \times g$ at 4 °C. Repeat once.
3. Resuspend cell pellets of the positive control and samples in 50 μL of 1× BD Perm/Wash™ buffer and of the negative control in 50 μL of flow buffer (*see* **Note 7**). Incubate cells for 30 min, at 4 °C.
4. Wash cells once by adding 1 mL of 1× BD Perm/Wash™ buffer to positive control and samples and 1 mL of flow buffer to the negative control. Centrifuge for 5 min, at $500 \times g$ at 4 °C.
5. Resuspend cell pellets of the positive control and samples in 50 μL of 1× BD Perm/Wash™ buffer and of the negative control in 50 μL of flow buffer. Add 10 μL of DAO-PE, 1 mg/mL (*see* **Note 8**). Incubate for 30 min, at 4 °C away from light.

6. Wash cells by adding 1 mL of 1× BD Perm/Wash™ buffer to the positive control and samples and 1 mL of flow buffer to the negative control. Centrifuge for 5 min, at 500 × *g* at 4 °C. Repeat once.
7. Resuspend cell pellets in 150 μL of flow buffer. Transfer cell suspension to the 1.2 mL cluster tubes for flow cytometric analysis.

3.2.4 Flow Cytometry Analysis

Set the DAO-PE$^+$ gate based on ≈2% DAO-PE-positive cells in the negative control. Apply the "DAO-PE$^+$ gate" to the positive control to determine DAO-PE-positive cells (cells which are DAO-PE$^+$ are positive for intracellular histamine) and to the samples to assess the degree of histamine release. An example of the flow analysis is shown in Fig. 3.

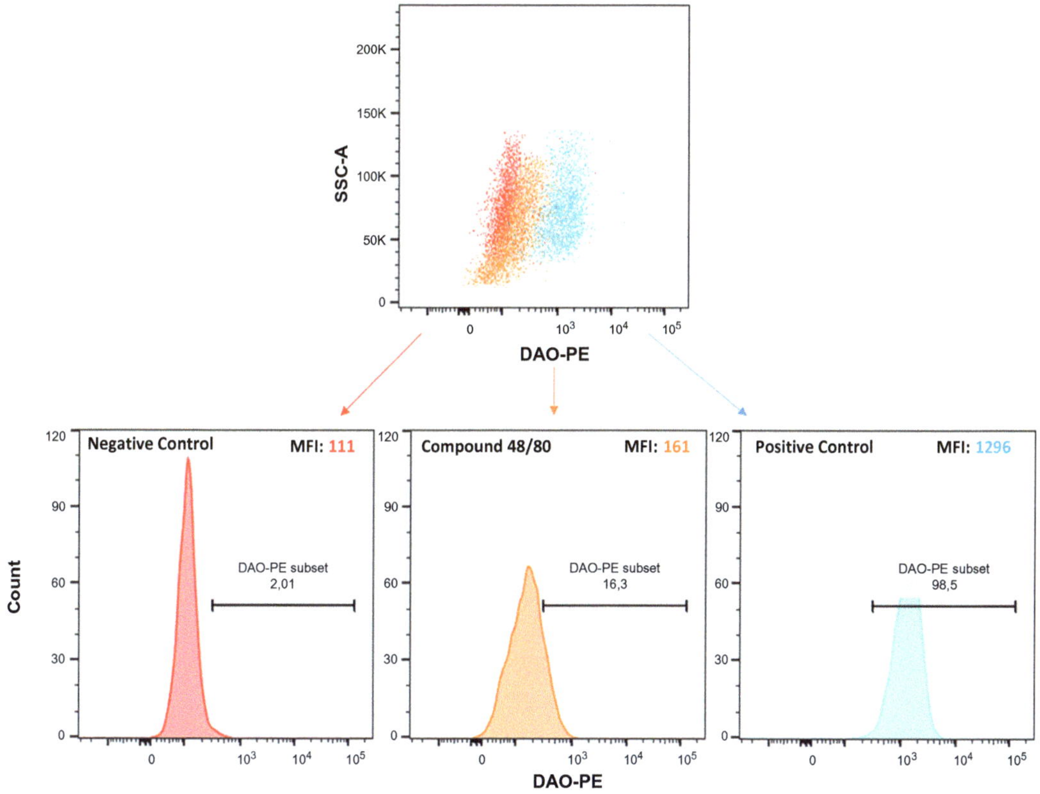

Fig. 3 Detection of intracellular histamine in LAD2 cells. The red, blue, and orange coloring indicates the negative and positive controls and the sample, respectively. The histamine release is induced by 30 min stimulation with compound 48/80 (1000 μg/mL). Red, blue, and orange values refer to the median fluorescence intensity (MFI), whereas black-colored values indicate the percentage of DAO-PE$^+$ cells

The percentage of histamine release can be calculated as follows:

$$\begin{aligned}&\%\text{ of DAO-PE}^{+}\text{cells in the positive control}\\&-\%\text{ of DAO-PE}^{+}\text{cells in the sample.}\end{aligned}$$

For the example given in Fig. 3, the percentage of release equals 82.2%.

3.3 Cytokine Release and Detection

Cytokine secretion experiments are often more time consuming than histamine release since many cytokines need to be produced before the cells can release them. However, the secretion rate can be different which is why studies have to be designed according to the specific cytokine of interest. Please note that if anti-IgE-mediated cytokine release is to be studied in in vitro generated human mast cells, it is necessary to sensitize the cells with IgE before activation (*see* Subheading 3.1.3). All the reagents used for cell activation are made up to 10× the final concentration (if cells will be stimulated with 100 ng/mL anti-IgE, make solutions of 1000 ng/mL anti-IgE). The volume of the cell samples depends on the following analysis, but cultures of 200 μL should be sufficient for most cytokine analysis that is the volume used in the description below.

3.3.1 Cytokine Release with Purified Basophils or In Vitro Cultured Human Mast Cells

Centrifuge the purified basophils or in vitro cultured human mast cells (300 × *g*, 10 min, room temp.) and resuspend in basophil or mast cell culture media, respectively, to a cell density of 1.11×10^6 cells/mL. This will result in a final density of 10^6 cells/mL in the cell sample. The culture medium for basophils is given in the Materials section. For human mast cells, the choice of culture medium depends on the culture protocol used.

Use a 48-well cell culture plate and add 20 μL of stimulant or medium to the wells. Add 180 μL cell suspension and place the culture plate in the CO_2 incubator (37 °C) for a time point relevant for the specific cytokine of interest (which can vary from 4 to 24 h).

After incubation, transfer the cell suspension to 0.5 mL tubes and centrifuge (300 × *g*, 10 min, 4 °C). Save the supernatant (containing released cytokines) and continue with cytokine detection or store the supernatant at −80 °C until further analysis. Avoid freeze/thaw cycles (*see* **Note 9**).

3.3.2 Cytokine Release from Skin Mast Cells

Microdialysis may also facilitate recovery of larger molecules when membranes with a high MWCO are used; however, sampling of cytokines is more complicated compared to sampling of histamine [26, 27]. As mentioned in Subheading 3.3, cytokine secretion from mast cells depends on the analyte in question; hence, kinetics must be considered, and the sampling profile should be adjusted accordingly to avoid misinterpretation of false-negative results. Furthermore, different parameters should be optimized to ensure optimal recovery of the cytokines in question, e.g., the flow rate can be lowered to increase the relative recovery of the analyte, and carrier proteins may be added to the perfusate, as this counteracts fluid leakage from the probes—a frequent problem with high-MWCO membranes [25, 28–30].

The procedure for cytokine sampling by microdialysis only varies in general terms from what is described in Subheading 3.1.4 on the following points:

(a) Use probes with a high MWCO (e.g., 3000 kDa).
(b) Larger guide cannulas must be used for intradermal insertion of the microdialysis membranes (21 G instead of 23 G due to the larger outer diameter of the membrane).
(c) An optimized perfusate should be used for probe perfusion (e.g., a physiological solution containing 1% human albumin).
(d) Probes can be primed for 1 h prior to cytokine sampling by perfusing the membranes with a solution containing a protein (e.g., human albumin) to minimize nonspecific adsorption of the analyte to probe components.
(e) A lower flow rate is usually used for cytokine recovery (e.g., 0.5–1.0 μL/min).

3.3.3 Cytokine Detection

Quantification of cytokines can be performed in several ways, and many commercial ELISA kits and multiplex technologies are available. Most of these kits are ready-to-use assays and accompanied by a clear protocol from the manufacturer.

The ELISA assays are typically offered as ready-coated 96-well plates with prepared reagents for a quick and convenient way to determine one cytokine per sample. Several companies also offer ELISA assays as kits consisting of a standard, HRP-streptavidin, capture, and detection antibodies. These kits require more preparation time and method optimization but once established can be a more economic and customizable assay when measuring larger quantities of samples. The ELISA method does not require specific laboratory equipment except from a plate reader, and the procedure takes only between 1 and 2 days, depending on the ELISA kit. The detection limit is generally good (down to 10–100 pg/mL), but since only a single cytokine can be measured per ELISA, you will need many cells in order to generate enough material for several cytokine tests.

The multiplex technology has added a new dimension to the cytokine detection. Several cytokines can now be measured in the same sample using either a bead-antibody-fluorometric-based platform (e.g., Luminex or Cytometric Bead Array from BD) or an electrochemiluminescence assay (from Meso Scale Discovery), and it appears that only the number of cytokines in the catalogue sets the limit of cytokines which can be detected in one sample. In comparison to the ELISA, most multiplex assay can be run in a 96-well plate format, but the multiplex platforms often require special equipment (e.g., a flow cytometer or a special plate reader), which some laboratories might not have access to. On the other hand, the detection limit is quite good (down to few pg/mL) and sometimes lower compared with an ELISA. Unfortunately, there might be variations between multiplex assays since difference in reproducibility and absolute quantities of detected cytokines has been reported [31–34].

4 Notes

1. The histamine release curve to IgE-dependent stimulation is often bell-shaped. Therefore, it is important to find the right concentration—too low or too high a concentration may result in no or very little histamine release.
2. If blood will be stored, keep it at 8–15 °C. Room temperature or below 4 °C will negatively affect the basophil histamine release.
3. IL-3 will often improve the basophil histamine release. Concentrations from 10 pg/mL to 1 ng/mL have been used to potentiate the anti-IgE-mediated histamine release. Higher concentrations might cause histamine release by IL-3 alone.
4. Between 10% and 20% of the population have non-releasing basophils [35]. This term is used when basophils do not release histamine when stimulated with anti-IgE. However, other stimuli, like fMLP, will activate these cells to release histamine [36].
5. Expression of % histamine release: The glass fiber plate contains blank wells and wells with 50 ng histamine which is used for calculating the histamine concentration in the unknown sample:

$$\text{ng/mL histamine in sample} = \left(\text{Sample}_{\text{FU}} - \text{Blank}_{\text{FU}}\right) \times 50/\left(50\text{ng}_{\text{FU}} - \text{Blank}_{\text{FU}}\right),$$

where Blank_{FU} is the mean fluorescence intensity of the blank wells (A1-D1) and 50ng_{FU} is the mean fluorescence intensity of the wells containing 50 ng histamine (well E1-H1). $\text{Sample}_{\text{FU}}$ is the fluorescence intensity of the sample.

The % histamine release (of maximal release) can be calculated from the PMA/ionomycin results:

$$\%\ \text{Histamine release} = (\text{ng/mL histamine in sample}/\text{ng/mL histamine in PMA/ionomycin}) \times 100\%.$$

6. Longer storage time (> month) increases the background signal and lowers the specific signal and is therefore not recommended.

7. An additional incubation, between the cell fixation/permeabilization and DAO-PE staining, improves the intracellular detection of histamine in LAD2 cells.
8. Titration of DAO-PE revealed that while the most optimal concentration for histamine staining in basophils was 0.1 mg/mL DAO-PE, detection of histamine in LAD2 cells required tenfold higher concentration—1 mg/mL.
9. When measuring mediator release from both mast cells and basophils, please be aware that the supernatants may contain large amounts of proteases that can degrade the cytokines present. If possible, keep your supernatants at −80 °C until measuring and avoid freeze-thaw cycles for the best possible outcome.

References

1. Ishizaka T, Dvorak AM, Conrad DH, Niebyl JR, Marquette JP, Ishizaka K (1985) Morphologic and immunologic characterization of human basophils developed in cultures of cord blood mononuclear cells. J Immunol 134:532–540
2. Nolte H, Spjeldnaes N, Kruse A, Windelborg B (1990) Histamine release from gut mast cells from patients with inflammatory bowel diseases. Gut 31:791–794
3. Okayama Y, Church MK (1992) IL-3 primes and evokes histamine release from human basophils but not mast cells. Int Arch Allergy Immunol 99:343–345
4. Hate KR, Schneider R (1966) The difference in behaviour of basophil leucocytes and mast cells towards compound 48/80. Br J Pharmacol Chemother 28:282–288
5. Lowman MA, Rees PH, Benyon RC, Church MK (1988) Human mast cell heterogeneity: histamine release from mast cells dispersed from skin, lung, adenoids, tonsils, and colon in response to IgE-dependent and nonimmunologic stimuli. J Allergy Clin Immunol 81:590–597
6. Oskeritzian CA, Zhao W, Min HK, Xia HZ, Pozez A, Kiev J, Schwartz LB (2005) Surface CD88 functionally distinguishes the MCTC from the MCT type of human lung mast cell. J Allergy Clin Immunol 115:1162–1168
7. Galli SJ, Nakae S, Tsai M (2005) Mast cells in the development of adaptive immune responses. Nat Immunol 6:135–142
8. Redegeld FA, Yu Y, Kumari S, Charles N, Blank U (2018) Non-IgE mediated mast cell activation. Immunol Rev 282:87–113
9. Sumbayev VV, Yasinska I, Oniku AE, Streatfield CL, Gibbs BF (2012) Involvement of hypoxia-inducible factor-1 in the inflammatory responses of human LAD2 mast cells and basophils. PLoS One 7:e34259
10. McCurdy JD, Olynych TJ, Maher LH, Marshall JS (2003) Cutting edge: distinct Toll-like receptor 2 activators selectively induce different classes of mediator production from human mast cells. J Immunol 170:1625–1629
11. Varadaradjalou S, Feger F, Thieblemont N, Hamouda NB, Pleau JM, Dy M, Arock M (2003) Toll-like receptor 2 (TLR2) and TLR4 differentially activate human mast cells. Eur J Immunol 33:899–906
12. Kulka M, Alexopoulou L, Flavell RA, Metcalfe DD (2004) Activation of mast cells by double-stranded RNA: evidence for activation through Toll-like receptor 3. J Allergy Clin Immunol 114:174–182
13. Supajatura V, Ushio H, Nakao A, Akira S, Okumura K, Ra C, Ogawa H (2002) Differential responses of mast cell Toll-like receptors 2 and 4 in allergy and innate immunity. J Clin Invest 109:1351–1359
14. Zhang B, Alysandratos KD, Angelidou A, Asadi S, Sismanopoulos N, Delivanis DA, Weng Z, Miniati A, Vasiadi M, Katsarou-Katsari A, Miao B, Leeman SE, Kalogeromitros D, Theoharides TC (2011) Human mast cell degranulation and preformed TNF secretion require mitochondrial translocation to exocytosis sites: relevance to atopic dermatitis. J Allergy Clin Immunol 127:1522–1531
15. Galli SJ, Tsai M (2012) IgE and mast cells in allergic disease. Nat Med 4:693–704
16. Falkencrone S, Poulsen LK, Bindslev-Jensen C, Woetmann A, Odum N, Poulsen BC, Blom L, Jensen BM, Gibbs BF, Yasinska IM, Sumbayev

VV, Skov PS (2013) IgE-mediated basophil tumour necrosis factor alpha induces matrix metalloproteinase-9 from monocytes. Allergy 68:614–620

17. Blakeley J, Portnow J (2010) Microdialysis for assessing intratumoral drug disposition in brain cancers: a tool for rational drug development. Expert Opin Drug Metab Toxicol 6:1477–1491
18. Boddu SH, Gunda S, Earla R, Mitra AK (2010) Ocular microdialysis: a continuous sampling technique to study pharmacokinetics and pharmacodynamics in the eye. Bioanalysis 2:487–507
19. Plock N, Kloft C (2005) Microdialysis—theoretical background and recent implementation in applied life-sciences. Eur J Pharm Sci 25:1–24
20. Kulka M, Metcalfe DD (2010) Isolation of tissue mast cells. Curr Protoc Immunol Chapter 7:Unit 7.25
21. Brazis P, Barandica L, Garcia F, Clough GF, Church MK, Puigdemont A (2006) Dermal microdialysis in the dog: in vivo assessment of the effect of cyclosporin A on cutaneous histamine and prostaglandin D2 release. Vet Dermatol 17:169–174
22. Groth L, Jorgensen A, Serup J (1998) Cutaneous microdialysis in the rat: insertion trauma and effect of anaesthesia studied by laser Doppler perfusion imaging and histamine release. Skin Pharmacol Appl Ski Physiol 11:125–132
23. Petersen LJ, Church MK, Skov PS (1997) Histamine is released in the wheal but not the flare following challenge of human skin in vivo: a microdialysis study. Clin Exp Allergy 27:284–295
24. Baumann KY, Church MK, Clough GF, Quist SR, Schmelz M, Skov PS, Anderson CD, Tannert LK, Giménez-Arnau AM, Frischbutter S, Scheffel J, Maurer M (2019) Skin microdialysis: methods, applications and future opportunities—an EAACI position paper. Clin Transl Allergy 9:24. https://doi.org/10.1186/s13601-019-0262-y
25. LYNX Rapid RPE Antibody Conjugation Kit® datasheet. https://images.bio-rad-antibodies.com/datasheets/datasheet-LNK023RPE.pdf. Accessed 21 Jun 2019
26. Ao X, Stenken JA (2006) Microdialysis sampling of cytokines. Methods 38:331–341. https://doi.org/10.1016/j.ymeth.2005.11.012
27. Clough GF (2005) Microdialysis of large molecules. AAPS J 7:E686–E692. https://doi.org/10.1208/aapsj070369
28. Wælgaard L, Pharo A, Tønnessen TI, Mollnes TE (2006) Microdialysis for monitoring inflammation: efficient recovery of cytokines and anaphylatoxins provided optimal catheter pore size and fluid velocity conditions. Scand J Immunol 64:345–352. https://doi.org/10.1111/j.1365-3083.2006.01826.x
29. Helmy A, Carpenter KLH, Skepper JN, Kirkpatrick PJ, Pickard JD, Hutchinson PJ (2009) Microdialysis of cytokines: methodological considerations, scanning electron microscopy, and determination of relative recovery. J Neurotrauma 26:549–561
30. Kirbs C, Kloft C (2014) In vitro microdialysis recovery and delivery investigation of cytokines as prerequisite for potential biomarker profiling. Eur J Pharm Sci 57:48–59. https://doi.org/10.1016/j.ejps.2013.11.006
31. Berthoud TK, Manaca MN, Quelhas D, Aguilar R, Guinovart C, Puyol L, Barbosa A, Alonso PL, Dobaño C (2011) Comparison of commercial kits to measure cytokine responses to *Plasmodium falciparum* by multiplex microsphere suspension array technology. Malar J 10:1–9
32. Breen EC, Reynolds SM, Cox C, Jacobson LP, Magpantay L, Mulder CB, Dibben O, Margolick JB, Bream JH, Sambrano E, Martínez-Maza O, Sinclair E, Borrow P, Landay AL, Rinaldo CR, Norris PJ (2011) Multisite comparison of high-sensitivity multiplex cytokine assays. Clin Vaccine Immunol 18:1229–1242
33. Moncunill G, Aponte JJ, Nhabomba AJ, Dobaño C (2013) Performance of multiplex commercial kits to quantify cytokine and chemokine responses in culture supernatants from *Plasmodium falciparum* stimulations. PLoS One 8:1–10
34. Dabitao D, Margolick JB, Lopez J, Bream JH (2011) Multiplex measurement of proinflammatory cytokines in human serum: comparison of the Meso Scale Discovery electrochemiluminescence assay and the cytometric bead array. J Immunol Methods 372:71–77. https://doi.org/10.1016/j.jim.2011.06.033
35. Kepley CL, Youssef L, Andrews RP, Wilson BS, Oliver JM (1999) Syk deficiency in nonreleaser basophils. J Allergy Clin Immunol 104:279–284
36. Knol EF, Mul FP, Kuijpers TW, Verhoeven AJ, Roos D (1992) Intracellular events in anti-IgE nonreleasing human basophils. J Allergy Clin Immunol 90:92–103

Chapter 22

Detection of Mast Cells and Basophils by Immunohistochemistry

Andrew F. Walls and Cornelia Amalinei

Abstract

Staining cells or tissues with basic dyes was the mainstay of mast cell and basophil detection methods for more than a century following the first identification of these cell types using such methods. These techniques have now been largely supplanted by immunohistochemical procedures with monoclonal antibodies directed against unique constituents of these cell types. Immunohistochemistry with antibodies specific for the granule protease tryptase provides a more sensitive and discriminating means for detecting mast cells than using the classical histochemical procedures, and using antibodies specific for products of basophils (2D7 antigen and basogranulin) has allowed detection of basophils that infiltrate into tissues. The application of immunohistochemistry to detect more than one marker in the same cell has underpinned concepts of mast cell heterogeneity based on differential expression of chymase and other proteases. The double labeling procedures employed have also provided a means for investigating the expression of cytokines and a range of other products. Protocols are here set out that have been used for immunohistochemical detection of mast cells and basophils and their subpopulations in human tissues. Consideration is given to pitfalls to avoid and to a range of alternative approaches.

Key words Basophil, Chymase, Heterogeneity, Immunohistochemistry, Mast cell, Tryptase

1 Introduction

Mast cells were first characterized by Ehrlich in the late nineteenth century as cells with prominent cytoplasmic granules that reacted with a basic aniline dye [1]. His application of the term *Mastzellen* (well-fed cells) derived from his assumption that the intracellular material had been ingested from the local tissue. Though mast cells do indeed have phagocytic properties [2], the distinctive granules are in fact produced by the cell itself and serve as stores for a range of preformed mediators of inflammation. These include histamine, various proteases, and heparin, which can be explosively secreted in response to allergens or other stimuli. The process is accompanied by the rapid generation of new mediators including prostaglandin D_2 and leukotriene C_4 and the release of multiple cytokines that can

Bernhard F. Gibbs and Franco H. Falcone (eds.), *Basophils and Mast Cells: Methods and Protocols*, Methods in Molecular Biology, vol. 2163, https://doi.org/10.1007/978-1-0716-0696-4_22,

be both preformed and newly generated. Methods for the identification of mast cells based on those of Ehrlich remain in use, but understanding of their constituents has led to development of immunocytochemical procedures of greater sensitivity and specificity.

Though not appreciated at the outset, the classical methods for identifying mast cells have depended on the binding of basic dyes to heparin or other proteoglycans in the granules. These procedures result also in staining of the granules of basophils, a leukocyte identified by Ehrlich in blood smears using the same aniline dyes that led to the discovery of mast cells. The most widely employed basic dyes were to become toluidine blue (which stains mast cells *metachromatically* a violet color) and Alcian blue (which stains the granules *orthochromatically* an intense blue). Other basic dyes employed in mast cell histochemistry have included safranin O, astra blue, azure A and B, and thionine, and differences in the performance of stains (possibly related to differing degrees of sulfation of proteoglycans in different granules) have led to combinations of these stains being applied in the investigation of mast cell heterogeneity (reviewed in [3]). Other approaches employed to detect mast cells have relied on the binding to proteoglycans of fluorescent dyes such as berberine sulfate or acridine orange or avidin conjugated to enzymes or fluorescent compounds.

While the dye-based procedures are inexpensive and rapid to perform, their application may lead to an underestimation of mast cell numbers. The use of formaldehyde-based fixatives can render mast cells difficult to stain with basic dyes, particularly those of mucosal tissues in humans and rodents [4, 5]. While formaldehyde-induced loss of dye binding can be overcome to some extent by protease treatment of fixed tissues or by prolonged staining periods (generally several days required), this may be to the detriment of cell morphology and can lead to artifactual staining of other cell types [4]. A further concern with dye-binding procedures is that they may fail to stain mast cells where there is extensive loss of granule contents, and unstained "phantom mast cells" have been postulated to occur in particular where there is mast cell activation [6].

The preparation of monoclonal antibodies specific for tryptase, a trypsin-like serine protease of mast cells, opened the way for methods offering important advantages in the detection of mast cells [7, 8]. Immunohistochemistry with these antibodies (*see* Fig. 1a) has now replaced the classical dye-binding techniques as the "gold standard" procedure for identifying mast cells in human tissue. This protease is largely unique to mast cells, and quantities as great as 30 pg per cell may be present [9]. Basophils, the only other cell type in which tryptase has been demonstrated convincingly, have been estimated to contain less than 1% of the amount within a mast cell [10]. Employing sensitive detection systems (e.g., with a

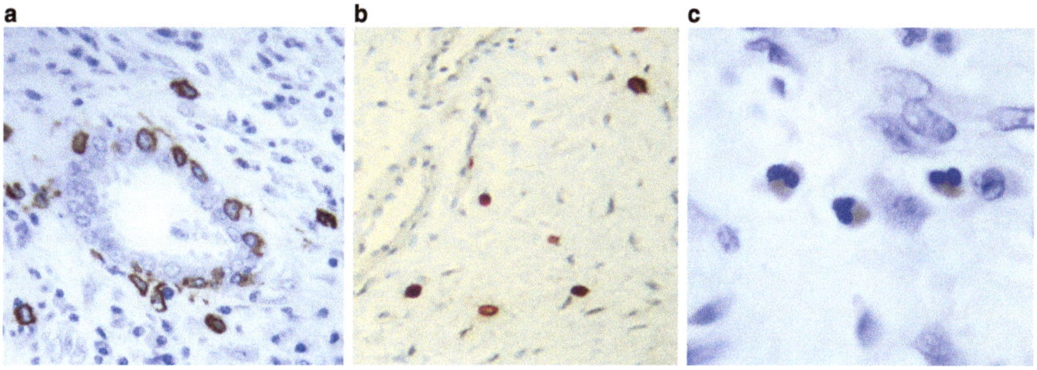

Fig. 1 (**a**) Mast cells detected using tryptase-specific monoclonal antibody AA1 in lung tissue from a case of bronchial carcinoma. Tryptase appears as reddish-brown staining and is localized predominantly within cytoplasmic granules. The presence of extracellular immunostaining would be consistent with partial degranulation of some of the cells. (**b**) Mast cells in synovial tissue from a patient with osteoarthritis detected using a double labeling procedure with tryptase-specific monoclonal antibody AA1 (reddish-brown immunostaining) and chymase-specific antibody CC1 (blue/black). (**c**) Basophils infiltrated into lung tissue in a case of bronchial carcinoma, detected using monoclonal antibody BB1 specific for the basophil marker basogranulin. Note the lobulated nuclei (blue counterstain) and the presence of basogranulin in the cytoplasm (brown)

secondary antibody and avidin biotin complex (ABC) amplification procedures), higher signal-to-noise ratios can be achieved than with conventional basic dye procedures [8] allowing more mast cells to be detected.

Immunohistochemistry with a tryptase-specific antibody can sometimes be forgiving of poor technique, but where there is potential for loss of tryptase from the granules (either as a consequence of mast cell activation in vivo or artifactual release from tissue ex vivo, e.g., as a consequence of crushing or decomposition), the use of amplification procedures and careful standardization is important. However, in most cases, mast cell degranulation in clinical disease may be partial rather than total, and tryptase will still be detected in mast cells albeit with reduced intensity. The presence of tryptase has been noted in the immediate vicinity of mast cells by immunohistochemistry (*see* Fig. 1a), occasionally as a "halo" around cells, or sometimes in discrete granules or as a "smear." Inadequate fixation can result in loss of this water-soluble protease, particularly in the case of sections of frozen tissues.

Though there is the theoretical risk that some mast cells may not be detected by immunohistochemistry with tryptase-specific antibodies on account of reduced quantities of tryptase being present following degranulation, other approaches have yet to supplant this method. A surface marker for mast cells which is not granule-associated is c-kit (CD117). Immunohistochemistry with the c-kit-specific monoclonal antibody YB5.B8 can be used for the identification of mast cells [11], but this antibody has generally proved unsuitable for use with paraffin-embedded tissues. Some success has been reported using an immunohistochemical procedure with an antibody specific to histamine [12], but histamine is present also in substantial quantities in basophils as well as being produced by diverse other cell types including neurons, neutrophils, monocytes/macrophages, dendritic cells, and platelets [13–15].

Certain mast cell markers are restricted to a subset of these cells, and their detection by immunohistochemistry has been used to investigate heterogeneity in mast cell populations. The first such marker to be studied was chymase, a chymotryptic serine protease. On the basis of double labeling procedures in immunohistochemistry (*see* Fig. 1b), mast cells have been categorized according to the presence of both tryptase and chymase (MC_{TC}) or of tryptase but not chymase (MC_T) [16, 17]. Relative numbers of mast cells distinguished in this way differ between anatomical sites, with MC_T cells most abundant at mucosal surfaces and cells of the MC_{TC} subset predominant in skin and the respiratory and gastrointestinal submucosa. The proportion of each subset can also be affected by disease, with cells of the MC_T population reported to be relatively sparse in the gut of AIDS patients but present in increased numbers in the affected tissues in a range of conditions associated with inflammation or tissue remodeling (including rhinitis, conjunctivitis, scleroderma, rheumatoid arthritis, and osteoarthritis (reviewed in [18]).

The significance of altered ratios of MC_T and MC_{TC} cells is unclear, and it remains to be determined to what extent each subset may differ in function other than that which would be conferred by the presence or absence of chymase. On the basis of fairly limited studies, it has been suggested that another mast cell-specific protease, carboxypeptidase A3, may be restricted to the MC_{TC} subset [19] and also cathepsin G, a protease better characterized as a neutrophil product [20]. Moreover, certain inflammatory cytokines may be differentially expressed in each of the major subsets defined, at least in the respiratory tract in allergic disease (reviewed in [21]). Immunohistochemical techniques have opened the way to investigate differential expression of protein products in mast cells.

Careful optimization and standardization of procedures is crucial when applying immunohistochemical procedures for the identification and characterization of mast cell subpopulations. Published reports of relative numbers of MC_T and MC_{TC} cells detected in similar tissues have varied greatly. Some of the variation may reflect differences between subjects studied or be a consequence of sampling too small a piece of tissue. However, it seems likely that the use of different antibodies and staining protocols will have contributed to the variation observed. When double labeling procedures are employed for tryptase and chymase, it is essential that there is effective detection of both these proteases. Different fixatives, different periods of incubation, or addition of different concentrations of primary antibodies on tissue sections can result in major differences in the proportions of MC_T and MC_{TC} subpopulations detected and even in the appearance of cells with chymase but not tryptase MC_C [22, 23]. Immunohistochemical procedures can be effective at determining the presence within cells of specific constituents but are less good for assessing relative amounts within each cell. The challenge is to use a limit of detection that is suitable for each study and to adhere closely to it. Mast cells are categorized according to the presence or absence of specific constituents by immunohistochemistry, but often it may be more appropriate to define these cells according to whether a defined constituent is abundant or present in only small quantities.

Considerations of detection sensitivity can be particularly important when applying immunohistochemistry to study mast cell subpopulations. Similar issues may arise also when seeking to distinguish mast cells from basophils and other cell types. It is generally assumed that the quantities of "mast cell proteases" present in a basophil are insufficient to be detected using standard immunohistochemical procedures, and basophils are undoubtedly less likely to be mistaken for mast cells using immunohistochemistry than using classical dye-binding techniques. However, quantities of tryptase present in the basophil can vary greatly between individuals and by more than two orders of magnitude [24], and this issue may require reappraisal at least for tissues from some allergic individuals.

Basophil numbers in human tissues may now be determined using antibodies specific for this cell type. Monoclonal antibodies 2D7 [25] and BB1 [26] bind to proteins of unknown function that are unique to the granules of basophils and which are secreted upon cell degranulation [27]. Antibody 2D7 binds to a protein of 72–76 kDa [25] that is produced at an early stage in basophil development from bone marrow progenitors [28], while BB1 antibody recognizes a basic protein, termed basogranulin, that is secreted as a large macromolecular complex of approximately 5000 kDa or more [29]. Elucidation of their mediator roles could one day provide important clues as to what may be the unique contribution of basophils in disease. In the meantime, immunohistochemistry with BB1 (*see* Fig. 1c) and 2D7 can provide a valuable means for the identification of basophils in tissues.

The protocols set out below have been used successfully to detect mast cells and basophils by immunohistochemistry. As histochemical staining methods have largely been replaced by the immunohistochemical methods, they are not provided here (though protocols for staining by dye binding are available elsewhere [30]). Mast cell and basophil markers are here described only for human cells. While the basic immunohistochemical procedures outlined may be suitable for use with cells and tissues from nonhuman species, such is the extent of mast cell heterogeneity between species that markers used in man may be inappropriate even if suitable antibodies were available for the nonhuman homologue. There are species-related differences not only in the distribution of mast cell proteases but also in the types of protease. Thus, for example, multiple chymases (with quite different patterns of expression) have been characterized in several species of small mammal, but there is just one chymase in man (reviewed in [18]). Simple categorization into MC_T and MC_{TC} populations is unlikely to be possible for many nonhuman species.

The techniques described are intended not as blueprints to be followed rigidly but as guides that may be helpful when developing suitable methods for a specific system being studied. It is important to validate and optimize methods for each source of material. Suggestions are made for antibodies to employ, for the preparation of some widely employed fixatives and reagents, and there is a list of basic equipment. The primary approach for which details are provided involves fixation and processing tissues in paraffin wax, though methods are indicated also for preparing and fixing cytocentrifuge preparations. Loss of antigenicity often occurs during fixation, and a heat-induced epitope retrieval procedure is described for use following de-waxing and rehydration of tissue sections. Protocols are provided for immunohistochemistry with one primary antibody or with two different antibodies. Pitfalls to avoid and possible alternative materials and strategies for immunohistochemistry are suggested in notes to the methods.

2 Materials

2.1 Antibodies

Monoclonal antibodies specific for mast cell tryptase (e.g., AA1), chymase (e.g., CC1), or basophils (e.g., BB1) or to an irrelevant antigen as a control (e.g., specific for *Aspergillus niger* glucose oxidase, X0931).

Biotinylated antiserum to the primary antibody (e.g., goat anti-mouse immunoglobulins).

2.2 Fixatives, Tissue Processing, and Slide Preparation

See **Note 1**.

1. 10% neutral buffered formalin: 100 mL 40% formaldehyde stock is diluted with 900 mL distilled water, and 4 g of sodium dihydrogen orthophosphate and 6.5 g of disodium hydrogen orthophosphate (anhydrous) are added and dissolved. The pH is adjusted to 7.0–7.6.
2. Aldehyde-Based Fixatives:

   Glutaraldehyde.

   Paraformaldehyde: A solution of 2.26% sodium dihydrogen orthophosphate (41.5 mL) is added to 2.52% sodium hydroxide (8.5 mL) and heated to 60 to 80 °C in a covered container. Paraformaldehyde (2 g) is added, and the mixture stirred until it becomes clear. This is filtered, cooled, and adjusted to pH 7.2 to 7.4. This fixative should be freshly prepared on each occasion.

3. Mercuric chloride fixatives (B5 fixative): A stock B5 solution is made with 60 g mercuric chloride, and 12.5 g sodium acetate dissolved in 1 L of distilled water. A working solution is prepared immediately before use by mixing 90 mL of the stock B5 solution with 10% neutral buffered formalin as prepared above.

   Zenker's fixative: Prepared by dissolving 25 g potassium dichromate and 50 g mercuric chloride in 950 mL distilled water and adding 50 mL glacial acetic acid. The solution should be mixed overnight to ensure solubilization of mercuric chloride crystals.

4. Alcohol fixatives—Carnoy's fluid: A mixture of 60 mL absolute ethanol, 30 mL chloroform, and 10 mL glacial acetic acid.

   Methacarn: A mixture of 60 mL absolute methanol, 30 mL chloroform, and 10 mL glacial acetic acid.

5. Susa fixative: A stock solution is prepared with 4.5 g mercuric chloride, 2 g trichloroacetic acid, and 0.5 g sodium chloride dissolved in 80 mL distilled water. To this is added a mixture of 20 mL 40% formaldehyde and 4 mL glacial acetic acid.

6. Acetone.

7. Methanol.

8. Absolute alcohol.

9. Histoclear II (or other xylene substitute).

10. Poly-L-lysine.

2.3 Reagents for Antigen Retrieval and Immunohistochemistry

1. Antigen retrieval solution: 0.01 M sodium citrate buffer, pH 6.0. Prepare stock solutions of 0.1 M citric acid solution (by dissolving 21 g of citric acid, monohydrate 1 L of distilled water) and 0.1 M sodium citrate solution (by dissolving 29.4 g trisodium citrate dihydrate in 1 L of distilled water). A working solution is prepared by adding 9 mL of 0.1 M citric acid and 41 mL of 0.1 M sodium citrate to 450 mL of distilled water and adjusting the pH to 6.0.

2. Peroxidase blocking solution: 0.5% hydrogen peroxide, 0.1% sodium azide in methanol (*see* **Note 2**).

3. Nonspecific protein-binding blocking buffer:

   For use with hydrogen peroxide reporting enzyme: 1% fraction V bovine serum albumin in phosphate-buffered saline (PBS-albumin).

   For use with alkaline phosphatase reporting enzyme (as in double labeling immunohistochemistry protocol):

1% fraction V bovine serum albumin in Tris-buffered saline (TBS-albumin).

4. Wash buffer:

   For use with hydrogen peroxide: phosphate-buffered saline (PBS), pH 7.4 (*see* **Note 3**).

   For use with alkaline phosphatase: Tris-buffered saline, pH 8.0.

5. ExtrAvidin®-peroxidase conjugate or ExtrAvidin®-alkaline phosphatase conjugate (*see* **Note 4**).

6. Substrates:

   For horseradish peroxidase: a stock solution is prepared with 10 mg/mL 3-amino-9-ethylcarbazole (AEC) in dimethylformamide (*see* **Note 5**). This can be stored in the dark for up to 1 week at 4 °C. The AEC substrate for applying to sections is comprised of 1 mL AEC stock, 19 mL 0.1 M acetate buffer, pH 5.0, and 3.3 μL 30% hydrogen peroxide. This should be filtered immediately before use. *See* **Note 6**.

   For alkaline phosphatase: Prepare stock solutions of (a) 50 mg/mL 5-bromo-4-chloro-3-indolyl phosphate (p-toluidine salt; BCIP) in dimethylformamide (BCIP stock), (b) 75 mg/mL p-nitrotetrazolium blue (NBT) in 70% dimethylformamide (BCIP stock), and (c) 0.1 M glycine, 1 mM magnesium chloride, and 1 mM zinc chloride, pH 10.4 (alkaline phosphatase buffer). These can be stored at 4 °C. Prior to use, prepare the alkaline phosphatase substrate solution by adding 75 μL BCIP stock and 100 μL NBT stock to 20 mL alkaline phosphatase buffer, and 0.25 mg/mL levamisole should be added (to block most endogenous alkaline phosphatases).

7. Counterstain: Mayer's hemalum. Filter each time before use.

2.4 Equipment

1. Glass microscope slides.
2. Microtome for cutting tissue sections.
3. Delimiting pen.
4. Access to automated tissue processing equipment is useful where there is high throughput of samples.
5. Water bath.
6. Microwave (if employed for fixation or for unmasking antigens prior to immunohistochemistry).
7. Slide tray which holds the slides above a pool of tap water. A close-fitting lid for the tray allows a humidified atmosphere to be maintained and prevents sections from drying out during immunohistochemistry.

8. Staining trough and slide racks.
9. Hemocytometer.
10. Cytocentrifuge.

3 Methods

3.1 Fixation and Processing of Tissues

1. Carefully cut tissues into pieces small enough to allow good penetration of the fixative (generally 1 cm^3 or less). *See* **Note 7**.
2. Fix specimens as soon as possible to preserve tissue architecture by incubating in 10% neutral buffered formalin for 6–24 h (*see* **Note 8**).
3. Decant the fixative and dispose according to local regulations. Add PBS and mix gently on a spiral or rotary mixer at 4 °C for 1–2 h to wash out the fixative.
4. At intervals of an hour or more, dehydrate the specimens through a series of graded alcohols, incubating progressively with 70% ethanol, 90% ethanol, and 100% ethanol (twice) with gentle mixing at 4 °C.
5. Embed the tissues in paraffin wax (*see* **Note 9**).
6. Cut into 4–6 μm sections (*see* **Note 10**) and float them in a tank of water.
7. Pick up cut sections from the water tank onto slides coated with poly-L-lysine (to provide an adherent surface) (*See* **Note 11**).
8. Maintain at 58 °C in a drying oven (with thermostat) for better adherence of tissue to the slides (*see* **Note 12**).

3.2 Preparation and Fixation of Cytocentrifuge Preparations

1. Establish the number of cells in suspension using a hemocytometer, and then adjust to 10^6 cells per mL using a physiological buffer.
2. Assemble slides in cytocentrifuge according to the manufacturer's instructions, then load 100 μL of cell solution per slide, and centrifuge for 5 min at 500 rpm (28 × *g*).
3. Remove the slides, taking care not to smear the cells when removing the filter cards. Allow the cells to air-dry, and then fix in methanol or acetone for 1 min (*See* **Note 12**).

3.3 De-waxing and Rehydration of Tissue Sections

1. Immerse slides with tissue sections in a trough of xylene and heat in the oven at 58 °C, for 45 min.
2. Incubate twice in Histoclear II (or other proprietary, less toxic substitute for xylene) at room temperature for 10 min each time.
3. Incubate three times in 100% ethanol for 10 min each time.
4. Incubate in 96% ethanol for 10 min.

5. Incubate in 80% ethanol for 10 min.
6. Incubate in 70% ethanol for 10 min.
7. Rinse in distilled water twice each for 10 min.
8. It may be helpful to use a delimiting pen to draw a circle around specimen on the slide (*see* **Note 13**). The trace is water-repellent and should help to prevent reagents added from running off the slide (*See* **Note 14**).

3.4 Heat-Induced Epitope Retrieval

See **Note 15**.

1. Transfer slides to a heat-proof vessel containing the sodium citrate buffer antigen retrieval solution—buffer solution and heat in a water bath at 97–99 °C for 20–30 min (*see* **Note 16**). If necessary, add more solution during this process to prevent the tissue sections from drying out.
2. While still submerged, allow slides to cool at room temperature for an additional 30–40 min.
3. Wash in distilled water for 5 min at room temperature, and proceed to immunohistochemistry.

3.5 Immuno-histochemistry with a Single Primary Antibody

See **Note 17**.

1. With the slides placed flat in a humid environment (e.g., in slide tray), add the peroxidase blocking solution for 5 min (to inhibit endogenous peroxidase activity).
2. Discard the peroxidase blocking solution and wash with distilled water for 5 min.
3. Block nonspecific protein-binding sites by adding the protein blocking solution for 10 min.
4. Discard the protein blocking solution and wipe around sections with a piece of tissue.
5. Apply 100 μL primary antibody at a pre-established working dilution (*see* **Note 18**) in PBS-albumin to the section, and incubate overnight at 4 °C (*see* **Note 19**).
6. Wash the slides three times with wash buffer, for 10 min each wash.
7. Wipe around the sections and apply 100 μL of biotinylated secondary antibody for 30 min, at room temperature.
8. Wash the slides as described in **step 6**.
9. Wipe around the sections and apply 100 μL ExtrAvidin®-peroxidase diluted 1:20 in PBS-albumin. Incubate the slides for 30 min at room temperature.
10. Wash the slides as described in **step 6**, and while sections are being washed, prepare the AEC substrate solution.

11. Apply the AEC substrate solution for 3–4 min or until a reddish-brown color develops (checking briefly under the microscope if necessary). In practice, this means starting to collect slides in a rack after 2 min as staining can continue with the slides in a vertical position (*See* **Note 6**).
12. Wash slides in a trough of running tap water for 5 min, avoiding excessive turbulence that could result in detachment of sections. Rinse briefly in distilled water.
13. Counterstain in Mayer's hemalum for 3 min, and then wash as described in **step 12**.
14. When the sections have dried completely, mount a coverslip using aqueous mounting medium (*see* **Note 20**).

3.6 Double Labeling Immunohistochemistry for MC_T and MC_{TC} Subpopulations

See **Note 21**.

1. With the slides in a humid environment, add the peroxidase blocking solution for 5 min (as in Subheading 3.5; *see* **Note 22**).
2. Discard the peroxidase blocking solution and wash with distilled water for 5 min.
3. Block nonspecific protein-binding sites by adding the protein blocking solution for 10 min.
4. Discard the protein blocking solution and wipe around sections with a piece of tissue.
5. Apply 100 μL primary antibodies at pre-established working dilutions (*see* **Note 18**) in TBS-albumin to the section, and incubate for 3 h (*see* **Note 19**). The antibodies added should comprise (a) peroxidase-conjugated anti-tryptase antibody, (b) biotinylated anti-chymase antibody, and (c) both peroxidase-conjugated anti-tryptase antibody and biotinylated anti-chymase antibody added together. Appropriate negative control antibodies (of the same isotype and quantity as each of the other antibodies) should also be included.
6. Wash the slides three times with wash buffer, for 10 min each wash, and meanwhile prepare the AEC substrate solution.
7. Apply the AEC substrate solution for 5 min. A reddish-brown color should develop.
8. Wash the slides as described in **step 6**.
9. Wipe around the sections and apply 1/20 ExtrAvidin®-alkaline phosphatase in TBS-albumin for 1 h.
10. Wash the slides as described in **step 6**, and while sections are being washed, prepare the BCIP/NBT substrate solution.
11. Apply the BCIP/NBT substrate solution for 10 min.

12. Wash slides in a trough of running tap water for 5 min avoiding excessive turbulence. Rinse briefly in distilled water.
13. Counterstain in Mayer's hemalum for 3 min, and then wash as described in **step 12**.
14. When the sections have dried completely, mount a coverslip using aqueous mounting medium. Using this protocol, there should be reddish-brown immunostaining for tryptase and blue/black for chymase (almost invariably co-localized with the reddish-brown color).

4 Notes

1. There are hazards associated with the constituents of many fixatives, and their preparation and use should be carried out in a well-ventilated area with personal protective equipment (PPE). Formaldehyde, paraformaldehyde, and mercuric chloride have irritant and corrosive effects on the eye and skin. Mercuric chloride is corrosive to metals. It should not be brought in contact with any metal object, and waste should not be poured down the drain.
2. Sodium azide is highly toxic, so PPE should be worn.
3. PBS may be adequate as a wash buffer in many situations, but where there is heavy background staining, there may be a need for a more extensive washing protocol and to add a suitable detergent, e.g., Tween 20® (Polysorbate 20), or increase the salt concentration of the wash buffer. A protocol that has been used in mast cell identification [30] has involved sequential washing with three separate wash buffers comprising:

   Buffer A: 0.4 M NaCl, 50 mM Tris, 0.05% Tween 20, pH 8.5.

   Buffer B: 0.15 M NaCl, 50 mM Tris, 0.05% Tween 20, pH 8.5.

   Buffer C: 50 mM Tris, 0.1% Tween 20, pH 8.5.

   As phosphate ions inhibit alkaline phosphatase activity, a TBS rather than a PBS-based wash buffer should be used where this is the reporter enzyme employed (as in the double labeling protocol here).
4. The protocols described involve an avidin-biotin complex (ABC) amplification step which relies on the high affinity of avidin for biotin. ExtrAvidin®, a modified form of avidin that retains affinity for biotin, is associated with less nonspecific binding at neutral pH, and its use is described here. Modified forms of avidin from other suppliers include NeutrAvidin®, and the bacterially derived streptavidin has also been employed in immunohistochemistry.

The principle of the ABC staining method is that a biotinylated secondary antibody is allowed to bind to the primary antibody once it has bound to the target antigen in tissue. This is in turn incubated with a mixture of biotinylated reporter enzyme (horseradish peroxidase or alkaline peroxidase) and free ExtrAvidin®. As avidin is tetravalent, a complex is formed that allows many more molecules of the reporter enzyme to be attached (via biotin) to the secondary antibody than would be possible by simply conjugating the enzyme directly to the immunoglobulin.

Various reporter enzymes are available, though the methods described here are restricted to horseradish peroxidase and to alkaline phosphatase (whose use is described in the double labeling protocol). The techniques set out could be adapted also to use fluorescent reporters, whose use is particularly suitable where several different primary antibodies are to be used on the same section, though the staining pattern can be short-lived (unlike with enzyme reporters). An alternative to avidin-biotin complex amplification of signal is the labeled avidin biotin (LAB) staining method (or LSAB where streptavidin is used instead of avidin) in which the enzyme reporter is conjugated directly to avidin (or streptavidin) molecules. Further amplification could be obtained by the use of a tertiary antibody directed against the reporter enzyme (peroxidase anti-peroxidase or phosphatase anti-phosphatase (PAP) method), though such a sensitive approach is unlikely to be required for mast cell or basophil identification in most situations.

5. When using dimethylformamide in preparing the substrate, avoid polystyrene vessels because these are attacked by the solvent. Avoid bright lights with AEC, and wear PPE.
6. A commonly used alternative substrate (to AEC) for horseradish peroxidase is 3,3′-diaminobenzidine (DAB) diluted 1:20 in substrate buffer, pH 7.5. DAB solution has to be prepared within 30 min before developing the reaction (being photosensitive strong light should be avoided when handling it). If using DAB as substrate, washing and counterstaining should proceed as described for AEC, though thereafter there should be dehydration in graded alcohols and clarification in Histoclear II (or other xylene substitute), before mounting a coverslip using a synthetic organic medium (e.g., DPX).
7. Prolonged fixation in formalin-based fixatives is likely to lead to antigens being masked so that they are not recognized by the primary antibody in immunohistochemistry. Careful optimization of the antigen retrieval stage may be required in such cases.

8. A number of other commonly employed fixatives are listed above, but selection of the most appropriate for each type of tissue is beyond the scope of this chapter. Fixatives differ in their suitability for different types of tissues and vary in their speed of penetration and in the degree of structural alterations they induce in the tissues (shrinkage, hardness rendering cutting difficult, masking of antigens, etc.).

   Tissue imprints, blood smears, cell cultures, and purified cells can be fixed with acetone (5 s and then 10 min and then rehydrated in buffer) or 10% neutral buffered formalin (as for tissue). Cells can also be centrifuged into a pellet (or cells resuspended in gelatin), and the pellet (or cells in gelatin) processed for histology as described for tissue specimens. Microwave treatment can be used for fixation of fresh cells or frozen tissues by heating or can be used to speed fixation by heating tissues in fixative.

9. A temperature higher than 60 °C should be avoided during processing of tissues in wax as this can result in loss of antigenicity.

10. In general, sections should be cut thinly, as when more than 5 μm thick there are likely to be multiple layers of cells present rendering interpretation more difficult.

11. Clean slides should be soaked in a 10% solution of poly-L-lysine for 10 min and then air-dried with the use of a fan or a drying oven (40 °C), prior to storage or use.

    When picking sections up onto the slides, it is important that sections are flat and that they adhere to the slides without forming ridges; otherwise, they may be torn off during processing.

12. Sectioned tissue on slides or cytocentrifuge preparations can be wrapped in aluminum foil and stored at 4 °C for up to 2 weeks before performing immunohistochemistry.

13. Drawing around the section with a delimiting pen is not universally practiced, and it may be sufficient (and less work) to simply rely on surface tension to maintain small volumes of liquid on top of the section in subsequent steps. In this case, more care may be required in wiping around the tissue sections at later stages when the surface of the slide has become wet.

14. After de-waxing and rehydration, sections can be stored for a short period with buffer to prevent drying out, but it is preferable to proceed directly to the next stage (antigen retrieval or immunohistochemistry) as soon as possible.

15. An antigen retrieval step may not be necessary for some fixatives but is often required where there has been prolonged incubation of the tissue with formalin. Masking of the epitope so that it does not bind to primary antibody can be a

consequence of formalin-induced cross-linking of peptides either within or adjacent to the epitope. This can be associated with conformational changes or alterations in charge at the epitope, such that the antibody no longer binds or fails to penetrate to the epitope.

Though a single technique is described here, a range of antigen retrieval methods are available, and for some tissues, fixation methods, and antibodies employed, it may be necessary to try more than one for best results, as well as varying conditions for each. Heat-induced epitope retrieval is thought to reverse some of the peptide cross-linkages that occur during fixation and allow restoration of secondary or tertiary structure at the epitope. Protease-induced epitope retrieval methods are also widely employed and involve addition of proteinase K, trypsin, pepsin, or other proteases to cleave peptides that may be masking the epitope. Protease-based methods require careful titration to avoid loss of tissue morphology or loss of the epitope of interest with prolonged incubation with protease or too high a concentration.

16. Heat-induced epitope retrieval can be performed also using microwave ovens, pressure cookers, vegetable steamers, and autoclaves. Microwave protocols can involve 5-min periods of heat with the antigen retrieval buffer added directly to the slide, followed by replacement of the buffer.

17. Sodium azide inhibits horseradish peroxidase activity and so should not be used as a preservative in buffers employed for immunohistochemistry with the method described using this enzyme for detection of antibody binding.

18. Antibodies should be titrated before using for the first time. As a rough guide, concentrations of monoclonal antibodies of around 1 μg/mL generally give good results, but the optimum concentration can differ between tissues processed in different ways as well as between different batches of antibody. If the type of tissue of interest is in short supply, an alternative that is more readily available can be used in antibody titration, ideally processed in exactly the same way. Staining patterns for serial dilutions of antibody should be carefully compared, and a concentration of antibody selected that gives a clear signal and minimal background staining.

    Appropriate control antibodies should be employed as well as the antibody of interest. Ideally, these should include a negative control antibody in the staining run and buffer-only controls. Anti-tryptase antibody AA1 can be suitable as a positive control antibody in many situations (even where mast cells are not the principal cell of interest) as it gives a strong signal and will detect mast cells clearly in most human tissues. Detection of basophils in tissues can be more problematic, as they can be sparse or absent altogether. Confirmation that the technique for immunohistochemistry is working appropriately in such situations can be provided by including sections of tissues in which basophil infiltration has been demonstrated previously or of a pellet of basophils or buffy coat cells that has been processed using similar histological procedures (greater volumes of such material can be produced by embedding the cells in gelatin prior to processing).

19. Incubation periods employed in immunohistochemistry are generally chosen to correspond to a point *after* the antibody-antigen reaction has achieved equilibrium. As such, there is generally scope for shortening incubation periods to speed up the staining process if required or to fit the procedure into the working day. Thus, for example, instead of incubating the primary antibody overnight at 4 °C, it may be acceptable for this incubation period to be reduced to 90 min at room temperature [30] or for a shorter period. However, once a protocol has been established in a study, it should be followed consistently to minimize variation in results.

20. A xylene-based mountant such as DPX should not be used with AEC as it will solubilize the stain (though it is appropriate with other substrates such as DAB; *see* **Note 6**).

21. The protocol here is for detection of tryptase and chymase in mast cells, but it could equally well be adapted for use with pairs of antibodies specific for other markers (e.g., to localize a cell product to mast cells or basophils or to determine the location of either of these cells in relation to other cells or structures). In order that the binding of two different primary antibodies can be visualized simultaneously, it is necessary to use different reporting enzymes for each and to select chromogenic substrates which will yield contrasting colors. The present method employs a biotinylated antibody specific for chymase (for which the signal is amplified by avidin-biotin complex formation with alkaline phosphatase) and an antibody specific for tryptase that is labeled with horseradish peroxidase. Using this protocol, the amplification step should be for the antibody that gives the weaker signal (either on account of the properties of the antibody or the quantity of antigen available to bind). In the present case, chymase is the less abundant in mast cells than tryptase [9]. It is crucial that care is exercised to get the correct balance of staining of the two target proteins, and it should be ensured that antigen retrieval steps if applied are appropriate for both antigens being studied. As with all immunohistochemistry, antibody concentrations should be titrated and incubation times selected carefully (*see* **Note 18**).

22. As alkaline phosphatase is inhibited by phosphate ions, PBS should not be employed as buffer or in washing steps in this protocol (*see* **Note 3**).

Acknowledgments

Special thanks to Dr. Mark Buckley, formerly of the Immunopharmacology Group, University of Southampton, UK, for fruitful discussions on immunohistochemical technique. Dr. Anne-Marie Skitt and Dr. Mark Buckley helped with provision of photographs. Financial support from the Medical Research Council, the National Institute for Health Research, and Asthma UK is gratefully acknowledged.

References

1. Ehrlich P (1879) Contributions to the theory and practice of histological staining. In: Himmelweit F (ed) The collected papers of Paul Ehrlich. Pergamon Press, New York, NY, pp 65–68
2. Wesolowski J, Caldwell V, Paumet F (2012) A novel function for SNAP29 (Synaptosomal-Associated Protein of 29 kDa) in mast cell phagocytosis. PLoS One 7(11):e49886
3. Heard BE (1986) Histochemical aspects of the staining of mast cells with particular reference to heterogeneity and quantification. In: Kay AB (ed) Asthma, clinical pharmacology and therapeutic progress. Blackwell, Oxford, pp 286–294
4. Walls AF, Roberts JA, Godfrey RC, Church MK, Holgate ST (1990) Histochemical heterogeneity of human mast cells: disease-related differences in mast cell subsets recovered by bronchoalveolar lavage. Int Arch Allergy Appl Immunol 92:233–241
5. Wingren U, Enerback L (1983) Mucosal mast cells of the rat intestine: a re-evaluation of fixation and staining properties, with special reference to protein blocking and solubility of the granular glycosaminoglycan. Histochem J 15:571–582
6. Claman HN, Choi KL, Sujansky W, Vatter AE (1986) Mast cell "disappearance" in chronic murine graft-vs-host disease (GVHD) – ultrastructural demonstration of "phantom mast cells". J Immunol 137:2009–2013
7. Craig SS, DeBlois G, Schwartz LB (1986) Mast cells in human keloid, small intestine, and lung by an immunoperoxidase technique using a murine monoclonal antibody against tryptase. Am J Pathol 124:427–435
8. Walls AF, Jones DB, Williams JH, Church MK, Holgate ST (1990) Immunohistochemical identification of mast cells in formaldehyde-fixed tissue using monoclonal antibodies specific for tryptase. J Pathol 162:119–126
9. Schwartz LB, Irani A-MA, Roller K, Castells MC, Schechter NM (1987) Quantitation of histamine, tryptase and chymase in dispersed human T and TC mast cells. J Immunol 138:2611–2615
10. Castells MC, Irani AM, Schwartz LB (1987) Evaluation of human peripheral blood leukocytes for mast cell tryptase. J Immunol 138:2184–2189
11. Mayrhofer G, Gadd SJ, Spargo LD, Ashman LK (1987) Specificity of a mouse monoclonal antibody raised against acute myeloid leukaemia cells for mast cells in human mucosal and connective tissues. Immunol Cell Biol 65:241–250
12. Johansson O, Virtanen M, Hilliges M, Yang Q (1994) Histamine immunohistochemistry is superior to the conventional heparin-based routine staining methodology for investigations of human skin mast cells. Histochem J 26:424–430
13. Xu X, Zhang D, Zhang H, Wolters PJ, Killeen NP, Sullivan BM, Locksley RM, Lowell CA, Caughey GH (2006) Neutrophil histamine contributes to inflammation in mycoplasma pneumonia. J Exp Med 203:2907–2917
14. Oh C, Suzuki S, Nakashima I, Yamashita K, Nakano K (1988) Histamine synthesis by non-mast cells through mitogen-dependent induction of histidine decarboxylase. Immunology 65:143–148

15. Alcañiz L, Vega A, Chacón P, El Bekay R, Ventura I, Aroca R, Blanca M, Bergstralh DT, Monteseirin J (2013) Histamine production by human neutrophils. FASEB J 27:2902–2910
16. Irani A-MA, Bradford TR, Kepley CL, Schechter NM, Schwartz LB (1989) Detection of MCT and MCTC types of human mast cells by immunohistochemistry using new monoclonal anti-tryptase and anti-chymase antibodies. J Histochem Cytochem 37:1509–1515
17. Buckley MG, McEuen AR, Walls AF (1999) The detection of mast cell subpopulations in formalin-fixed human tissues using a new monoclonal antibody specific for chymase. J Pathol 189:138–143
18. Walls AF (2000) The roles of neutral proteases in asthma and rhinitis. In: Busse WW, Holgate ST (eds) Asthma and rhinitis, 2nd edn. Blackwell, Boston, MA, pp 968–998
19. Irani A-MA, Bradford TR, Schwartz LB (1991) Human mast cell carboxypeptidase. Selective localization to MCTC cells. J Immunol 147:247–253
20. Schechter NM, Irani AM, Sprows JL, Abernethy J, Wintroub B, Schwartz LB (1990) Identification of a cathepsin G-like proteinase in the MCTC type of human mast cell. J Immunol 145:2652–2661
21. Bradding P, Walls AF, Holgate ST (2006) The role of the mast cell in the pathophysiology of asthma. J Allergy Clin Immunol 117:1277–1284
22. KleinJan A, Godthelp T, Blom HM, Fokkens WJ (1996) Fixation with Carnoy's fluid reduces the number of chymase-positive mast cells: not all chymase-positive mast cells are also positive for tryptase. Allergy 51:614–620
23. Beil WJ, Schulz M, McEuen AR, Buckley MG, Walls AF (1997) Number, fixation properties, dye-binding and protease expression of duodenal mast cells: comparisons between healthy subjects and patients with gastritis or Crohn's disease. Histochem J 29:759–773
24. Foster B, Schwartz LB, Devouassoux G, Metcalfe DD, Prussin C (2002) Characterization of mast-cell tryptase-expressing peripheral blood cells as basophils. J Allergy Clin Immunol 109:287–293
25. Kepley CL, Craig SS, Schwartz LB (1995) Identification and partial characterization of a unique marker for human basophils. J Immunol 154:6548–6555
26. McEuen AR, Buckley MG, Compton SJ, Walls AF (1999) Development and characterization of a monoclonal antibody specific for human basophils and the identification of a unique secretory product of basophil activation. Lab Investig 79:27–38
27. Mochizuki A, McEuen AR, Buckley MG, Walls AF (2003) The release of basogranulin in response to IgE-dependent and IgE-independent stimuli: validity of basogranulin measurement as an indicator of basophil activation. J Allergy Clin Immunol 112:102–108
28. Kepley CL, Pfeiffer JR, Schwartz LB, Wilson BS, Oliver JM (1998) The identification and characterization of umbilical cord blood-derived human basophils. J Leuk Biol 64:474–483
29. McEuen AR, Calafat J, Compton SJ, Easom NJ, Buckley MG, Knol EF, Walls AF (2001) Mass, charge, and subcellular localization of a unique secretory product identified by the basophil-specific antibody BB1. J Allergy Clin Immunol 107:842–848
30. Buckley MG, Walls AF (2008) Identification of mast cells and mast cell subpopulations. In: Jones MG, Lympany P (eds) Allergy methods and protocols; methods in molecular medicine vol. 138. Humana, Totowa, NJ, pp 285–297

Chapter 23

Detection of Basophils and Other Granulocytes in Induced Sputum by Flow Cytometry

Lionel Rohner and Michaela Fux

Abstract

Flow cytometry is one of the most widely used techniques for the detection of surface markers on various cells, particularly the cells of the immune system, at a single-cell resolution. Modern flow cytometers can identify rare cell population in highly heterogeneous samples. Here we present a protocol that allows a precise detection of basophils as well as eosinophils and neutrophils in induced sputum samples. The identification of sputum basophils and other granulocytes contributes to a better understanding of the cellular network that promotes and regulates inflammation of the lower respiratory tract.

Key words Basophils, Eosinophils, Neutrophils, Induced sputum, Allergy, Asthma, Airway inflammation, Flow cytometry

1 Introduction

The continuous advances in flow cytometry have enabled precise and highly standardized immunophenotyping of blood leukocytes, a discipline that has been adopted by many diagnostic laboratories. In contrast, flow cytometry analyses of samples from other body fluids such as induced sputum are less standardized. Here we present a protocol that allows detection of basophils as well as eosinophils and neutrophils in induced sputum samples using eight-color flow cytometry. Basophils are known to play a critical role as effector and immunoregulatory cells in protective immune responses against parasites (e.g., helminths and ticks) and during IgE-driven allergic reactions that, on the contrary, are destructive to the host. In allergic asthmatics, numbers of sputum basophils have been shown to correlate with the level of infiltrating eosinophils, which is associated with a more severe inflammatory response [1]. Consistent with this, basophils are found in high numbers in the airways of asthmatics where they accumulate in response to allergen challenge [2] or due to natural narrowing of the airway in asthma exacerbations [3].

Bernhard F. Gibbs and Franco H. Falcone (eds.), *Basophils and Mast Cells: Methods and Protocols*, Methods in Molecular Biology, vol. 2163, https://doi.org/10.1007/978-1-0716-0696-4_23,

The flow cytometric analysis of immune cells in induced sputum samples provides valuable information on the cellular composition of the inflammatory environment in the central and upper respiratory tract of patients suffering from airway inflammation. Importantly, induced sputum is a noninvasive procedure for immunophenotyping in the respiratory tract. However, analysis of sputum samples is not evident because of their highly heterogeneous composition. Induced sputum samples differ in their viscosity and exhibit a variable amount of mucus, squamous epithelial cells, and exogenous and cell-derived residues and thus require more intricate purification steps and data analysis as compared to detection of immune cells in blood or bronchioalveolar lavages (BAL) [4]. Using adequate purification and staining procedures, these hurdles can be overcome. Strict adherence to this protocol allows flow cytometry analysis of sputum basophils and other granulocytes and to study their involvement in airway inflammation.

2 Materials

All solutions for sputum purification have to be stored at 4 °C with the exception of the BD lysing solution, which should be kept at RT. For the preparation of the solutions, cold deionized water (dH_2O) or 1× phosphate-buffered saline (PBS) with a pH of 7.4 is recommended.

2.1 Sputum Induction

1. Nebulizer (PARI, Provocation Test II).
2. 0.9% isotonic saline solution.
3. 3% and 6% hypertonic saline solution.

2.2 Sputum Processing

1. 1 M 4-(2-hydroxyethyl)-1-piperazineethanesulfonic acid (HEPES) solution: Dissolve 11.915 g HEPES in 50 mL dH_2O.
2. 5 M NaCl solution: Dissolve 7.305 g NaCl in 25 mL dH_2O.
3. 1,4-Dithioerythritol (DTE) solution: Weigh 48 mg DTE and transfer to a graduated cylinder. Add 2.5 mL 1 M HEPES and 1.44 mL 5 M NaCl and top up to 50 mL with deionized water. Mix the solution until DTE is fully dissolved. The solution can be stored at 4 °C for at most one week (*see* **Note 1**).
4. Falcon 100 μm cell strainer, yellow, sterile (Corning).
5. EASYstrainer 70 μm for 50 mL tubes (Greiner Bio-One).

2.3 Cell Counting by Flow Cytometry

1. Anti-CD45 (APC/H7; clone 2D1, BD Bioscience).
2. CountBright Absolute Counting Beads (Invitrogen).

2.4 Surface Staining for Flow Cytometry

1. Intravenous immunoglobulin (IVIG) blocking solution: Prepare a 10% solution in 1× PBS. Prepare 100 μL per sample. Store at 4 °C.
2. Flow cytometry buffer (FCM): For the preparation of 250 mL of FCM, dissolve 186.1 mg EDTA (Fisher Scientific) in 200 mL 1× PBS in a graduated cylinder. Add 2.5 mL heat-inactivated fetal calf serum and top up to 250 mL. Store at 4 °C.
3. BD FACS lysing solution (BD Bioscience).
4. Propidium iodide (PI) solution: Dilute the stock solution to a final concentration of 50 μg/mL. Store at 4 °C in 2 mL black-tinted safe-lock tubes or alternatively protect a transparent tube with aluminum foil.
5. LIVE/DEAD™ Fixable Green Dead Cell Stain Kit (Invitrogen): Dissolve 1 μL of the stock solution in 100 μL 1× PBS (1:100 LD solution). If cell numbers vary between samples, scale PBS and LD solution accordingly. Cannot be stored (*see* **Note 2**).
6. The following antibodies can be used for this panel: Anti-CD123 (BV421; clone 6H6, Biolegend), anti-HLA-DR (BV510; clone G46-6, BD Bioscience), anti-CD193 (PE; clone 61,828, R&D Systems), anti-Siglec-8 (PerCp/Cy5.5; clone 7C9, Biolegend), anti-FcεRI (APC; clone AER-37, Invitrogen), anti-CD66b (PE/Cy7; clone G10F5, Biolegend), and anti-CD45 (APC/H7; clone 2D1, BD) (*see* **Note 3**).

3 Methods

3.1 General Information and Preparation for Sputum Induction

1. Sputum induction should be performed as early as possible in the morning since sputum accumulates during sleep at night.
2. Volunteers should avoid eating before sputum induction.
3. Ask volunteers to clean their nose and to perform a mouthwash by gargling water before starting the procedure.

3.2 Sputum Induction

1. Use the nebulizer as instructed by the study nurse. Each cycle should be performed for 7 min, whereby the concentration of the hypertonic saline solution is increased each time.
   (a) Healthy subjects: Start with 3% hypertonic saline solution and increase incrementally up to a maximum of 6% of NaCl.
   (b) Asthmatic subjects: Start with isotonic saline solution (0.9% NaCl) and if possible increase up to a maximum of 3% NaCl (*see* **Note 4**).

2. After each nebulizer cycle, a "huff coughing" should be performed while sitting:
   (a) Inhale deeply and slowly until 3/4 of the lungs are filled.
   (b) Hold the breath for 3 s.
   (c) Use strong but slow and even exhalation to carry the mucus out of the airways.
   (d) Repeat step c twice if necessary.
   (e) When the mucus dislodges, the test person should try to cough it out into a collection tube (*see* **Note 5**).
3. Either repeat the "huff coughing" technique or repeat inhalation using a higher content of NaCl with the nebulizer.
4. After three cycles, enough sputum should be produced.
5. Place sputum sample on ice (*see* **Note 6**).

3.3 Sputum Processing

1. Place the expectorated sputum in a Petri dish and try to spread the sample. Use a forceps to transfer the mucoid portion of the sputum sample into a 50 mL centrifugation tube (*see* **Note 7**).
2. Weigh the mucoid portion of the induced sputum sample in a 50 mL centrifugation tube.
3. Resuspend the sputum sample thoroughly in DTE solution. Add 4 mL of the DTE solution for each gram of sputum. Homogenize the sample by pipetting up and down.
4. Vortex the sample for 15 s and then place the tube on a bench rocker for 15 min at RT protected from light (*see* **Note 8**).
5. Add 20 mL of 1× PBS to reduce the viscosity of the samples before filtration.
6. Filter the DTE-processed induced sputum sample through 100 μm cell strainer that fits onto 50 mL centrifugation tubes. Rinse the cell strainer with 1× PBS (*see* **Note 9**).
7. Filter the DTE-processed induced sputum sample again through 70 μm cell strainer and rinse the filter with 1× PBS.
8. Centrifuge the filtrate at 400 × g for 15 min at 4 °C and carefully discard the supernatant (*see* **Note 10**).
9. Resuspend the cell pellet in 1 mL of 1× PBS (*see* **Note 11**).

3.4 Cell Counting by Flow Cytometry

1. Prepare an aliquot of the processed sputum sample for cell counting by flow cytometry by adding 10 μL of the cell suspension to 90 μL 1× PBS.
2. Stain the aliquot by using anti-CD45 for 15 min at RT in the dark.
3. Add 90 μL 1× PBS and 10 μL CountBright Absolute Counting Beads.

4. Add 2 μL of diluted PI solution (50 μg/mL) and stain for 30 s before flow cytometry analysis.
5. Acquire for 2 min at a high flow rate (based on BD FACS Canto II, BD Bioscience).
6. Count CD45-positive cells in Flow Jo and extrapolate to the total cell count in the sample using the number of analyzed Absolute Counting Beads according to the manufacturer's instructions. The fraction of CD45-positive events that are negative for PI represents the fraction of live leukocytes.
7. Spin down at 400 × *g* for 10 min at 4 °C, discard the supernatant, and seed CD45-positive cells at a density of 10 × 10^6 cells/mL.

3.5 Surface Staining for Flow Cytometry

1. Transfer cell suspension (10 × 10^6 cells/mL) into FACS tubes (*see* **Note 12**).
2. Rinse once with 1 mL 1 × PBS and centrifuge at 400 × *g* for 10 min at 4 °C.
3. Discard the supernatant and block nonspecific staining by resuspending the cell pellet in a 10% IVIG blocking solution. Use 100 μL of the blocking solution per 1 × 10^6 cells. Block the sample for 15 min at RT. Rinsing/washing of the blocking solution is not necessary.
4. Add 890 μL of 1× PBS and add 10 μL of the diluted (1:100) LD solution cells. Stain for 30 min on ice protected from light.
5. Rinse once in 1 mL 1× PBS and centrifuge at 400 × *g* for 10 min at 4 °C.
6. Discard the supernatant and resuspend the cell suspension in the antibody mix (*see* **Note 13**). Stain the sample for 25 min at 4 °C. Add the corresponding volume of buffer to unstained sample. From this point on, the samples should be protected from light.
7. Add 2 mL BD FACS lysing solution and incubate for 15 min at RT in the dark (*see* **Note 14**).
8. Rinse with 2 mL FCM buffer and centrifuge the sample at 400 × *g* for 10 min at 4 °C. Repeat this step twice to get rid of all the remaining BD FACS lysing solution.
9. Resuspend the cell suspension in 600 μL FCM buffer.
10. Acquire the processed and stained induced sputum samples with a flow cytometer equipped with an eight-color configuration matching the antibody fluorochromes (*see* **Note 15**).

3.6 Gating Strategy for Basophils, Eosinophils, and Neutrophils in Flow Jo or Equivalent

1. Exclude the debris from further analysis based on their light-scattering characteristics in the SSC-A and FSC-A channels (*see* Fig. 1a, R1) (*see* **Note 16**).
2. Exclude doublet events: Using the physical properties measured in the FSC-H and FSC-A channel, doublet events can be excluded from further analysis. In a pseudocolor plot, where the *x*- and *y*-axes are set to FSC-H or FSC-A, singlets will be arranged in a diagonal line (*see* Fig. 1b, R2).
3. Exclude dead cells based on a high staining intensity of the LD stain in the FITC-channel (*see* Fig. 1c, R3) (*see* **Note 17**).
4. Gate on CD45-positive events using the SSC-A and APC/H7 channel (*see* Fig. 1d, R4).
5. Display CD45-positive events of R4 with the *x*- and *y*-axis set to HLA-DR BV510 and CD66b PE/Cy7, respectively (*see* Fig. 1e).
6. Gate on basophils as follows (*see* Figs. 1e and 2):

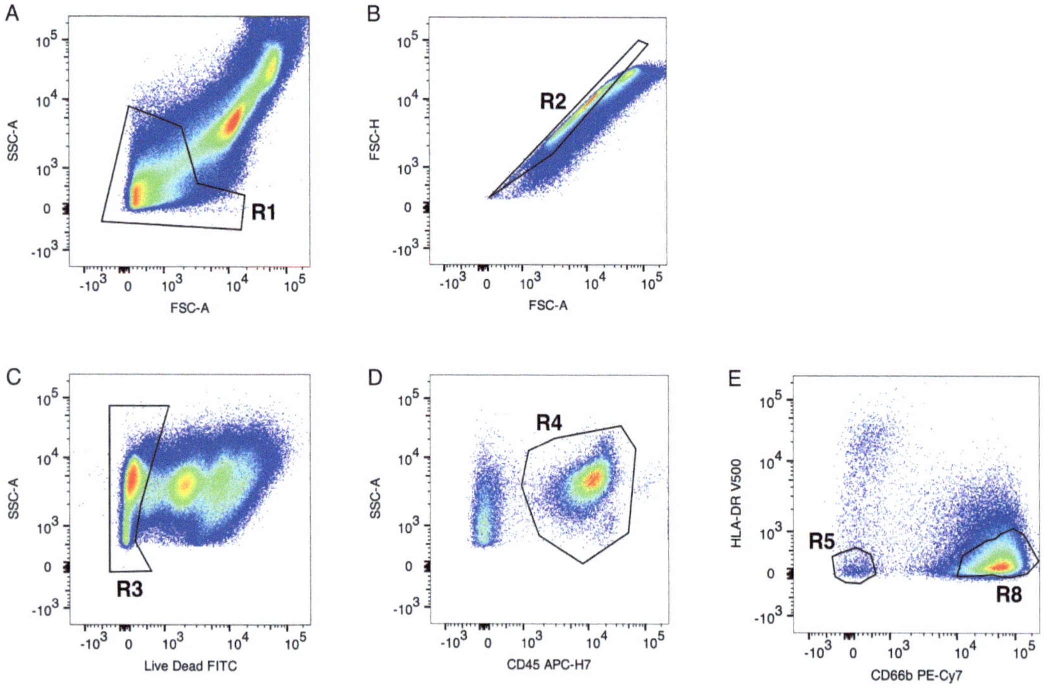

Fig. 1 Primordial gating strategy for all three types of granulocytes. (**a**) Exclusion of debris and cell fragments. R1 is a "NOT" Gate. (**b**) Gate used for the inclusion of singlets only (R2). (**c**) Live cells were selected (R3) based on negative staining for the LD solution. (**d**) Gating on CD45-positive events (R4) allows identification of leukocytes and discrimination of remaining cell debris or epithelial cells. (**e**) HLA-DR and CD66b double-negative events (R5) contain basophils. HLA-DR-negative and CD66b-positive events (R9) comprise eosinophils and neutrophils

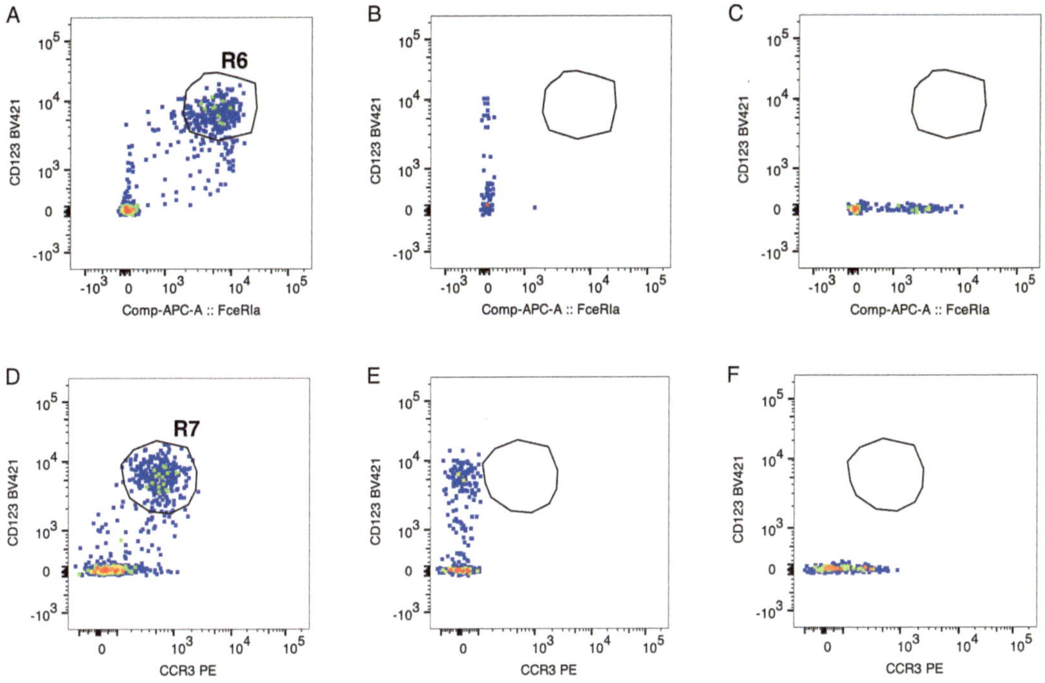

Fig. 2 Gating strategy for sputum basophils. (**a**) Basophils are identified based on high expression of CD123 and FcεRI (R6). (**b**, **c**) Depicted is the FMO control for the anti-FcεRI (APC) or the anti-CD123 (BV421) antibody, respectively. (**d**) In parallel to the gating strategy presented in (**a**), basophils are also identified based on the presence of CCR3 and CD123. (**e**, **f**) Shown is the FMO control for the anti-CD123 (BV421) or the anti-CCR3 (PE) antibody, respectively

(a) Display CD45-positive events of R4 (*see* Fig. 1e) with the *x*- and *y*-axis set to HLA-DR BV510 and CD66b PE/Cy7, respectively.

(b) Set R5 to HLA-DR/CD66b double-negative cells within CD45-positive cells (*see* Fig. 1e, R5).

(c) Display cells of R5 with the *x*- and *y*-axis set to FcεRI APC and CD123 BV421, respectively (*see* Fig. 2a).

(d) Set basophil gate to FcεRI/CD123 double-positive cells (R6). Use the corresponding FMOs to verify proper gating (*see* Fig. 2b, c) (*see* **Note 18**).

(e) Additionally, display cells of R5 with the *x*- and *y*-axis set to CCR3 PE and CD123 BV421, respectively (*see* Fig. 2d).

(f) Set basophil gate to CCR3/CD123 double-positive cells (R7). Use the corresponding FMOs to verify proper gating (*see* Fig. 2e, f) (*see* **Note 19**).

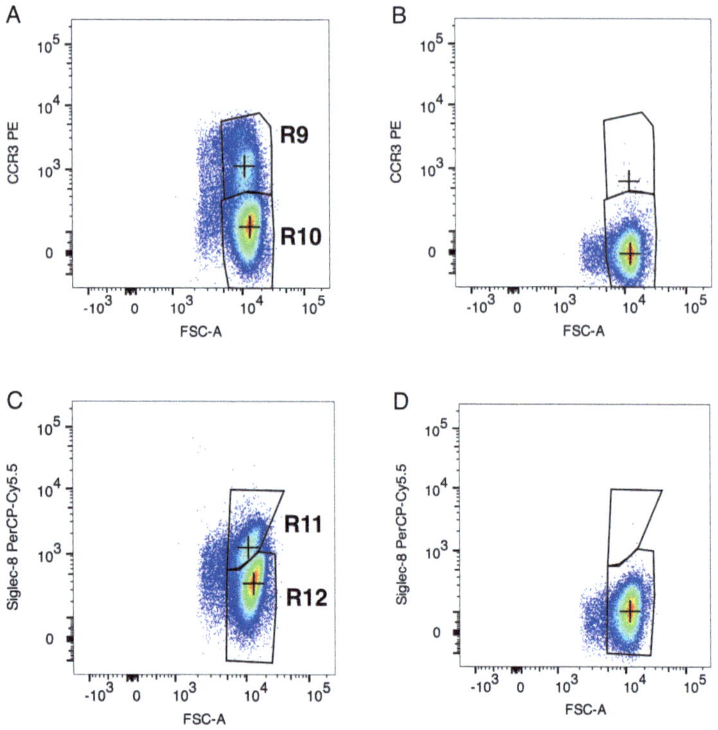

Fig. 3 Gating strategy for sputum eosinophils and neutrophils. (A) Within the CD66b-positive and HLA-DR-negative population, sputum eosinophils (R9) and neutrophils (R10) can be distinguished based on the presence or absence of CCR3. (**b**) Depicted is the FMO control for the anti-CCR3 antibody (PE), which was omitted from the antibody staining mix. (**c**) To clearly distinguish between the eosinophils and neutrophils, the variable expression intensity of Siglec-8 can be exploited. (**d**) Shown is the FMO control for the anti-Siglec-8 antibody (PerCP-Cy5.5). An exact determination of sputum eosinophils or neutrophils is obtained by using Boolean "AND" gates of R9 and R11 or R10 and R12, respectively

7. Gate on eosinophil and neutrophils as follows (*see* Figs. 1e and 3):
   (a) Display CD45-positive events of R4 (*see* Fig.1e) with the *x*- and *y*-axis set to HLA-DR BV510 and CD66b PE/Cy7, respectively.
   (b) Set R8 to HLA-DR-negative/CD66b-positive cells (*see* Fig. 1e, R8).
   (c) Display cells of R8 with the *x*- and *y*-axis set to FSC-A and CCR3 PE, respectively (*see* Fig. 3a).
   (d) Set R9 to CCR3 PE-positive/FSC-A cells for eosinophils and R10 to CCR3 PE-negative/FSC-A cells for neutrophils (*see* Fig. 3a). Use the corresponding FMOs to verify proper gating (*see* Fig. 3b).
   (e) Display cells of R8 with the *x*- and *y*-axis set to FSC-A and Siglec-8 PerCP/Cy5.5, respectively (*see* Fig. 3c).

(f) Set R11 to Siglec-8 PerCP/Cy5.5-positive/FSC-A cells for eosinophils and R12 to Siglec-8 PerCP/Cy5.5-negative/FSC-A cells for neutrophils (*see* Fig. 3c). Use the corresponding FMOs to verify proper gating (*see* Fig. 3d).

(g) Use a Boolean "AND" gate of Siglec-8-positive and CCR3-positive cells to determine eosinophils and a Boolean "AND" gate of Siglec-8- and CCR3-low/negative cells for neutrophils.

4 Notes

1. The DTE solution is photosensitive and should be kept on ice and protected from light in between usage. Inappropriate storage may impair its mucolytic activity. Seal the tube after usage as exposure to air can result in oxidization of DTE. Prepare DTE solution in advance as dissolving DTE can take some time. Always prepare fresh DTE solution if possible. Alternatively, the epimer of DTE, dithiothreitol (DTT), which has similar mucolytic activity may be used.
2. Since the diluted LD solution cannot be kept for further usage, it is recommended to prepare small aliquots of 1 μL of the stock solution, which is enough for the staining of 10×10^6 cells.
3. We highly recommend the titration of all antibodies to achieve an optimal resolution between the populations and to minimize nonspecific binding. The use of sputum cells is preferable for the titration of antibodies since the expression of certain surface markers (e.g., chemokine receptors) on immune cells originating either from sputum or blood can be quite different.
4. It is recommended to start with a physiological concentration of saline in patients with severe asthma as a higher concentration of the saline solution may cause irritations of the airways in asthmatics. Ask the patient whether the salt vapor negatively affects breathing before increasing the content of NaCl for the next cycle. In case the patient is very susceptible to the irritation from salt vapor, it is also advisable to perform FEV1 measurement after each cycle. If the FEV1 falls by over 20%, the procedure should be stopped.
5. If the cough remains dry, do not force it and do not harrumph. If saliva accumulates in the mouth due to the salt vapor, it should be disposed of in a separate collection cup to minimize squamous epithelial cell contamination.
6. Do not use cold packs kept at −20 °C.
7. Process the expectorated sputum as quickly as possible and avoid keeping the tube containing the sample for too long on ice, as the cold temperature makes the sample more viscous and

thus more difficult to separate the mucoid portion from saliva. Try to transfer as little saliva (clear liquid) as possible since it is rich in squamous epithelial cells from the mouth and throat region that will contaminate the sample and therefore make data analysis more complicated. Using a thin, curved forceps allows greater precision and minimizes the amount of saliva that is transferred. Try to pick the yellow, green, or gray mucoid plugs by squeezing the tip of the forceps (the curved end of the forceps should face down) underneath the plug, which will push the plug on top of the forceps. Place the mucoid plug into the 50 mL tube by swiping it to its edges.

8. Prolonged exposure to DTE or DTT should be avoided as it negatively affects cell viability and may change the surface density of certain antigens and thus negatively impact the fluorescence intensity of certain staining [5].
9. Depending on the viscosity of the induced sputum sample, more than one cell strainer might be required.
10. The supernatant should be discarded carefully. Ideally, pour the supernatant into the sink while aspirating with a laboratory liquid suction system; therefore, it is not necessary to get too close to the pellet. If the liquid is still opaque before discarding the supernatant, the induced sputum sample should be rinsed again with 50 mL 1× PBS and centrifuged at 400 × g for 15 min at 4 °C.
11. If the pellet is small, the volume in which the cells are resuspended can also be reduced. There is no reliable method to estimate the number of cells in the pellet before counting; however, we recommend using 1 mL if the pellet is visible to the naked eye.
12. At least 1.5×10^6 CD45 cells should be used for the fully stained samples. A lower number of cells may be used for the fluorescence minus one controls (FMO) if not otherwise possible.
13. We recommend using fluorescence minus one (FMO) samples to verify and confirm the gating strategies. In our experience, isotype controls are less powerful to assess the quality of the flow cytometry staining.
14. BD FACS lysing solution contains 15% of formaldehyde and thus also acts as a fixation solution. It is crucial to resuspend the cell pellet thoroughly before and after the fixation step, to prevent cells from clumping.
15. Cells in induced sputum samples exhibit quite different light-scattering characteristics compared to blood leukocytes. Make sure to adjust the forward (FSC) and side scatter channels (SSC) of the flow cytometer in such a way that all the relevant

cells are in scale (e.g., log scale). Furthermore, it is essential to activate the width and height of the FSC and SSC channels to exclude doublets.

16. Gating sputum granulocyte populations can be quite complex as populations are often small and overlap more in comparison to the typical blood granulocyte population.
17. It is recommended to use a pseudocolor plot of the FITC-channel and the SSC-A rather than a histogram since a clear threshold between live and dead cells is rarely observed in induced sputum samples.
18. Since lung mast cells exhibit very low levels of the IL-3 receptor α-chain (CD123), the combination of CD123 and the high-affinity IgE receptor (FcεRI) is a reliable gating strategy for sputum basophils.
19. In some cases, sputum basophils exhibit low surface levels of CCR3, and therefore this marker is not always suitable.

Acknowledgments

We thank the Center of Laboratory Medicine and the Clinical Cytomics Facility for their support.

References

1. Brooks CR, van Dalen CJ, Hermans IF et al (2017) Sputum basophils are increased in eosinophilic asthma compared with non-eosinophilic asthma phenotypes. Allergy 72:1583–1586. https://doi.org/10.1111/all.13185
2. Gauvreau GM, Lee JM, Watson RM et al (2000) Increased numbers of both airway basophils and mast cells in sputum after allergen inhalation challenge of atopic asthmatics. Am J Respir Crit Care Med 161:1473–1478. https://doi.org/10.1164/ajrccm.161.5.9908090
3. Maruyama N, Tamura G, Aizawa T et al (1994) Accumulation of basophils and their chemotactic activity in the airways during natural airway narrowing in asthmatic individuals. Am J Respir Crit Care Med 150:1086–1093. https://doi.org/10.1164/ajrccm.150.4.7921441
4. Lay JC, Peden DB, Alexis NE (2011) Flow cytometry of sputum: assessing inflammation and immune response elements in the bronchial airways. Inhal Toxicol 23:392–406. https://doi.org/10.3109/08958378.2011.575568
5. Qiu D, Tan WC (1999) Dithiothreitol has a dose-response effect on cell surface antigen expression. J Allergy Clin Immunol 103:873–876. https://doi.org/10.1016/s0091-6749(99)70432-x

Chapter 24

Mast Cell Migration and Chemotaxis Assayed by Microscopy

Monika Bambousková, Zuzana Rubíková, Lubica Dráberová, Pavel Dráber, and Petr Dráber

Abstract

A better understanding of the molecular mechanisms leading to mast cell migration and chemotaxis is the long-term goal in mast cell research and is essential for comprehension of mast cell function in health and disease. Various techniques have been developed in recent decades for in vitro and in vivo assessment of mast cell motility and chemotaxis. In this chapter, three microscopy assays facilitating real-time quantification of mast cell chemotaxis and migration are described, focusing on individual cell tracking and data analysis.

Key words Mast cells, Cell migration, Chemotaxis, Chemokine, Chemoattractant, Cell tracking

1 Introduction

Mast cells are best known for their role in mediating allergic diseases [1]. Recent studies have highlighted their important functions in innate immunity contributing to protection against bacterial infection, angiogenesis, and autoimmunity [2, 3]. Under pathological conditions, mast cells are involved in immediate hypersensitivity, chronic allergic reactions, asthma, and other inflammatory diseases [4]. As highly effective sentinel cells, they are localized close to blood vessels in common sites of potential infection, such as skin, airways, and the gastrointestinal tract. Chemoattractant-directed migration of mature mast cells or their progenitors might be one of the key mechanisms responsible for local accumulation of these cells under physiological or pathological conditions [5]. Chemotactic mechanisms can be, therefore, potentially targeted by therapeutic interventions.

$CD13^+/CD34^+/CD117$ (c-Kit)$^+$ mast cell progenitors arise in bone marrow where they start their differentiation under the

Monika Bambousková and Zuzana Rubíková contributed equally to this work.

Bernhard F. Gibbs and Franco H. Falcone (eds.), *Basophils and Mast Cells: Methods and Protocols*, Methods in Molecular Biology, vol. 2163, https://doi.org/10.1007/978-1-0716-0696-4_24, © Springer Science+Business Media, LLC, part of Springer Nature 2020

influence of chemokines and growth factors [6]. During this process, they leave the bone marrow and migrate into circulation to reach various body sites. The key signal for mast cell homing and recruitment into peripheral tissues is provided by the interaction of stem cell factor (SCF) with its receptor, Kit [7]. This receptor is also involved in mast cell priming [8]. In rodents, another critical factor, interleukin-3 (IL-3), which binds to its surface receptor is involved [9]. In peripheral tissues, mast cell progenitors mature and terminally differentiate under the influence of the local environment. Two predominant phenotypes of mature mast cells, the connective tissue and the mucosal mast cells, differ by the expression of various secretory granule proteins [10]. Migration of mast cell progenitors to the site of their residency is directed by various chemokines which bind to chemokine receptors [11]. The expression pattern of these molecules is different in progenitors, and mature mast cells and differences are also observed between particular mast cell subtypes [12]. Several other chemotactic ligands affecting mast cell migration have been identified; these include sphingosine-1-phosphate [13], arachidonic acid metabolites such as leukotriene B4 [14], and prostaglandin E_2 (PGE_2) [15–17], as well as antigen (Ag) that is recognized by IgE anchored to the high-affinity IgE receptor (FcεRI) [18, 19]. In target tissues, mast cells participate actively in immune responses against various pathogens by releasing a broad spectrum of mediators, either stored in secretory granules or rapidly synthesized after cell triggering. Mediators produced by mast cells include leukotrienes, prostaglandins, cytokines, and chemokines, which in turn can recruit other immune cells to sites of inflammation [20, 21]. Importantly, mature mast cells in sensitized individuals exposed to allergens produce mediators attracting mast cell-committed progenitors to the place of pathogen entry [14]. In healthy tissues, the number of mast cells is stable and can be regulated by proliferation, migration, and mast cell survival. On the other hand, inflamed tissues show an increased number of mast cells, and their enhanced accumulation is characteristic for a number of pathological states [22, 23].

Numerous methods have been developed for studies on mast cell migration and chemotaxis toward various chemoattractants. Most in vitro studies utilized various modifications of a Boyden chamber system where cells migrate through pores (usually 5 or 8 μm) in a polycarbonate membrane toward increasing concentrations of chemoattractants [24]. These techniques are well established for quantification of chemotactic responses [14, 25–27]. However, they do not allow recording the movement of individual cells at selected time intervals and analyzing cell trajectories. Therefore, new methods have been developed for analysis of chemotaxis using real-time imaging. The gradient of chemoattractant is usually created by arranging special chambers in various settings [28]. Alternatively, a gradient is produced by embedding chemoattractants in porous materials such as agarose [29–31]. Using a microscope equipped with a camera and proper software, the images are processed with image analysis software capable of obtaining coordinates of individual trajectories. They are then used for calculation of cell movement parameters (e.g., directionality, achieved distance, velocity, etc.) [28, 32, 33]. Since primary mast cells are nonadherent, the 2D or 3D tracking of their movement encounters some difficulties. To overcome these problems, mast cells are attached to the surface with some natural adhesive compounds, such as fibronectin. Furthermore, cell attachment has to be preserved during the whole experiment to prevent any artificial movement of cells during analysis. The combined efforts resulted in the development of new methods for analysis of migration and chemotaxis of nonadherent mast cells. In this chapter, we describe several experimental settings for real-time analysis of cell motility or directed chemotaxis optimized for nonadherent mast cells. The techniques utilize commercially available migration chambers or simple homemade devices.

2 Materials

2.1 Chemicals and Kits

1. Fibronectin from bovine plasma for cell cultures (Sigma-Aldrich; Cat. No. F1141).
2. Recombinant murine IL-3 (PeproTech EC, London; U.K.; Cat. No. 213-13).
3. Recombinant murine SCF (PeproTech; Cat. No. 250-03). Stock solution containing 1 mg/ml of SCF in H_2O is aliquoted to avoid repeated freeze-thaw cycles.
4. Agarose.
5. Collagen I (Advanced BioMatrix; Cat. No. 5005; 3.1 mg/ml).
6. PGE_2; 1 mM stock solution is prepared in DMSO.
7. Miltefosine; a 10 mM stock solution is prepared in H_2O.
8. Mouse monoclonal antibody SPE-7 (IgE) specific for dinitrophenyl (DNP; Sigma-Aldrich; Cat. No D8406); 1 mg/ml stock.
9. DNP-albumin, used as an antigen (Ag) (Sigma-Aldrich; Cat. No A6661). 0.1 mg/ml stock solution is prepared in H_2O.

All other chemicals of the highest purity were obtained from Sigma-Aldrich.

2.2 Buffers and Media

1. 1 M HEPES solution.
2. RPMI-1640.
3. Minimum Essential Medium (MEM) nonessential amino acids.

4. 10× concentrated MEM Eagle with Hanks' salts (H-MEM).
5. Migration buffer (MB): RPMI-1640, 0.1% BSA (w/v), 20 mM HEPES, pH 7.2.
6. StemPro-34® serum-free medium (SFM; Gibco; Cat. No. 10639).

2.3 Mast Cells

1. Bone marrow from femurs and tibias of 6- to 10-week-old mice is used as a source of mast cell precursors. The cells are cultured in RPMI-1640 supplemented with 20 mM HEPES, pH 7.5, 100 U/ml penicillin, 100 μg/ml streptomycin, 100 μM MEM nonessential amino acids, 1 mM sodium pyruvate, 41 μM 2-mercaptoethanol, 10% (v/v) fetal calf serum (FCS), IL-3 (20 ng/ml), and SCF (15 ng/ml). After 5–7 weeks, most of the cultured cells (up to 95%) express FcεRI and c-Kit, markers of mast cells. Such bone marrow-derived mast cells (BMMCs) have been extensively used in studies on cell signaling since they closely resemble normal mast cells in respect to their requirement of IL-3 and SCF for optimal growth, their ability to respond to various physiological stimuli, and particularly because they can be isolated from mouse strains deficient in genes which are possibly involved in a variety of signaling pathways.
2. A particular cell line derived from mouse BMMCs (BMMCL) has been isolated by M. Hibbs [34]. These cells are cultured in complete RPMI-1640 supplemented as described above but without SCF and with recombinant IL-3 replaced by 10% (v/v) WEHI-3 cell supernatant. These cells can be recovered after freezing in 10% (v/v) DMSO and storage at −70 °C or liquid nitrogen. BMMCL can be used instead of BMMCs for their easier culturing and fast growth. Furthermore, they react in almost the same way as BMMCs to a variety of treatments [35].
3. The human mast cell line (LUVA) was obtained by growing $CD34^+$-enriched mononuclear cells derived from peripheral blood of a human donor [36]. LUVA cells are maintained in StemPro-34 SFM in the absence of any exogenously added cytokines. Although they have morphology of mast cells, after several passages in vitro, they lost surface expression of FcεRI (our own unpublished data).

2.4 Other Materials

1. Test tubes and 1.5 ml Eppendorf tubes.
2. 96-well black plates with optical bottom suitable for microscopy.
3. Beveled pipette tips for filling ports on μ-Slide.
4. Slant tweezers for plug handling.
5. Cell culture dishes 40 × 11 mm.

6. Glass beads (2 mm in diameter).
7. μ-Slide Chemotaxis^{3D} chambers (Ibidi; Cat. No. 80326) containing μ-Slide^{3D} for evaluation of cell chemotaxis in three samples, a cultivation lid and 18 plugs.

2.5 Equipment

1. Laminar flow box.
2. CO_2 incubator.
3. Inverted microscope (Leica DMI6000 or equivalent) equipped with 10× bright field objective camera (Leica DFC350FX R2 or equivalent) for recording time-lapse video, motorized stage, and CO_2- and temperature-regulated chamber.
4. Scan^R system (Olympus, Hamburg, Germany) equipped with 10× and 20× bright field objective camera (Olympus DP80 or equivalent) for recording time-lapse video, motorized stage, CO_2- and temperature-regulated chamber, and platform for 96-well plate samples.
5. Milli-Q water purification system (Millipore S.A., Molsheim, France).

2.6 Software

1. MetaMorph Offline (Version 7.7.9.0; Molecular Devices).
2. ImageJ (Version 1.46; National Institutes of Health) [37] with Chemotaxis and Migration Tool plugin (Version 1.01; Ibidi).
3. LAS AF Lite (Version 1.7.0; Leica Microsystems).

3 Methods

For quantitative studies of mast cell migration and chemotaxis by microscopy, three assays are presented: simple motility assay, chemotaxis measured in μ-Slide chemotaxis chamber, and chemotaxis analyzed in homemade chamber with chemoattractants embedded in agarose cone.

3.1 Motility Assay

This technique is used to determine the general ability of mast cells to move. It is useful when cells of different properties are investigated or if activity of selected drugs on cell movement is tested. The method allows measurement in 96-well plate format. Mast cells are allowed to attach to fibronectin-coated surfaces and are treated with an activator such as PGE_2. The velocities of cell movement and/or the distances reached from the start point are automatically determined (Fig. 1). However, no information on directionality of chemotaxis is provided.

All solutions intended for handling live cells and the microscope heating chamber should be pre-warmed to 37 °C.

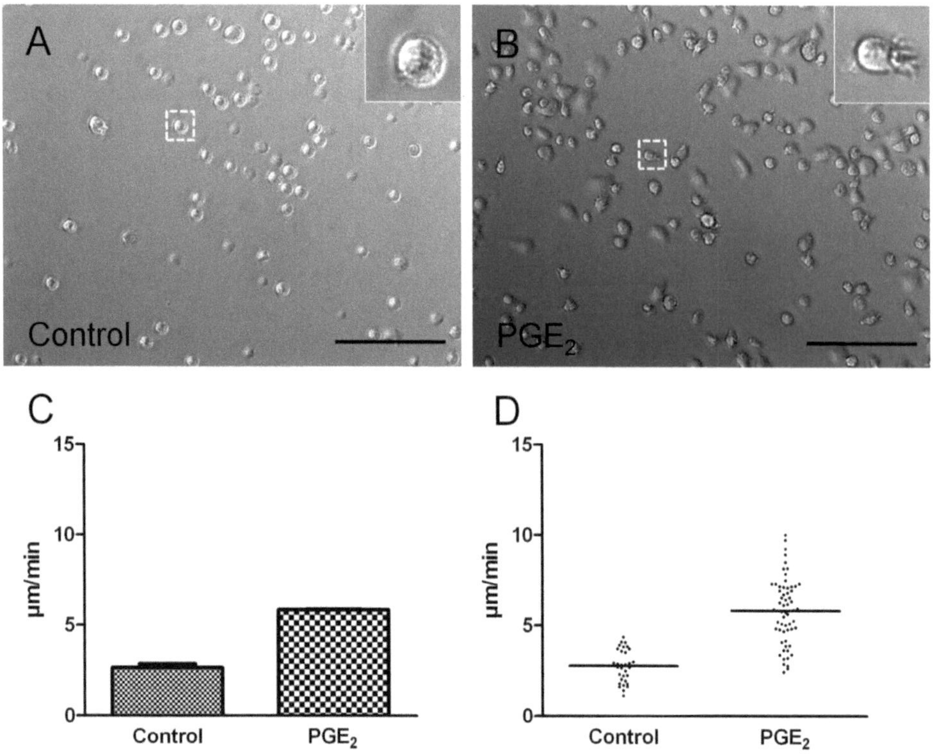

Fig. 1 Increased motility of BMMCs after PGE_2 activation. (**A**) Nonactivated (Control) or (**B**) activated (PGE_2) BMMCs attached to fibronectin-coated surface were monitored with $Scan^R$ system with a 20× objective. Time-lapse images were taken for 60 min with the frequency of 1.27 min/frame. Representative cells magnified and shown in inserts (**A**, **B**; right upper corners). (**C**) Individual trajectories of 30–50 cells were processed, and the average velocity ± SD was calculated from duplicates in one representative experiment. (**D**) Distribution of velocities of individual cells is presented with vertical scatter plot; average values of all analyzed cells are also shown. Bars in **A** and **B**, 100 μm

1. Prepare fibronectin-coated 96-well optical-bottom plate by adding 50 μl of 50 μg/ml fibronectin in PBS into each well of the plate and incubate for 14–16 h at 4 °C (*see* **Note 1**).
2. Grow BMMCs at concentration 2×10^6/ml without SCF and IL-3 for 12–16 h before experiment (*see* **Note 2**).
3. Wash the cells twice with MB, resuspend in MB, and adjust the concentration to $5 \times 10^5\ ml^{-1}$.
4. Wash the fibronectin-coated surfaces briefly with PBS and pipet 100 μl of cell suspension into each well.
5. Allow the cells to attach for 30 min at 37 °C and then gently wash the wells once with warm MB (*see* **Note 3**).
6. Add carefully 100 μl of MB or activator (e.g., PGE_2) diluted in MB into each well (*see* **Note 4**).

7. Monitor the cell motility on Scan^R system with 10× or 20× objective and define the number of examined sites per well and their *XYZ* positions. Take time-lapse images for 1–2 h at 1-min intervals (*see* **Note 5**).
8. Analyze the data using the procedure described in Subheading 3.4.

3.2 Mast Cell Chemotaxis in μ-Slide Chemotaxis^{3D} Chambers

The μ-Slide Chemotaxis^{3D} chambers make it possible to study directed migration of nonadherent cells toward the chemoattractant. Each μ-Slide^{3D} contains three identical chambers for parallel monitoring of samples. Cells are kept in a chamber using 3D collagen I gel. The unique working arrangement of μ-Slide^{3D} enables to create the required chemoattractant gradient. Mast cell movement is monitored by time-lapse microscopy, followed by data analysis (*see* Subheadings 3.4 and 3.5). Here we describe protocols to analyze chemotaxis toward SCF, PGE_2, or Ag. Protocol for the evaluation of inhibition of chemotaxis to Ag by miltefosine, a new candidate for the treatment of mast-cell-driven diseases [38], is also provided.

All procedures with μ-Slide Chemotaxis^{3D} chambers are performed in laminar flow box to maintain sterility.

1. One day before experiment, place the culture medium for BMMCs, BMMCL, or LUVA cells, and sterile H_2O for tissue culture and the μ-Slide^{3D} (Fig. 2a) into incubator for gas equilibration. Aqueous solutions should be transferred into semi-closed test tubes, while the μ-Slide^{3D} should stay packaged to maintain sterility (*see* **Note 6**).

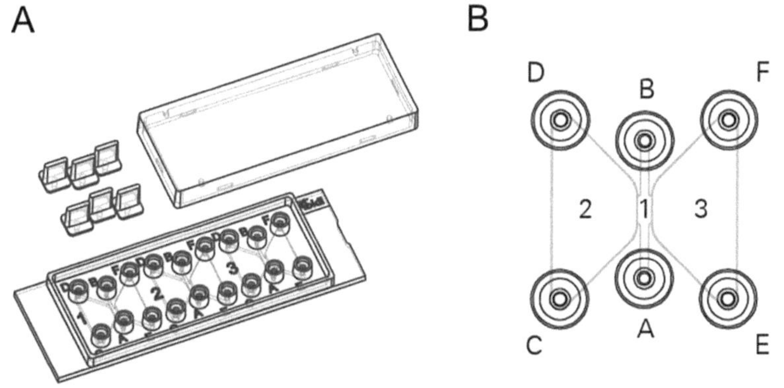

Fig. 2 Components of μ-Slide Chemotaxis^{3D} and configuration of a chamber for evaluation of cell motility in one tested sample. (**A**) μ-Slide Chemotaxis^{3D} contains a slide with chambers for simultaneous evaluation of three samples (Areas 1–3), six plugs, and a lid. (**B**) Chamber configuration. 1, channel (observation area); 2, left reservoir; 3, right reservoir; A–F in Figure **B**, filling ports

2. For chemotaxis toward Ag, dilute DNP-specific IgE stock 1:1000 in culture medium without 10% WEHI-3 cell supernatant to get 1 μg/ml IgE. Sensitize cells at a concentration of 5–10 × 10^6 cells/ml in this solution for 4 h in the CO_2 incubator.
3. Prepare mast cell suspension at a concentration of 9 × 10^6 cells/ml in proper equilibrated culture medium (*see* **Note 7**). From this step, use culture medium with 0.5% FCS, without 10% WEHI-3 cell supernatant, and without IgE to analyze chemotaxis toward Ag.
4. Mix 6 μl of 10× H-MEM, 21.9 μl of sterile H_2O for tissue culture, 2.1 μl of 7.5% (w/v) $NaHCO_3$, and 15 μl of proper culture medium in sterile Eppendorf tube (premix; total volume 45 μl). Pipet 30 μl of collagen I into a second Eppendorf tube (*see* **Note 8**).
5. Using tweezers, close the filling ports C, D, E, and F (Fig. 2b) with the plugs in μ-Slide^{3D}. Prepare a humid chamber by placing moistened tissue into a sterile 10 cm Petri dish.
6. Pipet 45 μl of the premix into Eppendorf tube containing 30 μl of collagen I, and immediately add 15 μl of cell suspension. In the case of treating cells with miltefosine (final concentration of 15 μM), add 2.7 μl of miltefosine pre-diluted 1:20. Mix well with a 100 μl pipette tip (*see* **Note 9**).
7. Apply 6 μl of the mixture to the top of filling port A (use beveled pipette tip). Immediately after loading, aspirate the air from the opposite port B to fill the channel (#1 in Fig. 2b) homogenously with the mixture (*see* **Note 10**). Repeat this step in all three chambers using proper premix (with or without the drug).
8. Remove all plugs from the slide and close filling ports A and B with plugs (filling ports A and B will stay closed till the end of the experiment). Insert the slide into the humid chamber and put into CO_2 incubator for 30–40 min (*see* **Note 11**).
9. Meanwhile, dilute chemoattractants (SCF, PGE_2, or Ag) in proper culture medium to get 200 ng/ml SCF, 0.2 μM PGE_2, or 200 ng/ml Ag, respectively. These concentrations correspond to 2× concentrated chemoattractants used during imaging. In drug-treated cells, add 15 μM miltefosine (final concentration; 1:667 dilution). Prepare 40 μl of the chemoattractants for one chamber (*see* **Note 12**).
10. After collagen polymerization, fill the reservoirs (#2 and #3 in Fig. 2b). First, close reservoir #2 with plugs (close filling ports C, D) and directly pipet 65 μl of chemoattractant-free medium through one of the two opened filling ports (E or F). Then remove plugs from filling ports C, D and close filling

ports E, F. Fill reservoir #2 with 65 μl of chemoattractant-free medium. Repeat this step twice to fill reservoirs in all three chambers. For drug treatment, add miltefosine at a final concentration 15 μM into chemoattractant-free medium before filling the reservoirs.

11. To create the chemoattractant gradient, apply 15 μl of 2× concentrated chemoattractant (e.g., SCF, PGE_2, or Ag) with or without miltefosine to the top of the filling port C, and immediately aspirate 15 μl of the medium from the filling port D. This time press the pipette tip directly inside the port. Repeating this step leads to filling of 30 μl of 2× concentrated chemoattractant into the reservoir. An equilibrium gradient of chemoattractant is created during 1–2 h (*see* **Note 13**).
12. Monitor cell chemotaxis with Leica DMI6000 inverted microscope immediately after chemoattractants are added. Define *XYZ* positions of channels (observation areas) of all three chambers. Take time-lapse images for 7–10 h at 1- to 2-min intervals with the intensity of 71 and exposure time 6 ms. Autofocus can be used during the time-lapse (*see* **Note 14**).
13. Analyze the data using procedure described in Subheading 3.4.

Further information about this method is available online (*see* **Note 15**).

3.3 Chemotaxis in Homemade Chambers

This method allows characterization of the directional movement of cells toward a chemoattractant embedded in an agarose cone. After attachment of cells to a fibronectin-coated surface, the chemoattractant-embedded agarose cone is added. Movement of cells starts after an initial 15-min period of sensing the chemoattractant. As a relatively simple and cheap alternative to commercial systems, this method can be used for observation of cell behavior and changes in morphology during chemotaxis.

1. Prepare fibronectin-coated culture dish by adding 1 ml of fibronectin (50 μg/ml) in PBS and incubate it for 12–16 h at 4 °C (*see* **Note 1**).
2. Grow BMMCs at concentration 2×10^6/ml without SCF and IL-3 for 12–16 h (*see* **Note 2**).
3. For preparation of agarose cone with embedded chemoattractant, dissolve 1% (w/v) agarose in PBS by short heating at 80 °C, followed by slow chilling to approximately 40 °C. Keep the agarose solution at the same temperature and dilute PGE_2 in agarose solution to final 1 μM concentration. Dispense 15 μl aliquots of this solution to the bottom of 1.5 ml Eppendorf tubes (tubes should be pre-warmed to 40 °C). Immediately insert a glass bead into the agarose solution at

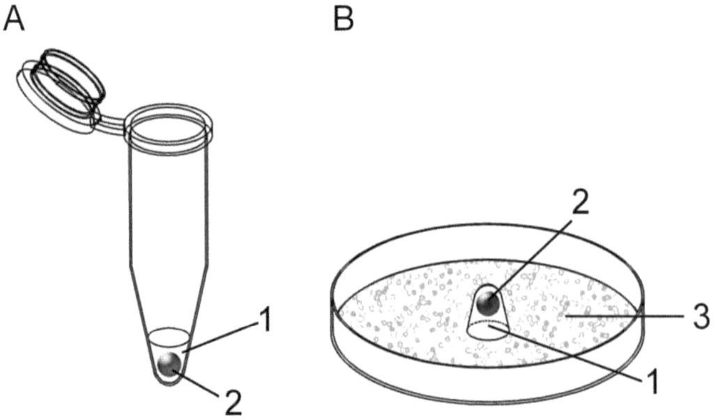

Fig. 3 Agarose cone with embedded chemoattractant. (**A**) Agarose cone with embedded chemoattractant and glass bead in an Eppendorf tube; (**B**) position of the agarose cone in a culture dish with cells and medium; 1, chemoattractant in 1% agarose; 2, glass bead; 3, cell layer

the bottom of each tube (Fig. 3a). The glass bead acts as a weight and will help stabilize the cone at the same position in culture dish during the test. Prepare control cones without chemoattractant in the same way. Let the agarose solidify in the tubes at 4 °C for at least 15 min and keep it so until use (*see* **Note 16**).

4. Wash the cells with MB and adjust their concentration to 5×10^5/ml.
5. Wash the fibronectin-coated dish with 1 ml of PBS before attaching the cells.
6. Dispense 2 ml aliquots of cell suspension (1×10^6 cells) into each dish and incubate undisturbed for 30 min at 37 °C.
7. Gently wash the dish with 1 ml of MB and add 5 ml of new MB.
8. Fix the dish in a proper position in microscope. Carefully remove the agarose cone containing the glass bead from the Eppendorf tube by using tweezers, and place it slowly in the middle of the dish with the flat bottom facing down the cell layer (Fig. 3b). Cover the dish with a lid.
9. Find the edge of agarose cone in the field of view of an inverted microscope with a 10× objective (Fig. 4a) and monitor the cells in time-lapse images for 2–3 h at 1-min intervals (Fig. 4b).
10. Analyze the data as described in Subheading 3.4. Start the analysis at frame 60 and end with 120–180 frames (*see* **Note 17**).

3.4 Tracking Cells

Automatic tracking in the MetaMorph program with optimized parameters for mast cells is described hereafter. However, there are also other options how to track cells in bright field time-lapse movies (*see* **Note 18**).

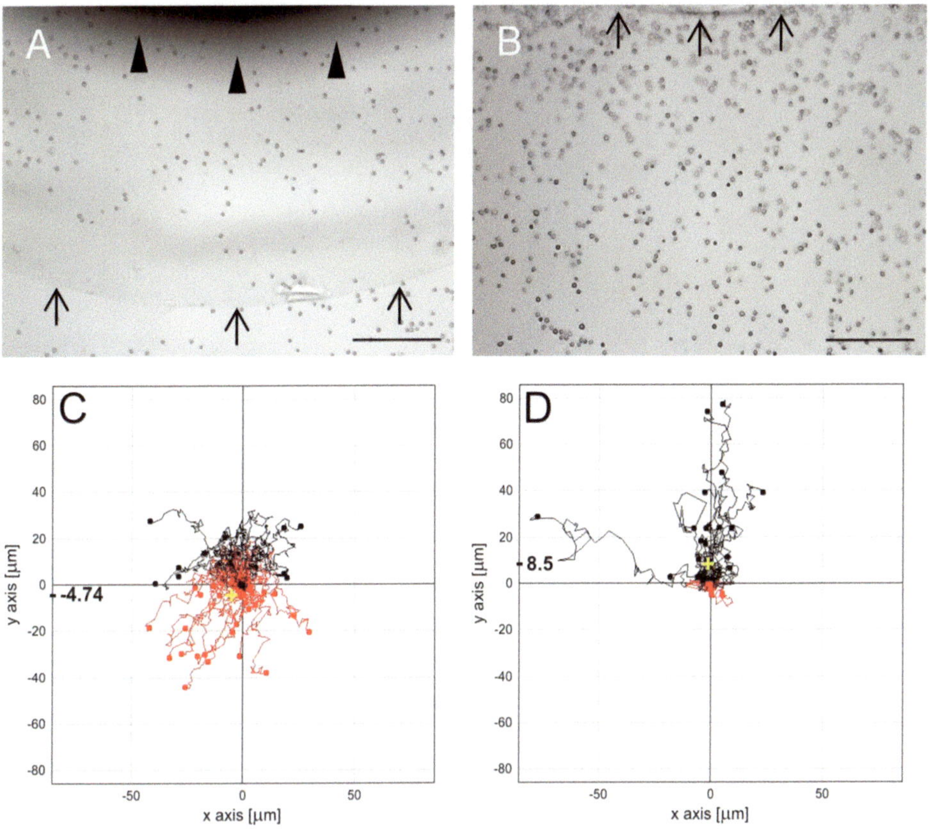

Fig. 4 Mast cell chemotaxis toward PGE_2 embedded in an agarose cone. (**A**) The edge of the agarose cone is marked by arrows and the glass bead by triangles. (**B**) BMMCs were attached to a fibronectin-coated surface, and the agarose cone with embedded PGE_2 was inserted into dish with cells. The cone edge is indicated by arrows. The image was captured 15 min after insertion of the cone (when the first changes in cell morphology were observed in the vicinity of the agarose cone). The images were taken with a Leica inverted microscope equipped with a 10× objective. (**C**, **D**) Representative images from a single experiment showing migration trajectory graphs expressing directionality of BMMC movement. Chemotaxis toward PGE_2 is reflected by a shift of a center of cell mass in *x* axis (yellow crosses). Bars represent 250 µm

[*Italic font style is used for buttons, radio buttons, and headlines. Underlined text refers to dialog boxes.*]

1. Open the time-lapse sequence in the MetaMorph program. Choose *Threshold* → *Threshold Image*. Mark off *Inclusive* and use the slider to adjust threshold so as only cells are marked (Fig. 5a) (*see* **Note 19**).
2. Select *Apps* → *Track Objects* and set the following parameters in Track Objects dialog box:

   In *Source for images*, mark off *Stack*.

   (a) Using the *Search Options* button, open the Search Options dialog box and in *Algorithm* set *Threshold Results*. Then choose what the program should do in *If object not*

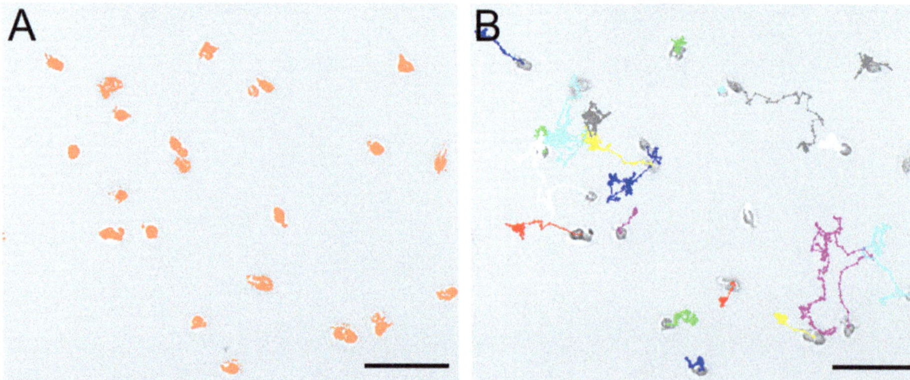

Fig. 5 Analysis of cell migration with MetaMorph. (**A**) Representative threshold image of time-lapse sequence. (**B**) Trajectories of individual cells. Chemotaxis of BMMCs toward 100 μM PGE_2 loaded into the left reservoir of a μ-Slide Chemotaxis^{3D} chamber. Images were taken with a Leica inverted microscope with 10× objective. Bars, 100 μm

found (e.g., *Select position*). Also set *Delay* (e.g., 0 s/object) and *Object size match requirement (as %)* (e.g., [38]).

(b) Set other parameters in Track Objects Interval Options, Track Overlay Options, and Origin Options dialog boxes by choosing *Set Interval*, *Set Overlay*, and *Set Origin* buttons, respectively.

(c) Choose *Track* to select tracked cells.

3. Select Objects dialog box appears. To choose objects to be tracked, hold down the [Ctrl] and click on each object using left mouse button. Adjust the size of the search regions (*see* **Note 20**). Then choose *OK* and object tracking will proceed automatically.

4. It is recommended to follow tracking during whole analysis. In case MetaMorph tracks in the wrong way (e.g., skip to another cell), press [Esc] to stop tracking. Tracking dialog box opens. If necessary, press the *Step Back* to return the analysis to the last correctly tracked plane. Then do a correction by hand (move the search region to the correct position using left mouse button) and continue in the analysis by choosing *Update Position And Continue* in Tracking dialog box. In such cases, it can be worth to mark off the *Continue one step and pause* in the Tracking dialog box.

5. If MetaMorph loses an object during the tracking, it proceeds according to the previous selection in the *If object not found* options in Search Options dialog box. In case *Select position* is marked off, it is necessary to do the correction by hand as described in **step 4**.

6. After the tracking is finished, all trajectories of individually tracked cells are marked in the time-lapse movie (Fig. 5b), and Track Objects dialog box appears again.
   (a) Select *Open Log*, define where the data should be saved, and blank Excel file opens. Thereafter, choose *Log Data* to save all the data from tracking. Sometimes it is easier to track only several cells at one time and each time log newly tracked data to the same Excel document.

3.5 Presenting Chemotaxis Data

Chemotaxis is analyzed by means of the Chemotaxis and Migration Tool. This is free software for ImageJ that can be downloaded at https://ibidi.com/chemotaxis-analysis/171-chemotaxis-and-migration-tool.html.

[Italic font style is used for buttons, check boxes, and headlines. Underlined text refers to bookmarks.]

Download and open the Chemotaxis and Migration Tool in ImageJ. Import the data table from MetaMorph by choosing *Import Data* in Import dataset bookmark (*see* **Note 21**).

1. Choose which cell tracks from the data table will be analyzed by filling in the *Number of slices*. Only specified tracks with exact number or within an entered range of slices will be used. Then choose *Add dataset* and mark off *Selected Dataset* in the main panel (*see* **Note 22**).
2. Calibrate the software in Settings bookmark. *XY* coordinates from MetaMorph are in pixels; it is therefore necessary to define the size of one pixel in *X/Y Calibration*. Also fill in *Time interval* that was used during time-lapse monitoring. Afterward, choose *Apply settings* in the main panel.
3. To analyze only selected time intervals from an imported time-lapse, choose *Open restrictions* in the main panel, mark off *Split dataset* check box, and fill in the required slices (*see* **Note 23**). Then choose *Apply settings* in the main panel.
4. Now the tool is prepared to plot graphs and diagrams and export various statistic values. One can plot Chemotaxis Plot in Plot feature bookmark → *Plot graph* (Figs. 6a, b and 7a, b). Sector feature and Diagram feature bookmarks allow plotting other graphs (e.g., Rose diagram; Fig. 6c, d). Statistic feature bookmark provides statistic values that can be used for further statistical analysis.

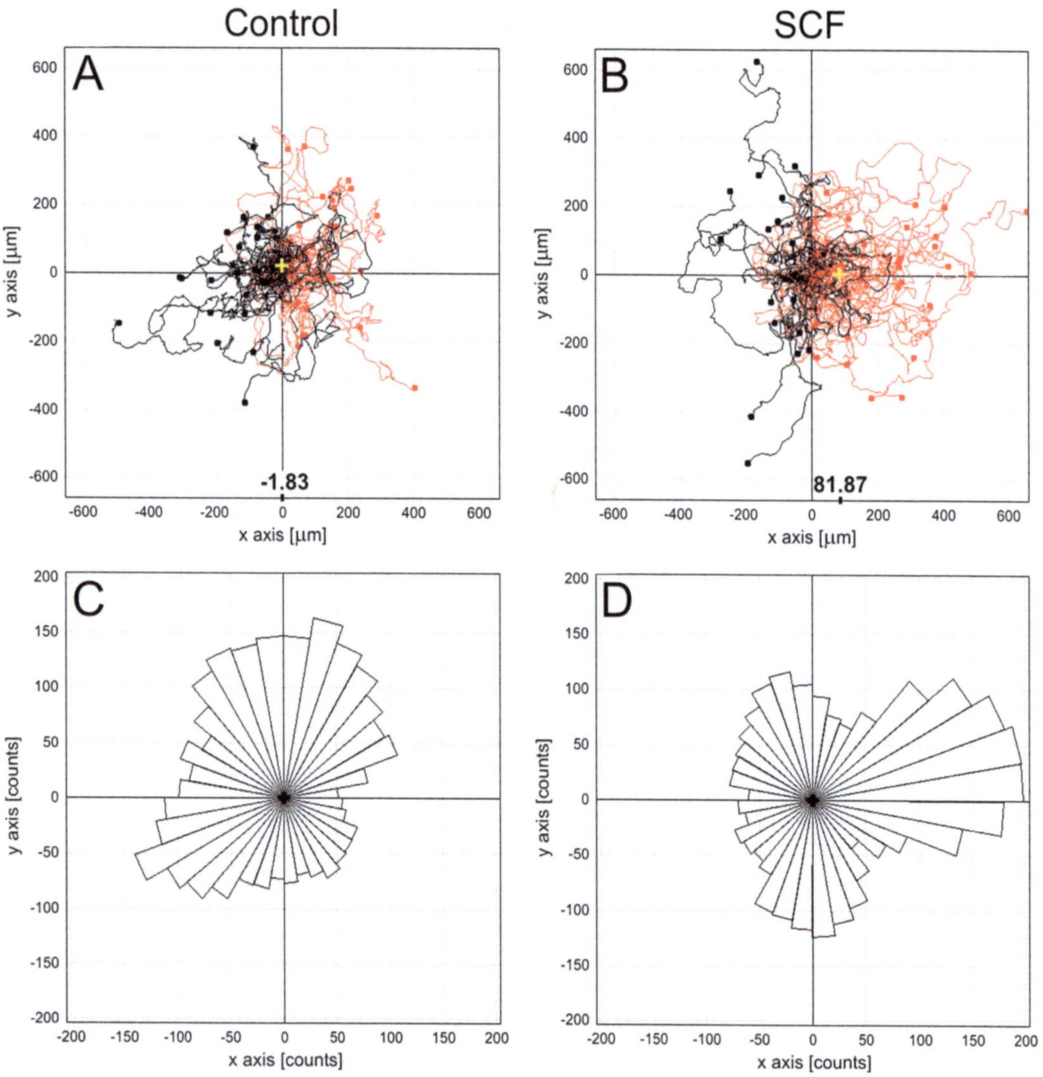

Fig. 6 Chemotaxis of LUVA cells. Representative images from a single experiment showing migration trajectory graphs (**A**, **B**) and rose diagrams expressing direction tendencies of LUVA cells (**C**, **D**). Cells without (**A**, **C**; Control) and with 100 ng/ml SCF in the right reservoir (**B**, **D**; SCF). Chemotaxis of LUVA cells toward SCF is reflected by a shift of the center of cell mass in the x axis (yellow crosses in **a**, **b**). LUVA cells in collagen I matrix were monitored for 10 h in a Leica inverted microscope equipped with a 10× objective

4 Notes

1. Alternatively, fibronectin-coated surface can be prepared by 1 h incubation at 37 °C.
2. BMMC starvation of IL-3 should not exceed 14–16 h. In case of suboptimal results, overnight culturing with a small amount of IL-3 can be optionally tested.

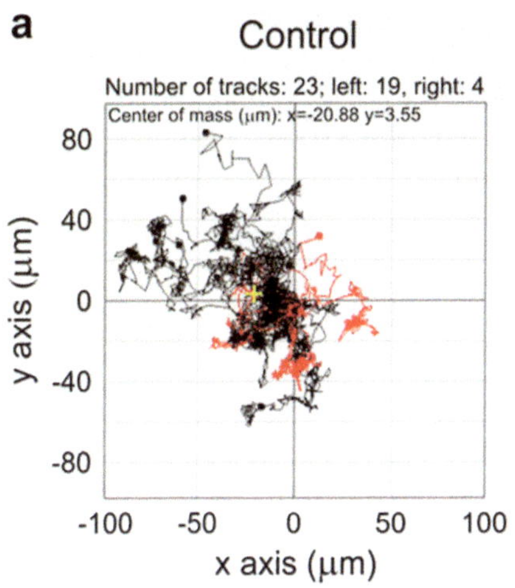

Fig. 7 Chemotaxis of BMMCL cells toward Ag in the presence or absence of miltefosine. Representative images are shown from a single experiment showing migration trajectory graphs. Control cells (**a**) and cells treated with 15 μM miltefosine (**b**) were attracted to Ag (100 ng/ml) in the left reservoir. BMMCL in collagen I matrix were monitored for 7 h in Leica inverted microscope equipped with 10× objective. Chemotaxis of BMMCL toward Ag is reflected by a shift of the center of cell mass in *x* axis (yellow cross) (Reproduced from [38] with permission from Frontiers Media SA)

3. Removal of fibronectin-unbound cells is an important step and should be carried out with maximal caution to avoid any washing out or damaging of the fibronectin-attached cell layer.
4. Best results are achieved with PGE_2 concentrations between 50 and 100 nM.
5. The length of time intervals depends on the number of monitored sites. If many sites are being examined, it is advisable to choose as short a time interval as possible. Cells become most active after the initial 15 min. An optimal time interval for subsequent analysis can be defined for individual experiments.
6. Equilibration prevents formation of air bubbles during loading of the μ-Slide^{3D}. Packaged μ-Slide^{3D} enables gas equilibration. Do not equilibrate collagen I.
7. The cell suspension will be consequently diluted into a final concentration of 1.5×10^6 cells/ml. Cell concentration could vary according to individual experimental preferences—more cells for obtaining extensive statistic data and less cells for easier cell tracking (cells are isolated and trajectories do not interfere).
8. This amount corresponds to a collagen I concentration of 1 mg/ml.
9. Mix well but avoid generating bubbles. Polymerization starts approximately 5 min after mixing collagen I with premix.

10. Do not inject the mixture directly; rather, maintain a small gap between the tip and the filling port A. Insert then the pipette tip directly into the filling port B and aspirate until the gel level reaches the pipette tip. Both ports should be filled but not overfilled.
11. Cell concentration and morphology should be checked with a microscope. The remaining gelation mixture in the Eppendorf tube helps determine the moment when the process is completed.
12. To generate a chemoattractant gradient, only half of the reservoir will be filled. Therefore, to achieve the desired final concentration, double the chemoattractant concentrations.
13. Reservoirs must be filled, but not overfilled. This will prevent unwanted medium fluxes.
14. For later easier cell tracking, it is better to contrast the cells against the background rather than perfectly focus on the cells themselves.
15. Further information about working with μ-Slide Chemotaxis^{3D} chambers can be found on the Ibidi website, https://ibidi.com/content/category/23-chemotaxis.
16. Manipulation with viscous agarose solution might be facilitated if pipette tips with cutoff narrow ends are used. Low-melting-point agarose with gelling point around or below 40 °C is not recommended.
17. Do not include data obtained during the first 60 min into analysis because of possible irregularities caused by addition of the agarose cone.
18. Cells can be also tracked with Manual Tracking plugin for ImageJ, which is freely available. Furthermore, several commercially available platforms offer automated cell tracking online (e.g., Wimasis Image Analysis).
19. If the time-lapse displays dark borders or periphery, it is advisable to cut it off from the time-lapse sequence. The goal is to obtain as many marked cells as possible without marked background. Cells that are not properly marked would not be tracked easily. On the other hand, too much background signal would slow down the tracking. It is possible to modify the threshold during automated tracking.
20. For this type of tracking, it is best to use search regions of the size similar to that occupied by the cell.
21. Data loaded from MetaMorph must be in the same order and format as those from the Manual Tracking from ImageJ. It is therefore necessary to alter .xlsx file from MetaMorph. The first column should contain numbers of each tracked slice (from

1 to the very last line). The second column should indicate the number of tracks (Track n°). The third column should show the number of each slice in every track (Slice n°). Two other columns indicate *X* and *Y* coordinates. Save data table as a .txt file.

22. This tool enables to work with more imported datasets at one time.
23. This option enables to analyze cell chemotaxis from the same time after adding the chemoattractant in every experiment.

Acknowledgments

This work was supported in part by Grants 17-20255S, 17-20915S, 18-18521S, 18-27197S, and 19-20716S from the Czech Science Foundation, project LTAUSA17052 from Ministry of Education, Youth and Sport of the Czech Republic and the institutional support RVO 68378050.

References

1. Galli SJ, Tsai M, Piliponsky AM (2008) The development of allergic inflammation. Nature 454:445–454
2. Abraham SN, St John AL (2010) Mast cell-orchestrated immunity to pathogens. Nat Rev Immunol 10:440–452
3. Maltby S, Khazaie K, McNagny KM (2009) Mast cells in tumor growth: angiogenesis, tissue remodelling and immune-modulation. Biochim Biophys Acta 1796:19–26
4. Galli SJ, Tsai M (2012) IgE and mast cells in allergic disease. Nat Med 18:693–704
5. Halova I, Draberova L, Draber P (2012) Mast cell chemotaxis – chemoattractants and signaling pathways. Front Immunol 3:119
6. Kirshenbaum AS, Metcalfe DD (2006) Growth of human mast cells from bone marrow and peripheral blood-derived CD34$^+$ pluripotent progenitor cells. Methods Mol Biol 315:105–112
7. Ashman LK (1999) The biology of stem cell factor and its receptor C-kit. Int J Biochem Cell Biol 31:1037–1051
8. Halova I, Ronnberg E, Draberova L, Vliagoftis H, Nilsson GP, Draber P (2018) Changing the threshold-signals and mechanisms of mast cell priming. Immunol Rev 282:73–86
9. Lantz CS, Boesiger J, Song CH, Mach N, Kobayashi T, Mulligan RC, Nawa Y, Dranoff G, Galli SJ (1998) Role for interleukin-3 in mast-cell and basophil development and in immunity to parasites. Nature 392:90–93
10. Welle M (1997) Development, significance, and heterogeneity of mast cells with particular regard to the mast cell-specific proteases chymase and tryptase. J Leukoc Biol 61:233–245
11. Taub D, Dastych J, Inamura N, Upton J, Kelvin D, Metcalfe D, Oppenheim J (1995) Bone marrow-derived murine mast cells migrate, but do not degranulate, in response to chemokines. J Immunol 154:2393–2402
12. Okayama Y, Kawakami T (2006) Development, migration, and survival of mast cells. Immunol Res 34:97–115
13. Jolly PS, Bektas M, Olivera A, Gonzalez-Espinosa C, Proia RL, Rivera J, Milstien S, Spiegel S (2004) Transactivation of sphingosine-1-phosphate receptors by FcεRI triggering is required for normal mast cell degranulation and chemotaxis. J Exp Med 199:959–970
14. Weller CL, Collington SJ, Brown JK, Miller HR, Al-Kashi A, Clark P, Jose PJ, Hartnell A, Williams TJ (2005) Leukotriene B4, an activation product of mast cells, is a chemoattractant for their progenitors. J Exp Med 201:1961–1971
15. Weller CL, Collington SJ, Hartnell A, Conroy DM, Kaise T, Barker JE, Wilson MS, Taylor GW, Jose PJ, Williams TJ (2007) Chemotactic

action of prostaglandin E_2 on mouse mast cells acting via the PGE_2 receptor 3. Proc Natl Acad Sci U S A 104:11712–11717

16. Kuehn HS, Jung MY, Beaven MA, Metcalfe DD, Gilfillan AM (2011) Prostaglandin E_2 activates and utilizes mTORC2 as a central signaling locus for the regulation of mast cell chemotaxis and mediator release. J Biol Chem 286:391–402
17. Halova I, Bambouskova M, Draberova L, Bugajev V, Draber P (2018) The transmembrane adaptor protein NTAL limits mast cell chemotaxis toward prostaglandin E_2. Sci Signal 11:556
18. Ishizuka T, Okajima F, Ishiwara M, Iizuka K, Ichimonji I, Kawata T, Tsukagoshi H, Dobashi K, Nakazawa T, Mori M (2001) Sensitized mast cells migrate toward the antigen: a response regulated by p38 mitogen-activated protein kinase and Rho-associated coiled-coil-forming protein kinase. J Immunol 167:2298–2304
19. Kitaura J, Kinoshita T, Matsumoto M, Chung S, Kawakami Y, Leitges M, Wu D, Lowell CA, Kawakami T (2005) IgE– and IgE+Ag-mediated mast cell migration in an autocrine/paracrine fashion. Blood 105:3222–3229
20. Echtenacher B, Mannel DN, Hultner L (1996) Critical protective role of mast cells in a model of acute septic peritonitis. Nature 381:75–77
21. Malaviya R, Ikeda T, Ross E, Abraham SN (1996) Mast cell modulation of neutrophil influx and bacterial clearance at sites of infection through TNF-α. Nature 381:77–80
22. Brightling CE, Bradding P, Symon FA, Holgate ST, Wardlaw AJ, Pavord ID (2002) Mast-cell infiltration of airway smooth muscle in asthma. N Engl J Med 346:1699–1705
23. Quintas-Cardama A, Jain N, Verstovsek S (2011) Advances and controversies in the diagnosis, pathogenesis, and treatment of systemic mastocytosis. Cancer 117:5439–5449
24. Boyden S (1962) The chemotactic effect of mixtures of antibody and antigen on polymorphonuclear leucocytes. J Exp Med 115:453–466
25. Hálová I, Dráberová L, Bambousková M, Machyna M, Stegurová L, Smrž D, Dráber P (2013) Crosstalk between tetraspanin CD9 and transmembrane adaptor protein non-T cell activation linker (NTAL) in mast cell activation and chemotaxis. J Biol Chem 288:9801–9814
26. Draberova L, Bugajev V, Potuckova L, Halova I, Bambouskova M, Polakovicova I, Xavier RJ, Seed B, Draber P (2014) Transmembrane adaptor protein PAG/CBP is involved in both positive and negative regulation of mast cell signaling. Mol Cell Biol 34:4285–4300
27. Potuckova L, Draberova L, Halova I, Paulenda T, Draber P (2018) Positive and negative regulatory roles of C-terminal Src kinase (CSK) in FcεRI-mediated mast cell activation, independent of the transmembrane adaptor PAG/CSK-binding protein. Front Immunol 9:1771
28. Lee J, Veatch SL, Baird B, Holowka D (2012) Molecular mechanisms of spontaneous and directed mast cell motility. J Leukoc Biol 92:1029–1041
29. Poole TJ, Zetter BR (1983) Stimulation of rat peritoneal mast cell migration by tumor-derived peptides. Cancer Res 43:5857–5861
30. Heit B, Kubes P (2003) Measuring chemotaxis and chemokinesis: the under-agarose cell migration assay. Sci STKE 2003:L5
31. Wiggins H, Rappoport J (2010) An agarose spot assay for chemotactic invasion. BioTechniques 48:121–124
32. Zengel P, Nguyen-Hoang A, Schildhammer C, Zantl R, Kahl V, Horn E (2011) μ-Slide chemotaxis: a new chamber for long-term chemotaxis studies. BMC Cell Biol 12:21
33. Kuiper JW, Sun C, Magalhaes MA, Glogauer M (2011) Rac regulates PtdInsP(3) signaling and the chemotactic compass through a redox-mediated feedback loop. Blood 118:6164–6171
34. Hibbs ML, Tarlinton DM, Armes J, Grail D, Hodgson G, Maglitto R, Stacker SA, Dunn AR (1995) Multiple defects in the immune system of *Lyn*-deficient mice, culminating in autoimmune disease. Cell 83:301–311
35. Hájková Z, Bugajev V, Dráberová E, Vinopal S, Dráberová L, Janáček J, Dráber P, Dráber Pa (2011) STIM1-directed reorganization of microtubules in activated mast cells. J Immunol 186:913–923
36. Laidlaw TM, Steinke JW, Tinana AM, Feng C, Xing W, Lam BK, Paruchuri S, Boyce JA, Borish L (2011) Characterization of a novel human mast cell line that responds to stem cell factor and expresses functional FcεRI. J Allergy Clin Immunol 127:815–822
37. Schneider CA, Rasband WS, Eliceiri KW (2012) NIH Image to ImageJ: 25 years of image analysis. Nat Methods 9:671–675
38. Rubíková Z, Sulimenko V, Paulenda T, Dráber P (2018) Mast cell activation and microtubule organization are modulated by miltefosine through protein kinase C inhibition. Front Immunol 9:1563

Chapter 25

Basophil Stimulation and Signaling Pathways

Edward F. Knol and Bernhard F. Gibbs

Abstract

Despite the growing use of flow cytometry to analyze the functional characteristics of basophils, the intracellular signaling cascades that control their ability to elaborate various pro-allergic and inflammatory mediators and cytokines remain comparatively obscure. Additionally, some studies require the analysis of pro-allergic and inflammatory mediators, such as histamine, LTC_4, and various basophil-derived cytokines (e.g., IL-4 and IL-13). Elucidation of intracellular signaling proteins by Western blotting, cytosolic free calcium concentration by spectrofluorophotometry, and detection of mediator releases, as well as analysis of gene expressions by RT-PCR, generally requires relatively large numbers of purified basophils. In selected assays, flow cytometry enables the analysis of relatively low cell numbers and purity for the expression of intracellular signaling proteins or measurement of cytosolic free calcium concentrations by basophil-specific gating strategies. Unfortunately, many aspects of signal transduction relevant to human basophils cannot be readily extrapolated from the use of basophil or mast cell lines. This chapter therefore focuses on how to employ primary human basophils for studying mediator releases and signaling characteristics.

Key words Basophils, Signaling, IgE receptors, Western blotting, Flow cytometry, Inhibitors

1 Introduction

Human basophilic granulocytes (basophils) are inflammatory cells that are involved in the pathogenesis of a variety of allergic diseases (reviewed in [1]). Upon stimulation with various stimuli, they release several biologically active mediators, such as histamine and leukotriene C_4, as well as cytokines such as IL-4 (*see* Chapter 21). Due to the limited availability of human basophils of high purity, precise knowledge of the signal transduction leading to the release reactions is still sparse. Activation of basophils through FcεRI signaling is mostly studied *via* anti-IgE or allergen-induced activation. Other receptor-mediated stimuli for human basophils include the bacterial peptide formyl-methionyl-leucyl-phenylalanine (fMLP or fMLF) and the complement fragment C5a. Calcium ionophores A23187 or ionomycin as well as phorbol myristate acetate (PMA) activate basophils independent of cell membrane receptors. Priming of basophils can be achieved by several

Bernhard F. Gibbs and Franco H. Falcone (eds.), *Basophils and Mast Cells: Methods and Protocols*, Methods in Molecular Biology, vol. 2163, https://doi.org/10.1007/978-1-0716-0696-4_25,

cytokines, particularly interleukin (IL)-3, IL-5, IL-33, GM-CSF, and NGF as well as platelet-activating factor which markedly potentiate mediator release caused by various stimuli.

It is important to note that although basophils share certain functional properties with mast cells (e.g., FcεRI-induced degranulation), these cells differ markedly to their mast cell counterparts in terms of priming factors and certain other stimuli (e.g., fMLP) (reviewed in [2]). It is clear that the receptor-mediated stimulation of basophils results in a strong donor-dependent activation determined, for instance, by histamine release. For FcεRI-induced degranulation, the strength of activation of Syk has been hypothesized to be causal [3]. For many other stimuli, this is not clear yet, and we need better insights in basophil signal transduction cascades that control degranulation or the release of cytokines and other mediators.

2 Materials

2.1 Chemicals and Kits

1. Recombinant human IL-3. Prepare aliquots of a stock solution containing 10 μg/mL of IL-3 in Ca^{2+}-free HEPES-buffered Tyrode's solution to avoid repeated freeze-thaw cycles. Use at 0.1–1 ng/mL final concentration as a priming factor to enhance IgE-dependent basophil stimulation.
2. Anti-human IgE. Polyclonal IgG antibodies are most often employed. These need to be titrated in a pilot experiment to determine optimum basophil mediator release and concentrated aliquots prepared accordingly and frozen at −20 °C for subsequent experiments. Note that some basophil mediators, e.g., IL-4, are released at five- to tenfold lower concentrations than those required for histamine release or CD63 expression. It is therefore advisable to use concentrations of anti-IgE (or other IgE-dependent stimuli) that are slightly suboptimal with respect to histamine release or CD63 expressions.
3. IgE-independent stimuli. 1 mM stock solutions in DMSO to avoid freeze-thaw issues (which are particularly of concern regarding calcium ionophores) and perform pilot basophil stimulation experiments to determine optimum (or suboptimum) basophil stimulation. For A23187, this is usually between 1 and 10 μM (note that there is a sharp bell-shaped response for this stimulus), for fMLP 100–1000 nM, for C5a 1–10 nM, and for PMA 10–100 ng/mL. For C5a and IL-3, use either buffer or 0.9% NaCl with 1 mg/mL BSA to prepare stock solutions.
4. ELISA kits for histamine, LTC_4, and pro-allergic cytokines (e.g., IL-4). Note that for detection of cytokine release, the most sensitive ELISA kits or multiplex platforms such as xMAP

Luminex are usually required (i.e., minimum detection at ca. 1–3 pg/mL). Histamine releases also are mostly determined spectrofluorometrically based on the method described by Shore et al. [4].

5. Laemmli lysis buffer for SDS-PAGE/Western blotting: Prepare lysis buffer consisting of 50 mM Tris–HCl pH 7.5, 5 mM EDTA, 10 mM EGTA, 5 mM DTT, 1% Nonidet P-40, 1 mM PMSF, 100 μg/mL aprotinin, 20 μg/mL leupeptin, and 10 mM benzamidine and then mix with equal volumes of 2× concentrated Laemmli sample buffer (obtained from commercial sources). Freeze aliquots at −20 °C. Ensure complete dissolution of contents prior to use.
6. SDS-PAGE reagents: Tris pH 8.8, Tris pH 6.8, 30% acrylamide-bis solution, glycerol, glycine, SDS, TEMED, and APS. Blotting and transfer buffers.
7. Western blotting reagents: Tris-buffered saline and Tween 20 (TBST) solution (containing 20 mM Tris–HCl, pH 7.5, 137 mM NaCl, 0.1% Tween 20), skimmed milk powder, antibodies to signaling proteins, secondary antibodies (e.g., labeled with either HRP or dyes if using LiCor or similar fluorescence-based imaging systems).
8. Calcium indicators: The free Ca^{2+} indicators indo-1 and fura-2 are acetoxy-methyl ester-labeled. Basophils are incubated with approximately 1 μM in a Ca^{2+}-containing buffer. Within basophils, the acetoxy-methyl ester, which can diffuse freely through the membrane, is cleaved by cytosolic esterases, resulting in a charged, cell impermeable compound form of indo-1 and fura-2.

2.2 Buffers for Cell Culture and Stains

1. N-2-hydroxyethylpiperazine-N-2-ethanesulfonic acid (HEPES)-buffered Tyrode's solution (pH 7.4): 137.0 mM (8.0 g/L) NaCl, 5.6 mM (1.0 g/L) glucose, 2.7 mM (201 mg/L) KCL, 1.0 mM (147 mg/L) $CaCl_2{\cdot}2H_2O$, 1.0 mM (203 mg/L) $MgCl_2{\cdot}6H_2O$, 10.0 mM (2.9 g/L) HEPES; 0.4 mM (62.5 mg/L) $NaH_2PO_4{\cdot}2H_2O$. The pH is adjusted to 7.4 with 1 M NaOH. This buffer will keep for 1 week at 4 °C and may be used for brief periods (<1 h) of basophil stimulation.
2. HEPES-buffered Tyrode's solution (pH 7.4) without $MgCl_2$ and $CaCl_2$ (Ca^{2+}-free buffer). Used for washing basophils before experimentation and for terminating reactions at the end of incubation periods (though this is not usually done following long-term incubation periods, especially where ELISAs for LTC_4 and cytokine releases are required and dilution of samples must be avoided).

3. RPMI 1640 with L-glutamine, 1% penicillin-streptomycin, and 1 mg/mL BSA. Used for experiments requiring long incubation periods (>1 h).
4. Alcian blue staining of basophils (adapted from [5]): Dissolve 76 mg cetylpyridinium chloride, 700 mg lanthanum chloride, 900 mg NaCl, 21 μL Tween 20, and 143 mg Alcian blue in 100 mL distilled H_2O and stir for at least 2 h at room temperature. Filter using a 1 μm filter or, alternatively, centrifuge at 1000 × *g* for 15 min and carefully collect the supernatant. The solution is stable for at least 1 year at 4 °C. To 10 μL of cell suspension, add 65 μL of NaCl solution (1 mg/mL containing 0.1% EDTA) followed by 65 μL of Alcian blue solution (described above). Finally, add 5 μL of 1 M HCl, gently resuspend, and determine the total number and purity of basophils using a hemocytometer.

2.3 Basophils

1. Enrichment and purification of primary human basophils is described in detail in Chapter 2.
2. Ensure sufficient basophil purity and number using Alcian blue stain and a hemocytometer (*see* Chapter 2, **Note 9**) or flow cytometer with basophil discriminating antibodies (*see* Chapter 9, Subheading 1). For determining mediator release, at least 100,000 basophils/vial are usually necessary. For RT-PCR and Western blotting, use at least 200,000–300,000 basophils/vial.

2.4 Equipment and Other Materials

1. Water purification system.
2. Microscope and hemocytometer.
3. CO_2 incubator or water bath set to 37 °C.
4. Centrifuges (400–1000 × *g*).
5. Heating block for 1.5 mL tubes.
6. ELISA plate reader.
7. LiCor Odyssey or comparable imaging system or dark room for ECL detection using photographic film.
8. Flow cytometer.
9. Spectrofluorophotometer.
10. 50 and 10 mL test tubes (polypropylene) and 1.5 mL Eppendorf tubes.
11. Pipette tips.

3 Methods

The following methods describe the basic procedure for analyzing basophil responses as well as variations depending on whether the release of various mediator types is being assessed or signal transduction studies conducted. It is vital that cell handling and preparation of purified basophils is kept as short as possible.

3.1 Basophil Stimulation Assay: General Procedure

1. Wash basophils and resuspend in either Tyrode's buffer (short incubations) or RPMI 1640 medium (long incubations to determine cytokine release), at room temperature, and aliquot (400 μL) into test tubes at a density of *ca.* 50,000–150,000 basophils/mL. Include tubes for positive control (e.g., anti-IgE and/or IgE-independent stimulus) and negative control (buffer with cells alone).
2. Place all tubes in a water bath or culture cabinets set at 37 °C.
3. Add 50 μL of prewarmed buffer, priming factors (e.g., IL-3), or signal transduction inhibitors to the tubes and preincubate for 15 min. Avoid cooling.
4. Stimulate all tubes except negative control with 50 μL anti-IgE (or IgE-independent stimulus) for the required times (the total volume for each vial should now be 500 μL).

3.2 Histamine Release Analysis

1. Prepare and treat basophils as described above (Subheading 3.1; *see* **Note 1**) and stimulate for 30 min (unless release kinetics are being addressed; *see* **Note 2**).
2. Quench reactions by adding 1 mL of ice-cold Ca^{2+}-free HEPES-buffered Tyrode's solution into each tube and centrifuge immediately (850 × *g*, 2 min, 4 °C).
3. Decant supernatants into fresh tubes and add 1.45 mL Ca^{2+}-free HEPES-buffered Tyrode's solution to the pellet tubes. For automated histamine analysis, deplete proteins, including enzymes, in the supernatant by adding 2% w/v perchloric acid. Lyse the pellet tubes by adding 50 μL perchloric acid (70%) for automated analysis of histamine release according to the method described by Shore et al. [4]. Alternatively, freeze-thaw cell pellet sampled (without perchloric acid) three times.
4. Briefly vortex pellet tubes and centrifuge (850 × *g*, 10 min, 4 °C) to remove debris.
5. The tubes can be stored at −20 °C until ready for assay.
6. Assay for histamine (either using commercial kits or fluorometric analysis).

7. Histamine release is calculated as a percentage of the total content (present in the cell pellet and corresponding supernatant). Results are typically expressed as either relative percent releases (the % release into the supernatant for each condition/vial) or corrected percent release (the net % release after subtracting spontaneous releases obtained from the negative controls). Spontaneous release should be below 10%, preferably below 5%. A high spontaneous release may indicate nonspecific activation caused by cell handling issues or an in vivo activation [6].

3.3 Eicosanoid and Cytokine Release Analysis

1. Experimental conditions are the same as described in Subheading 3.1 except for the variable incubation periods and termination of reactions.
2. Unless the basophil density is very high (e.g., 200,000 basophils in 500 μL total reaction volume), it is not recommended to stop reactions using ice-cold Ca^{2+}-free HEPES-buffered Tyrode's solution since the mediators of interest may be diluted too much to detect. Instead, briefly place all tubes in ice water and then centrifuge (850 × *g*, 2 min, 4 °C).
3. Decant supernatants into fresh vials and freeze immediately (−20 °C or lower) until ready for assay.
4. Eicosanoid (e.g., LTC_4) release is usually complete after 30 min stimulation and is measured only in the supernatants.
5. Optimum cytokine production from basophils varies from 4 h, for the detection of IL-4 release, to over 16 h for IL-13 releases. At the end of each stimulation period, harvest the supernatants as described above and freeze immediately for storage prior to assaying cytokine content (*see* **Note 3**).

3.4 Measurement of Cytosolic Free Calcium Concentrations by Spectro-fluorophotometer or Flow Cytometer

1. Prepare and treat basophils as described above (Subheadings 2.1, **item 8** and 3.1).
2. For the spectrofluorophotometer and indo-1-loaded basophils, excitation and emission wavelengths should be taken at 340 and 390 nm, respectively.
3. Before each measurement, the cells are diluted in medium with 1 mM $CaCl_2$ to a density of approximately 1×10^6 cells/mL.
4. Calibration of the signal is obtained by permeabilization of the cells with 5 μM digitonin to obtain maximum fluorescence and then quenching with 0.5 mM $MnCl_2$ to obtain minimal fluorescence (*see* Fig. 1).
5. Cytoplasmic free calcium concentration is calculated by the equation $[Ca^{2+}] = Kd \times ((F - F_{min})/(F_{max} - F))$, using 250 nM as the Kd for the Ca^{2+}/indo-1 complex.

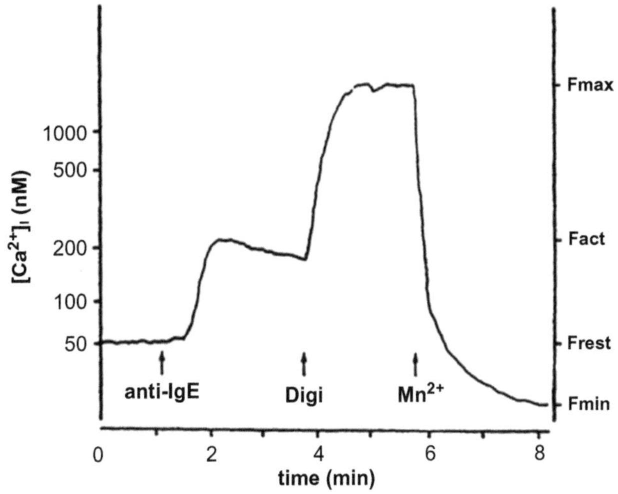

Fig. 1 Increase in $[Ca^{2+}]_i$ in a purified basophil suspension after anti-IgE stimulation. Basophils were loaded with indo-1, and fluorescence measurements were performed using a spectrofluorophotometer, as described in Subheading 3.4. In this recording, the calibration of indo-1 fluorescence as a function of $[Ca^{2+}]_i$ was also carried out. All trapped indo-1 was saturated by permeabilization of the cells with 5 μm digitonin (Digi), resulting in a maximal fluorescence (F_{max}). Subsequently, the indo-1 signal was quenched by addition of 0.5 mM $MnCl_2$ (Mn^{2+}) resulting in minimal fluorescence (F_{min}). Fluorescence of the unstimulated population is marked as F_{rest} and the anti-IgE-stimulated population as F_{act} [8]

6. For flow cytometry, it is not possible to determine the precise $[Ca^{2+}]$ using the calibration strategy described above; only the relative ratio of $[Ca^{2+}]$ can be obtained (*see* Fig. 2).
7. Flow cytometry emission and excitation wavelengths are determined somewhat by the filters available but should be within the range described above.
8. For flow cytometry, a sampling chamber with low magnetic stirring and heating at 37 °C is required.
9. Addition of 1.5 mM EGTA or buffering Ca^{2+} concentrations in incubation buffers to about 100 nM selectively will prevent Ca^{2+} influx and will only show Ca^{2+} increase due to release from intracellular stores (*see* Fig. 3).

3.5 Determination of Activation Pathways by Flow Cytometry

1. Prepare and treat basophils as described above (Subheading 3.1). Depending on the system, unprocessed blood, peripheral blood mononuclear cell preparation, or purified (partly or fully) basophils can be activated (*see* **Note 4**).
2. Stop reactions by adding ice-cold buffer with EDTA.

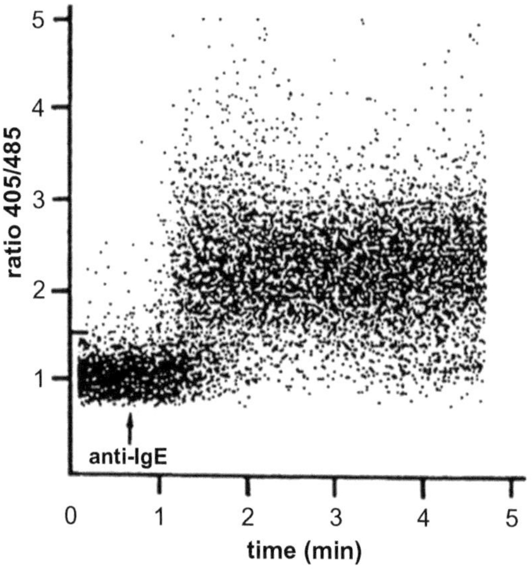

Fig. 2 Increase in $[Ca^{2+}]_i$ determined by flow cytometer in a purified basophil suspension after anti-IgE stimulation. Basophils were loaded with indo-1, and fluorescence measurements were performed in a flow cytometer, as described in Subheading 3.4. In contrast to the data shown in Fig. 1, the $[Ca^{2+}]_i$ in individual basophils is depicted [9]

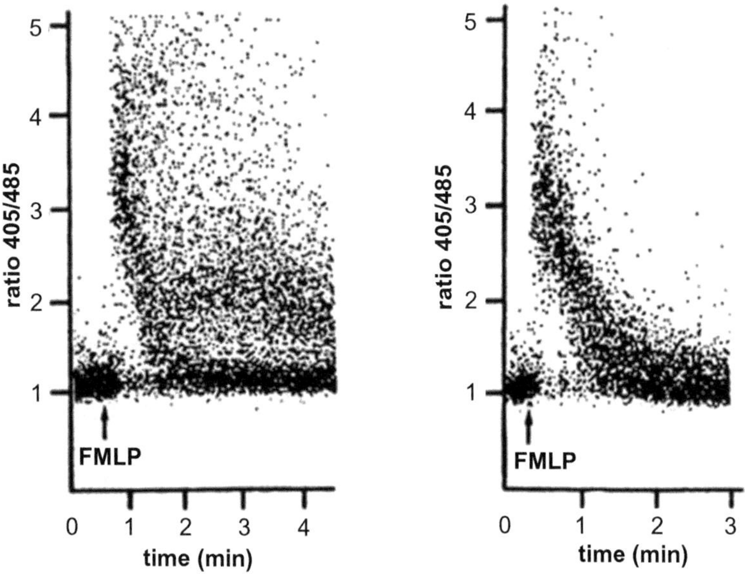

Fig. 3 Changes in $[Ca^{2+}]_i$ determined by flow cytometer in a purified basophil suspension following the addition of fMLP. Experiments were performed as described above. In the right panel, cells were stimulated without Ca^{2+} but with 1.5 mM EGTA to deplete extracellular Ca^{2+}. The right panel depicts the increase in $[Ca^{2+}]_i$ from intracellular stores only [8]

3. Antibodies to characterize basophils, such as $CD123^+/CD203C^+/HLA\text{-}DR^-$ or CCR3, are added together with CD63 (in order to estimate activated basophils) for about 30 min at 4 °C.
4. If unprocessed blood was used, residual erythrocytes should be lysed.
5. To stain for intracellular proteins, cells should be fixed and permeabilized by resuspending them in 300 μL of 0.3% saponin and 3.7% formaldehyde in PBS for 30 min at 4 °C.
6. Staining of phosphorylated (activation) proteins, such as phospho-p38 mitogen-activated protein kinase (p38 MAPK), is performed on the permeabilized cells at relatively early time points (e.g., 1, 3, 5, 10 min) following stimulation (*see* **Note 5**).
7. Control cells that are stored in ice-cold buffer are used to determine baseline phosphorylation.
8. Flow cytometric analysis should be performed on either labeled basophils or purified basophils.

3.6 Signal Transduction Experiments and Preparation of Samples for SDS-PAGE and Western Blotting

1. Prepare and treat basophils as described above (Subheading 3.1) and stimulate for the required times (*see* **Note 6**).
2. Stop cell incubations either by briefly placing tubes on ice water (if concentrated supernatants are required for mediator release assays) or by adding 1 mL of ice-cold Ca^{2+}-free HEPES-buffered Tyrode's solution.
3. Centrifuge immediately (850–1000 × *g*; 2 min).
4. Remove supernatants and lyse cell pellets on ice by vigorous mixing in 15–30 μL of Laemmli lysis buffer (*see* **Note 7**).
5. Immediately heat all tubes to over 95 °C for 3 min and either store at −70 °C for later analysis or proceed to SDS-PAGE.
6. Separate proteins by SDS-PAGE and transfer to nitrocellulose membranes. Include a molecular weight (MW) marker, e.g., pre-stained MW rainbow markers for each SDS-PAGE run.
7. Block membranes for 2 h in 5% skimmed milk in TBST buffer (*see* **Note 8**).
8. Briefly wash twice with TBST and incubate membranes overnight at 4 °C with specific antibodies to either phosphorylated, total signaling proteins or beta-actin on a swirling platform.
9. Wash membranes four to five times in TBST for a total of 15 min.
10. Incubate with secondary antibodies (e.g., anti-mouse or anti-rabbit Ig horse radish peroxidase or LiCor-labeled antibodies for Odyssey imaging) for 1 h.

11. Remove unbound secondary antibodies by washing membranes 4–5 times in TBST for a total of 15 min.
12. Visualize proteins either by chemiluminescence according to the manufacturer's instructions (e.g., ECL plus) or using a LiCor imaging system.
13. After detection, if additional signaling proteins require analysis, strip membranes for 15–25 min using commercial stripping reagents, wash in TBST (4 × 5 min), re-block and wash depending on the manufacturer's instructions, and re-probe.

4 Notes

1. If histamine release is being determined alone, i.e., without Western blotting or cytokine/LTC_4 releases, basophil preparations of lower purity may be used, especially for pilot experiments to gauge the optimum concentrations of stimuli required for subsequent experiments. In this case, even Ficoll-density enrichment may suffice for determining basophil histamine releases since, other than platelets in some donors, most of the blood histamine content is basophil derived.
2. Optimum kinetics for IgE-dependent histamine releases is usually around 30-min stimulation. However, this may vary if highly suboptimal or supraoptimal stimulus concentrations are used. IgE-independent stimuli, such as fMLP, display faster release kinetics. This is not usually a problem if a 30-min incubation period is employed (e.g., when fMLP or A23187 are being used as a positive control compared to anti-IgE or allergens) since the released histamine will not usually be broken down under the conditions described in this chapter. When basophils are stimulated with PMA, at least 60-min stimulation is required for maximal histamine release.
3. Optimum cytokine production from basophils varies from 4 h, for the detection of IL-4 release, to over 16 h for IL-13 release. However, since some preformed cytokines may be released very rapidly (within the first 30 min), it is recommended to initially use several stimulation periods, e.g., 30 min, 4 h, 8 h, 16 h, and more, in order to gauge the kinetics of both preformed and de novo generated cytokines as well as for assessing gene expressions by RT-PCR.
4. Analysis of basophil activation by flow cytometry enables to use basophil preparation of low purity or even whole blood. In this case, selective basophil markers ($CD203c^+$, $CD123^+$/$HLA\text{-}DR^-$, $CCR3^+$/low side scatter) should be used. Further information regarding the advantages and disadvantages of basophil testing by flow cytometry has recently been published [7].

5. Many signaling proteins are optimally phosphorylated within the first 15 min after basophil stimulation (e.g., Syk, p38 MAPK, ERK1&2). It is strongly advised to determine the optimal activity of the signal of interest by performing a kinetics experiment using basophils stimulated, for example, for 1, 5, 15, and 30 min. Other signaling proteins, e.g., HIF-1α, are maximally expressed at far greater periods (hours). Note that the concentrations of the stimulus and the degree of priming substantially affect signal transduction kinetics.
6. IgE-independent stimuli can be differentiated in receptor-dependent (e.g., fMLP, C5a) and receptor-independent (e.g., A23187) stimuli. The bacterial peptide fMLP serves as important positive control for basophil degranulation (especially in determining so-called nonresponders to IgE-dependent triggers; if the IgE-independent controls fail to elicit a response, this may indicate cell handling issues). C5a can be alternative for fMLP for degranulation but requires preceding stimulation with IL-3 for release of LTC_4. Calcium ionophores (e.g., A23187) or activators of protein kinase C (PKC), such as PMA, can be informative if receptor-mediated stimuli have no effect to check for status of basophil function.
7. Care must be taken to immediately disrupt cell pellets with Laemmli lysis buffer. On some occasions, cell pellets are not clearly visible, but cells are deposited on the sides of 1.5 mL tubes on fixed-angle rotors. Ensure, therefore, that Laemmli lysis buffer comes into contact with this material. If pellet debris does not immediately disrupt (i.e., becomes turbid and difficult to resuspend), proceed to the next step of heating the vials and vortex after 30 s of the heating/boiling step. If protein concentrations need to be determined prior to SDS-PAGE and Western blotting, disrupt the cells in lysis buffer alone on ice for 30 min before assessing protein concentrations (Bradford Assay). The remaining lysed cells are then treated with Laemmli for SDS-PAGE and Western blotting. These procedures can also be carried out at a later date by freezing the lysed cells at −80 °C.
8. There are several new commercial Western blotting systems (e.g., iBind) which are semiautomated and allow for multiplexing approaches. **Steps 7–10** may vary accordingly depending on the Western blotting system employed.

Acknowledgments

This work was originally supported by the COST Action BM1007 Mast Cells and Basophils—Targets for Innovative Therapies.

References

1. Falcone FH, Knol EF, Gibbs BF (2011) The role of basophils in the pathogenesis of allergic disease. Clin Exp Allergy 41:939–947
2. Varricchi G, Raap U, Rivellese F, Marone G, Gibbs BF (2018) Human mast cells and basophils – how are they similar how are they different? Immunol Rev 282:8–34
3. Miura K, Lavens-Phillips S, MacGlashan DW (2001) Piceatannol is an effective inhibitor of IgE-mediated secretion from human basophils but is neither selective for this receptor nor acts on Syk kinase at concentrations where mediator release inhibition occurs. Clin Exp Allergy 31:1732–1739
4. Shore PA, Burkhalter A, Cohn VH (1959) The detection of histamine, a new approach. J Pharmacol Exp Ther 127:182–186
5. Gilbert HS, Ornstein L (1975) Basophil counting with a new staining method using alcian blue. Blood 46:279–286
6. Sampson HA, Broadbent KR, Bernhisel-Broadbent J (1989) Spontaneous release of histamine from basophils and histamine-releasing factor in patients with atopic dermatitis and food hypersensitivity. N Engl J Med 321:228–232
7. Hoffmann HJ, Knol EF, Ferrer M, Mayorga L, Sabato V, Santos AF, Eberlein B, Nopp A, MacGlashan D (2016) Pros and cons of clinical basophil testing (BAT). Curr Allergy Asthma Rep 16:56
8. Knol E.F. (1991) Cellular mechanisms in the activation of human basophilic granulocytes. Chapter 5: Characterization of the role of Ca^{2+} in the degranulation of human basophilic granulocytes. Academic thesis, University of Amsterdam
9. Knol EF, Mul FPJ, Kuijpers TW, Verhoeven AJ, Roos D (1992) Intracellular events in anti-IgE non-releasing human basophils. J Allergy Clin Immunol 90:92–103

Chapter 26

Gene Silencing Approaches in Mast Cells and Primary Human Basophils

Vadim V. Sumbayev and Bernhard F. Gibbs

Abstract

The ability to silence gene expression is an invaluable tool for elucidating the importance of intracellular signaling proteins which contribute to the effector functions of mast cells and basophils. However, primary mast cells and their terminally differentiated blood counterpart, basophils, pose a difficult challenge for gene silencing approaches given not only their state of maturation and difficulty to transfect but also because their functions are readily altered by cell handling conditions. Here, we describe a method using lipofection which has been successfully employed to silence gene expression using siRNA in human LAD2 mast cells as well as primary human basophils.

Key words Transfection, siRNA, Lipofection, Electroporation, Nanofection, DOTAP

1 Introduction

Gene silencing is an epigenetic process which allows regulation of gene expression. The term "gene silencing" normally means "turning the gene off" without genetic modification. In other words, a gene which is expressed under normal circumstances is switched off by the cellular machinery. Gene silencing occurs when it is impossible to translate mRNA and can be performed on a transcriptional or posttranscriptional level [1].

Transcriptional gene silencing is normally achieved by histone modifications. This involves creating an environment of heterochromatin around the target gene making it inaccessible for transcription factors or other components of the transcriptional machinery [1]. However, in most cases, gene silencing is performed at the posttranscriptional level. This is achieved by destruction or blocking of the mRNA transcribed from a target gene that normally encodes a specific target protein. The blocking of the target mRNA normally occurs through the activity of silencers which interact with repressor regions. However, a more reliable effect could be obtained upon destruction of the target mRNA [1]. This is

Bernhard F. Gibbs and Franco H. Falcone (eds.), *Basophils and Mast Cells: Methods and Protocols*, Methods in Molecular Biology, vol. 2163, https://doi.org/10.1007/978-1-0716-0696-4_26,

normally performed using RNA interference (RNAi), which is also known as posttranscriptional gene silencing (PTGS). During RNAi, specific RNA molecules inhibit the expression of the target gene, typically by inducing the destruction of specific mRNA molecules [2].

This could be achieved by two ways:

(a) Transfection of the cells with small interfering RNA (siRNA). Small interfering or silencing RNAs form a class of small double-stranded RNA molecules (max 20–25 bp in length) [2–4].

(b) Use of small hairpin RNA or short hairpin RNA (shRNA) which makes a tight hairpin turn that can be used to silence target gene expression. shRNA is transfected into the cells in the form of plasmid DNA where it then undergoes transcription into shRNA [5].

In both cases, gene expression is ultimately silenced using siRNA. In the first case, the siRNA duplex is directly transfected into the cells; in the second, shRNA is produced on the matrix of transfected DNA and then processed by the intracellular machinery [2–5]. We will therefore consider the mechanisms of shRNA processing/action, since this will address both types of posttranscriptional mRNA destruction.

Plasmid DNA acts as a matrix for shRNA production. Generated shRNA is processed by the Dicer enzyme, an endoribonuclease of the RNase III family that cleaves double-stranded RNA (dsRNA) into short double-stranded RNA fragments of about 20–25 bp in length (duplexes), with a two-base overhang on the 3′ terminus. One of these duplexes is the siRNA of interest. This siRNA binds the target mRNA and induces its degradation which is achieved by activation of the RNA-induced silencing complex (RISC) of endonucleases called "argonaute proteins," cleaving the target mRNA strand complementary to the bound siRNA. As the RNA fragments produced by Dicer are double-stranded, each of them could, in theory, become converted into functional siRNA. However, only one of the two siRNA strands (guide strand) binds the argonaute protein and directs gene silencing [1, 2, 5]. These processes are summarized in Fig. 1.

Intracellular signaling networks occurring in human mast cells and basophils are poorly understood. In order to study the biological/biochemical activity of any signaling component, it is necessary to specifically switch it off and monitor the affected processes. This would then allow conclusions to be made on the contribution of the signaling protein of interest to specific intracellular biochemical cascades or cycles as well as changes in overall cell function, such as mediator production and release. Such studies could be performed by gene silencing since pharmacological

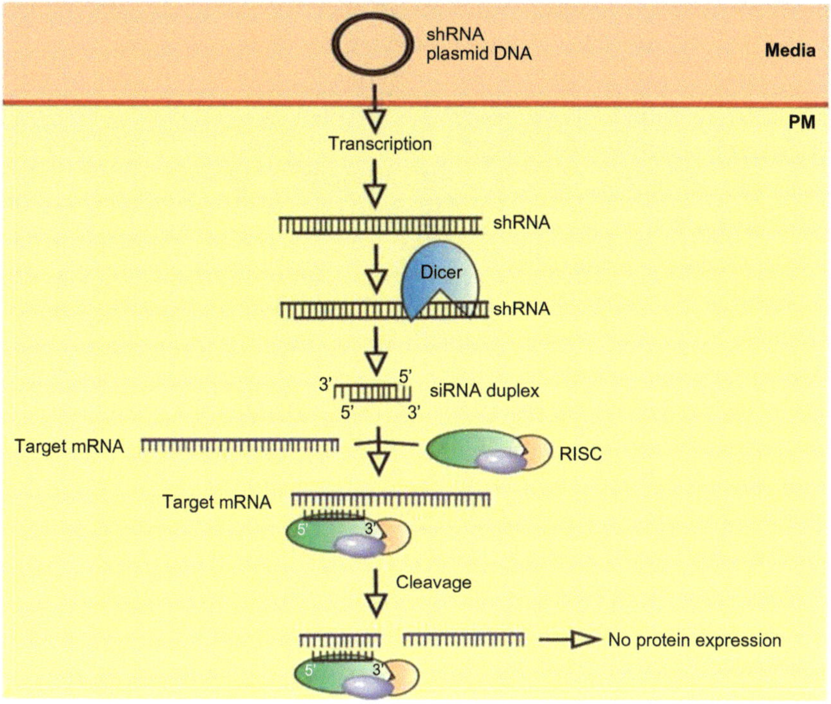

Fig. 1 Mechanism of siRNA activation and biochemical action. *Media* cell culture medium, *PM* plasma membrane

inhibitors of various signaling proteins often do not display absolute specificity. Additionally, we face the problem that both mast cells and basophils are hematopoietic cells of myeloid lineage. These cells (myeloid hematopoietic cells) are notoriously difficult to transfect [6, 7] because of the following reasons:

(a) They are highly differentiated, and therefore these cells are not geared toward overexpressing plasmids.
(b) They can often engulf and destroy foreign biologically active substances, including plasmids.
(c) Their membrane is much less permeable compared to other cells and has substantially more proteins integrated into it, which hinders the ready uptake and diffusion of transfection material.

Currently, there are several methods of transfection which have been employed for human mast cells/basophils and other myeloid cells.

The first is *electroporation*. In this case, an electric current is used to deliver genetic material. The membrane becomes relatively conductive, thus allowing current to pass through the cell when the electric fields are intensified above a certain threshold. This leads to the transient formation of pores by electrical breakdown. Only the genetic material bound to the plasma membrane is transferred into the cell. Another possible mechanism is that the electric pulse affects membrane stability where the appearance of unstable lipid bilayers results in the formation of vesicles containing genetic material which are then endocytosed. The potential problem with this technique is that biological responses of mast cells and basophils are regulated through plasma membrane-associated receptors. Electroporation could therefore affect the membrane potential by negatively influencing the ability of the inflammatory/immune receptors to recognize their ligands, which are vital in order to subsequently study their normal function [6, 7].

Another possible approach is *lipofection* (a contraction of "liposome transfection"). This is a technique which allows delivery of genetic material into a cell by the use of liposomes. Liposomes are lipid-derived vesicles that can easily interact with the cell membrane since both binding partners are made of a phospholipid bilayer. Lipofection generally assumes the use of positively charged (cationic) lipid to form an aggregate (complex) with negatively charged (anionic) genetic material. A net positive charge of this aggregate increases the effectiveness of transfection through the negatively charged phospholipid bilayer of the plasma membrane. The main advantages of lipofection are the relatively high transfection efficiencies, the possibility of transfection of all types of nucleic acids into a wide range of cell types, and the relatively low toxicity of the transfection reagents [3, 4, 8].

Currently, a new approach is under development which uses nanomaterials/nanoparticles for the delivery of genetic materials. This approach, known as *nanofection*, can result in very high transfection efficiencies, but in terms of its application to mast cells or basophils, it could be less effective since, upon endocytosis, nanomaterials often affect cytoskeleton elements. Upon degranulation, mast cells and basophils undergo major cytoskeletal alterations, and this type of transfection could therefore affect the ability of mast cells/basophils to respond to stimuli normally and give rise to observations that are unrelated to specific silencing of a target gene [3, 9].

Recent developments [10, 11] suggest that gold nanoconjugates, carrying antibody directed against unique target cell-specific receptor/surface protein, may also be considered for transfection. However, this technology has so far only been used successfully in the delivery of small molecular weight drugs and needs to be further developed for siRNA.

At present, lipofection is currently deemed to be the best transfection approach which one could use to silence genes in human mast cells and basophils.

There are several commercial transfection reagents which could be used in this case: DOTAP, Lipofectamine, FuGENE 6, etc. These reagents allow for relatively high transfection efficiencies

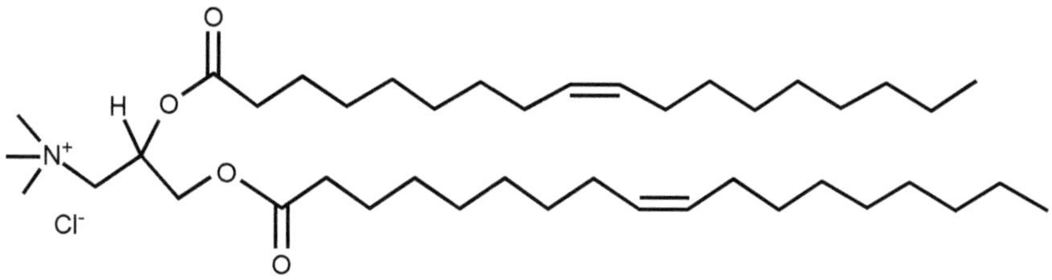

Fig. 2 Chemical structure of DOTAP chloride

(up to 40%) to be obtained while avoiding major cell damage or toxic effects on the target cells.

In our lab, we have successfully used DOTAP transfection reagent to deliver siRNA into LAD2 mast cells [4], primary human basophils [3], the monocyte cell line THP-1 [12], and monocyte macrophage 6 cell line (MM6) as well as human myeloid leukemia cells [13]. We will now present a protocol which describes the procedure for primary human basophils and LAD2 mast cells. The method is based on the ability of DOTAP chloride (*N*-[1-(2,3-dioleoyloxy)propyl]-*N*,*N*,*N*-trimethylammonium chloride; *see* Fig. 2), a cationic liposome-forming compound, to interact with genetic material forming aggregates which are endocytosed by target cells followed by a release of the genetic material inside the cell.

2 Materials

2.1 Cells

1. LAD2 human mast cells (*see* **Note 1**). These cells are available from Prof. Dean Metcalfe and Dr. Arnold Kirshenbaum (NAID, NIH, USA). Cells should be cultured in the StemPro-34 serum-free medium with supplements containing 2 mM L-glutamine, penicillin (100 units/mL), streptomycin (100 μg/mL), and recombinant human stem cell factor (SCF) (100 ng/mL) [14].
2. Primary human basophils (purified and handled as described in Chapter 2 of this book; *see* **Note 2**).

2.2 Reagents

1. Prepare HEPES (4-(2-hydroxyethyl)-1-piperazineethanesulfonic acid)-buffered saline (HBS): 20 mM HEPES, 150 mM NaCl, pH 7.4.
2. Solution A. Dissolve 5 μg of siRNA (or DNA) in 50 μL of HBS (*see* **Note 3**).
3. Solution B. Gently mix 30 μL of DOTAP chloride reagent (obtained from commercial sources) with 70 μL of HBS.
4. Ultrapure water suitable for molecular biology work.

3 Methods

3.1 Transfection

1. Gently mix solutions A and B (*see* **Note 4**; avoid vortexing or centrifugation) and leave for 30 min at 15 ± 10 °C. This is your working reagent (WR).
2. Add 150 μL of the WR to 5 mL of the culture medium containing your cells and mix gently, for example, by a careful swirling of the Petri dish (*see* **Note 5**).
3. Leave for 2–6 h for primary human basophils or overnight (~16 h for LAD2 mast cells or other types of myeloid cell lines).
4. Upon completion of the transfection, change the culture medium. Your cells are then ready for experimentation (*see* **Note 6**).

3.2 Analysis of Gene Silencing

Gene silencing should always be confirmed using two approaches:

1. Measurement of the mRNA of the target gene (protein) using the method which is most established in your laboratory. This could be reverse-transcription PCR or quantitative real-time PCR.
2. Detection of the actual amounts of your target protein. This should normally be done by Western blot analysis. However, if you are trying to silence expression of plasma membrane-associated proteins (e.g., Toll-like receptors or IgE receptors), then flow cytometry could also be an excellent approach to test protein levels (*see* **Note 7**).

4 Notes

1. LAD2 mast cells are currently the best human mast cell line for studying IgE-dependent as well as IgE-independent mast cell responses (*see* Chapter 10). However, we have in the past noticed variations in responses (e.g., changes in the percentage histamine release to anti-IgE stimulation after sensitization as well as very high spontaneous degranulation) while maintaining LAD2 cultures. It is advised that pilot experiments are first conducted to assess for acceptable functional responses before performing gene silencing studies.
2. It is essential to isolate and purify basophils in the shortest possible time in order to obtain viable cells after transfection. The current gene silencing methods, including the one described, can mainly be used in basophils for studying intracellular signaling processes and effects on de novo mediator generation. However, degranulation processes (e.g., histamine release), in our hands, appear to be more affected by the transfection reagent than we have seen with LAD2 mast cells.

3. It is recommended to optimize experimentally the amount of siRNA (or DNA) supplied, bearing in mind that DOTAP is normally nontoxic at concentrations below 150 μg/mL.
4. Use sterile tubes to make both solutions.
5. DOTAP allows for equal transfection quality in culture medium both in the presence and absence of serum. Therefore, for transfection, it is recommended to use the medium which you normally use to culture your cells [3, 4, 11].
6. It is essential to perform a cell viability check in parallel to the main experiment or upon completion of the treatment. In this regard, an MTS (3-(4,5-dimethylthiazol-2-yl)-5-(3-carboxymethoxyphenyl)-2-(4-sulfophenyl)-2H-tetrazolium) test would be optimal. Additionally, ensure that you use a random or mutated (scrambled—same overall nucleotide composition but different sequence) siRNA control to confirm that the effect you observe is specific and is a result of gene silencing. In addition, it is recommended to add a DOTAP reagent control, thus making sure that it does not affect the pathways you study.
7. Gene silencing affects mRNA, but it does not destroy the protein molecules which were already expressed. Therefore, the decrease in the constitutive levels of the silenced protein will depend not only on the mRNA levels but also on the stability of the actual protein in the cells.

References

1. Norata GD, Tibolla G, Catapano AL (2013) Gene silencing approaches for the management of dyslipidaemia. Trends Pharmacol Sci 34:198–205
2. Macrae I, Zhou K, Li F et al (2006) Structural basis for double-stranded RNA processing by Dicer. Science 311:195–198
3. Sumbayev VV, Nicholas SA, Streatfield CL et al (2009) Involvement of hypoxia-inducible factor-1 in IgE-mediated primary human basophil responses. Eur J Immunol 39:3511–3519
4. Sumbayev VV, Yasinska IM, Oniku AE et al (2012) Involvement of hypoxia-inducible factor-1 in the inflammatory responses of human LAD2 mast cells and basophils. PLoS One 7: e34259. https://doi.org/10.1371/journal.pone.0034259
5. Wang Z, Rao DD, Senzer N et al (2011) RNA interference and cancer therapy. Pharm Res 28:2983–2995
6. Imai E, Isaka Y (2002) Gene electrotransfer: potential for gene therapy of renal diseases. Kidney Int 61:S37–S41
7. Vilarino N, MacGlashan D Jr (2005) Transient transfection of human peripheral blood basophils. J Immunol Methods 296:11–18
8. Burke B (2003) Macrophages as novel cellular vehicles for gene therapy. Expert Opin Biol Ther 3:919–924
9. Sumbayev VV, Yasinska IM, Gibbs BF (2013) Biomedical applications of gold nanoparticles. In: Recent advances in circuits, communications and signal processing. Springer, New York, NY, pp 342–348
10. Yasinska IM, Calzolai L, Raap U et al (2019) Targeting of basophil and mast cell pro-allergic reactivity using functionalised gold nanoparticles. Front Pharmacol 10:333. https://doi.org/10.3389/fphar.2019.00333
11. Yasinska IM, Ceccone G, Ojea-Jimenes I et al (2018) Highly specific targeting of human acute myeloid leukaemia cells using pharmacologically active nanoconjugates. Nanoscale 10:5827–5833
12. Nicholas SA, Sumbayev VV (2009) The involvement of hypoxia-inducible factor

1 alpha in Toll-like receptor 7/8-mediated inflammatory response. Cell Res 19:973–983

13. Shatrov VA, Sumbayev VV, Zhou J et al (2003) Oxidized low density lipoprotein triggers HIF-1alpha protein via redox-dependent mechanism. Blood 101:4847–4849

14. Kirshenbaum AS, Akin C, Wu Y et al (2003) Characterization of novel stem cell factor responsive human mast cell lines LAD 1 and 2 established from a patient with mast cell sarcoma/leukemia; activation following aggregation of FcepsilonRI or FcgammaRI. Leuk Res 27:677–682

Chapter 27

Identification and Immunophenotypic Characterization of Normal and Pathological Mast Cells

José Mário Morgado, Laura Sánchez-Muñoz, Cristina Teodósio, and Luis María Escribano Mora

Abstract

Mast cells (MCs) are secretory cells that are central players in human allergic disease and immune responses. With the exception of a few pathological situations, MCs are usually present at relatively low frequencies in most tissues. Since their first description, MCs in tissues were identified mostly using their morphological characteristics and their typical coloration when stained with aniline dyes. However, increasing availability of highly specific antibodies now permits the use of fluorescence-based flow cytometry as the method of choice for the quantification, characterization, and purification of cells in suspension. This technique allows for a rapid analysis of thousands of events and for the identification of cells present at frequencies as low as one event in 10^6 unwanted cells. This method also permits for simultaneous characterization of multiple antigens at a single cell level, which is ideal in order to study rare populations of cells like MCs. Here we describe the basis of flow cytometry-based immunophenotyping applied to the study of MC. The protocol focuses on the study of human MCs present in body fluids (mainly bone marrow) but can easily be adapted to studying MCs from other tissues and species.

Key words Mast cell, Mastocytosis, Flow cytometry, Immunophenotype

1 Introduction

Mast cells (MCs) are unique secretory cells which have gained notoriety over the years for their role as central players in human allergic disease and immune responses mediated by high-affinity IgE receptor (FcεRI) [1, 2]. Nowadays, MCs are known to participate in various cell-mediated reactions and in tissue repair processes after injury (reviewed in [3]). MCs develop from $CD34^+$ progenitors in the bone marrow (BM) [4, 5], but, unlike other leukocytes, they migrate through the blood stream as precursors and are recruited to peripheral tissues where they begin their final maturation under the influence of cytokines and growth factors in the local microenvironment [6, 7].

Bernhard F. Gibbs and Franco H. Falcone (eds.), *Basophils and Mast Cells: Methods and Protocols*, Methods in Molecular Biology, vol. 2163, https://doi.org/10.1007/978-1-0716-0696-4_27,

Before the onset of immunophenotyping techniques, the distinction between different cell types was dependent on morphological or functional differences between them. Since then, quantitative evaluation of MCs has been attempted in several tissues and pathological conditions using different techniques. The rise of antibodies highly specific for many different proteins potentiated the use of fluorescence-based flow cytometry; this technique became the method of choice for the quantification, characterization, and purification of cells in suspension, allowing the rapid analysis of thousands of cells [8–11]. Moreover, the use of a double-step acquisition procedure, in which an appropriate live-gate acquisition is performed, has largely contributed to the detection and study of small populations of cells present in single cell suspension samples. This method allows identification of cells even when their frequency is as low as that of one event in up to a total of 10^6 unwanted cells [12].

In 1987, it was reported that a mouse monoclonal antibody (mAb) (YB5.B8) reacted with MCs without distinguishing mucosal from connective tissue MCs and did not bind to basophils [13]. This antibody was later demonstrated to be specific for the c-kit proto-oncogene product (KIT or CD117) [14] which was expressed not only by MCs but also on hematopoietic precursors in the bone marrow (BM) [15].

In 1996, Orfao et al. demonstrated that BMMCs may be clearly identified by flow cytometry on the basis of their light scatter properties and strong CD117 expression [16]. In addition, these cells were shown to be negative for CD34, CD38, and BB4 antigens and displayed a high reactivity for anti-IgE and anti-CD33 (Siglec-3) mAb's. This technical advance was revealed to be of major value in the study of MC-related disorders like mastocytosis and other clonal mast cell diseases. Applying the same technical principles, Escribano et al. reported that BMMCs from patients with systemic mastocytosis (SM) display unique aberrant immunophenotypic characteristics, as compared to normal MCs [17]. Among other features, pathological MCs show aberrant expression of CD25 and CD2 together with abnormally high levels of CD11c and CD35 complement receptors, CD59 complement regulatory molecule, CD63 lysosomal membrane antigen, and CD69 early-activation antigen [18–20]. More recently, three clearly different maturation-associated immunophenotypic profiles were found for BMMCs in SM [21] and related to both genetic markers of the disease and its clinical behavior.

Even though, in situations like MC hyperplasia or SM, a great number of MCs may be observed in tissues like BM or skin [22], under normal circumstances and in most SM cases, the number of MCs present in different tissues is, with few exceptions, considerably low [23]. For that reason, it is mandatory to analyze a high number of cells (1–2 × 10^6 cells] in order to obtain information from sufficient MCs to correctly interpret their phenotypic characteristics.

Although flow cytometry is highly indicated for the study of cells and other particles in suspension, like those from peripheral blood (PB) and BM, when appropriate methods are employed to create cell suspensions from solid tissues, these may also be efficiently studied using this technique. The protocols described in this chapter allow for the identification and enumeration of MCs using multiparametric flow cytometry on the basis of technical approaches used for the identification of cells present at low frequencies. This approach can be applied to virtually any cell suspension prepared from a wide range of human specimens, including BM, PB, organic fluids (e.g., ascitic fluid), and solid tissues, among others. In addition, the protocols described here allow for the study of surface membrane antigens and intracellular molecules and are therefore particularly useful to study the typical MC granule contents stored within these cells. These protocols are mainly directed to study human MCs, but the same principles may be easily applied to study other sources of MCs like animal model tissues or cell cultures.

2 Materials

1. 14–8 G biopsy needle for BM aspiration.
2. Vacutainer® tubes containing either tripotassium EDTA or heparin.
3. 25 G needle and syringe.
4. 50 mL conical bottom centrifuge tubes.
5. Phosphate-buffered saline (PBS, 1×): 137 mM NaCl, 2.7 mM KCl, 10 mM Na_2HPO_4, 1.8 mM KH_2PO_4, pH 7.4.
6. Hematological cell analyzer and/or other cell counting system.
7. Cytospin centrifuge (optional).
8. Microscopy crystals and coverslips.
9. Toluidine blue 1% (w/v) in methanol, pH 2.0–2.5.
10. Bright field light microscope.
11. 12 × 75 mm polystyrene tubes.
12. Monoclonal antibodies directly conjugated with fluorochromes suitable to be detected by your flow cytometer (*see* **Note 1**) (*see* Tables 1 and 2).

Table 1
Four color combinations recommended for immunophenotypic analysis of mast cells in the diagnosis of systemic mastocytosis

Tube #	FITC	PE	PerCP-Cy5.5	APC
Screening tubes[a]				
1	Control	Control	CD45	CD117
2	CD2	CD25	CD45	CD117
3	FcεRI[b]	CD30	CD45	CD117
Further mast cell characterization x[c]				
4	CD35	CD69	CD45	CD117
5	CD63	CD32	CD45	CD117
6	HLA-DR	CD123	CD45	CD117
7	cy Control	cy Control	CD45	CD117
8	cy Total tryptase (B12)[d]	CD34	CD45	CD117
9	cy CPA	CD203c	CD45	CD117

Control, unstained negative control; CPA, carboxypeptidase A3; cytoplasmic markers are preceded by "cy." Otherwise, we refer to membrane markers

FITC fluorescein isothiocyanate, *PE* phycoerythrin, *PerCP* peridinin-chlorophyll-protein complex, *Cy* cyanin, *APC* allophycocyanin

[a]Usually these combinations are enough to characterize the mast cell phenotype in the process of systemic mastocytosis diagnosis

[b]Indirectly observed by anti-IgE staining

[c]These combinations are merely suggestive and may be adapted according to the objective of the study

[d]This particular clone is not available commercially

13. Red cell lysing solution [e.g., FACS lysing solution 1× (Becton Dickinson): stock commercial 10× solution diluted 1/10 (v/v) in distilled water (dH_2O)] (*see* **Note 2**).
14. 1× PBS containing 0.5% (w/v) of paraformaldehyde (optional).
15. Flow cytometer with at least four fluorescence detectors and appropriate filter sets.
16. Software for analysis of flow cytometry data.
17. Cell fixation and permeabilization reagents to assess the cytoplasmic expression of proteins (e.g., Fix&Perm (An der Grub, Bio Research GmbH, Vienna, Austria)) (*see* **Note 3**).

Table 2
Eight color combinations recommended for immunophenotypic analysis of mast cells in the diagnosis of systemic mastocytosis

Tube #	Pacific blue	Pacific orange	FITC	PE	PerCP-Cy5.5	PE-Cy7	APC	APC-H7
Screening tubes[a]								
1	Control	CD45	Control	Control	CD34	CD117	Control	Control
2	CD2	CD45	CD35	CD25	CD34	CD117	CD203c	CD69
3	HLA-DR	CD45	FcεRI[b]	CD30	CD34	CD117	CD123	CD38
Further mast cell characterization[c]								
3	HLA-DR	CD45	CD35	CD59	CD34	CD117	CD123	CD25[d]
4	HLA-DR	CD45	CD63	CD32	CD34	CD117	CD123	CD25[d]
5	HLA-DR	CD45	cy Control	cy Control	CD34	CD117	cy Control	CD25[d]
6	HLA-DR	CD45	cy Total tryptase (B12)[e]	cy Chymase	CD34	CD117	cy CPA	CD25[d]

Control, unstained negative control; CPA, carboxypeptidase A3; cytoplasmic markers are preceded by "cy." Otherwise, we refer to membrane markers

FITC fluorescein isothiocyanate, *PE* phycoerythrin, *PerCP* peridinin chlorophyll protein complex, *Cy* cyanin, *APC* allophycocyanin, *H7* analogue to Cy7

[a]Usually these combinations are enough to characterize the mast cell phenotype in the process of systemic mastocytosis diagnosis

[b]Indirectly observed by anti-IgE staining

[c]These combinations are merely suggestive and may be adapted according to the objective of the study

[d]To be included when a double population (CD25+ and CD25−) is observed in screening tubes

[e]This particular clone is not available commercially

3 Methods

3.1 Cell Suspension Preparation

3.1.1 Specific Procedures for Bone Marrow Samples

1. Collect a minimum of 1.5–2 mL of BM aspirate using tubes containing EDTA or heparin as anticoagulant.
2. Prepare smears according to conventional methods taking special attention to prepare the smears with BM particles.
3. Immerse air-dried preparations in 0.5% toluidine blue solution during 4 min and then wash in water.
4. Mount with a coverslip and observe under the microscope to get an overall impression on whether MC numbers are low, normal, or high (*see* Fig. 1) (*see* **Note 4**).
5. Pass the BM aspirate two or three times through a 25 G gauge needle in order to disaggregate the particles.

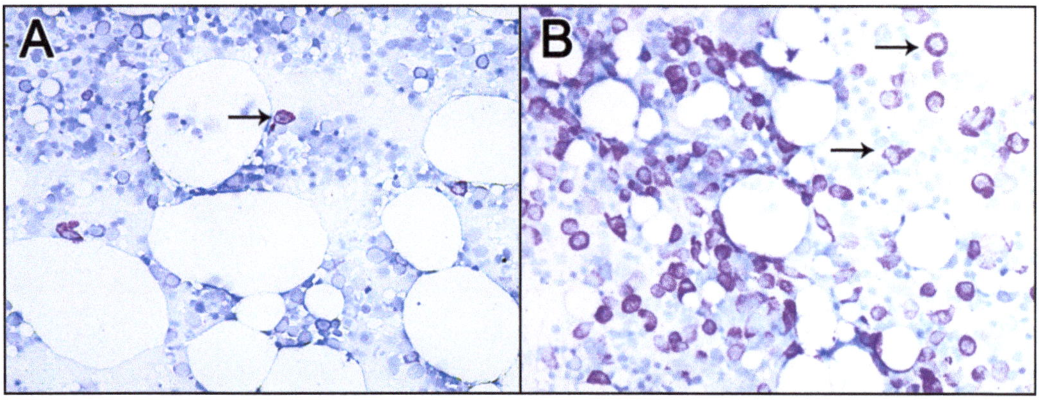

Fig. 1 Toluidine blue staining of two different BM smears. MCs may be identified based on their characteristic reddish violet color (arrows). The visual evaluation of a BM smear using this stain is of great help to determine the amount of sample to be stained for immunophenotyping by flow cytometry

6. Quantify the amount of cells in the sample using a conventional hematological analyzer or other methods (e.g., Neubauer counting chamber).

3.1.2 Specific Procedures for High Diluted Samples (Organic Fluids)

1. Divide the sample into 40–45 mL aliquots (in 50 mL conic bottom tubes) and centrifuge 10 min at 540 × *g* at RT.
2. Discard the supernatant and resuspend the pellet in 1 mL of 1× PBS.
3. Transfer all the samples to a single 50 mL conical bottom centrifuge tube.
4. Quantify the amount of cells in the sample using a conventional hematologic analyzer or other methods (e.g., Neubauer counting chamber).

3.1.3 Other Samples

As a general rule, any tissue susceptible of being turned into a cell suspension may be analyzed by flow cytometry. The process is relatively straightforward for some tissues which are, by definition, cell suspensions, like PB and BM blood. In the case of PB, the number of MC is usually very low, and it is advisable to stain triplicates of each tube.

For solid tissues, mechanical and/or enzymatic disaggregation must be employed in a process that may vary depending on the nature of the tissue.

3.2 Antigen Staining

Flow cytometry allows the detection of membrane-bound antigens and/or cytoplasmic contents of cells. After a cell suspension is obtained, the samples may be studied following the same general procedure regardless the origin of the sample. However, depending on the antigens to be assessed, some variations of the protocol must be applied (i.e., membrane vs. cytoplasmic antigens) as detailed below.

If only membrane antigens are to be studied, continue with the protocol in Subheading 3.2.1. If cytoplasmic antigens are meant to be studied, proceed with the protocol in Subheading 3.2.2.

3.2.1 Staining Procedure for Surface Antigens Only

1. Label 12 × 75 mm polystyrene tubes according to the panel to be employed (*see* Tables 1 and 2).
2. Pipette the volume of sample containing approximately 2.5×10^6 cells into each tube.
3. Add the appropriate volume of antibodies directed against cell surface markers to the tubes containing sample aliquots.
4. Incubate for 15 min at room temperature (RT) protected from light.
5. Add 2 mL of 1× red cell lysing solution and mix well by vigorous vortexing.
6. Incubate for 10 min at RT protected from light (*see* **Note 5**).
7. Centrifuge for 5 min at 540 × *g* and discard the supernatant using a Pasteur pipette or vacuum system without disturbing the cell pellet, leaving approximately 50 μL residual volume in each tube.
8. Add 2 mL of 1× PBS to the cell pellet and mix well.
9. Centrifuge for 5 min at 540 × *g* and discard the supernatant using a Pasteur pipette or vacuum system without disturbing the cell pellet, leaving approximately 50 μL residual volume in each tube.
10. Resuspend the cell pellet in 300 μL of 1× PBS.
11. Acquire the cells after staining or store at 4 °C (in the dark) for a maximum of 3 h until measurement in the flow cytometer.

3.2.2 Combined Staining Protocol for Intracellular and Surface Membrane Antigens

1. Label 12 × 75 mm polystyrene tubes according to the panel to be employed (*see* Tables 1 and 2).
2. Pipette the volume of sample containing approximately 2.5×10^6 cells into each tube.
3. Add the appropriate volume of antibodies directed against cell surface markers to the tubes containing sample aliquots.
4. Incubate for 15 min at RT protected from light.
5. Add 2 mL of 1× PBS to the cell pellet to rinse the unbound antibody and mix well.
6. Centrifuge for 5 min at 540 × *g* and discard the supernatant using a Pasteur pipette or vacuum system without disturbing the cell pellet, leaving approximately 50 μL residual volume in each tube.
7. Resuspend the cell pellet by mixing gently.
8. Add 100 μL of Reagent A (fixation medium; Fix&Perm).

9. Incubate for 15 min at RT protected from light.
10. Add 2 mL of 1× PBS to the cell pellet and mix well.
11. Centrifuge for 5 min at 540 × *g* and discard the supernatant using a Pasteur pipette or vacuum system without disturbing the cell pellet, leaving approximately 50 μL residual volume in each tube.
12. Resuspend the cell pellet (*see* **Note 6**).
13. Add 100 μL of Reagent B (permeabilizing solution; Fix&-Perm) and mix well.
14. Add the appropriate volume of the antibodies directed against intracellular antigens and mix well.
15. Incubate for 15 min at RT protected from light.
16. Add 2 mL of 1× PBS to the cell pellet and mix well.
17. Centrifuge for 5 min at 540 × *g* and discard the supernatant using a Pasteur pipette or vacuum system without disturbing the cell pellet, leaving approximately 50 μL residual volume in each tube.
18. Resuspend the cell pellet in 500 μL of 1× PBS.
19. Acquire the cells after staining or store at 4 °C for a maximum of 3 h until measurement in the flow cytometer.

3.3 Instrument Setup and Data Acquisition

The sample acquisition should be performed using a two-step procedure. A first file is created with the information on enough total cells to have a general representation of the sample (i.e., 1×10^5 cells). For reference, we will name these files as "Step 1 files." Then, a second file is created, and only the information is stored on those events falling inside a live gate drawn to include $CD117^+$ events. This will allow the analysis of a higher amount of cells while saving disc space and make the manipulation of the files easier. For reference, we will name these files as "Step 2 files":

1. Run the quality control and compensation setup according to specific instructions from the manufacturer of the cytometer. Make sure that forward scatter (FSC) width (FSC-W), height (FSC-H), and area (FSC-A) parameters are enabled (*see* **Note 7**).
2. Create bivariate dot plots as represented in Fig. 2:
   (a) FSC versus SSC.
   (b) CD117 versus SSC.
   (c) CD117 versus CD45.
3. In the dot plot "CD117 versus SSC," display all events and draw a gate (R1) to include all $CD117^+$ events (*see* Fig. 2b).
4. In the dot plot "CD117 versus CD45," display R1 events and draw a gate (R2) to include all $CD117^{high}$ events (*see* Fig. 2c).

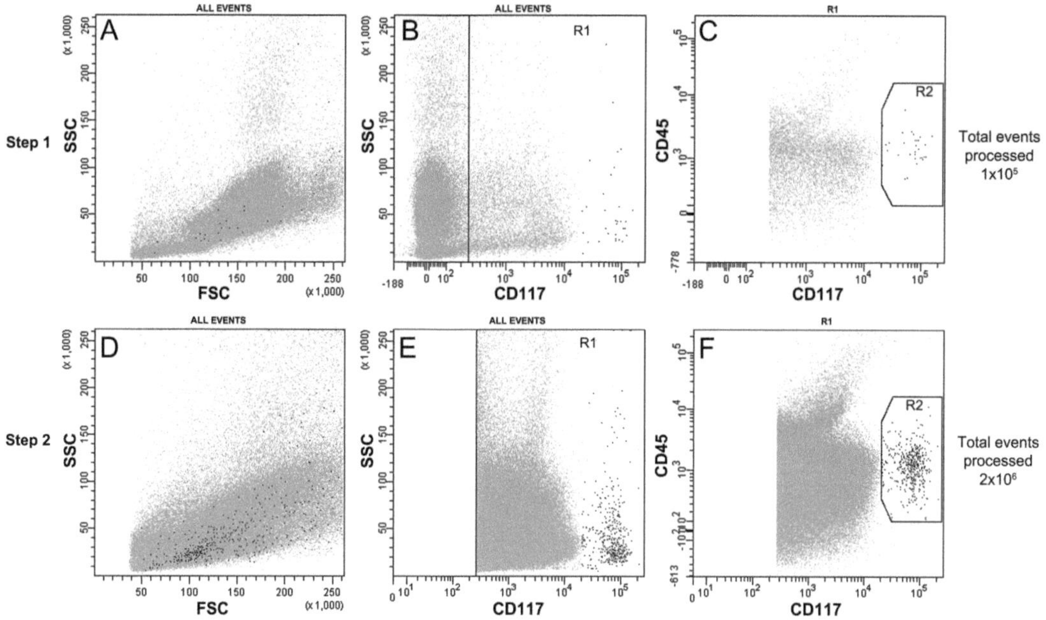

Fig. 2 Two-step acquisition procedure in a BM sample to study MCs. In Step 1, the information on all the events processed is stored to have general information on the sample studied. In Step 2, only the information on CD117^{+} events is stored; however, it is important to register the number of total events processed in order to calculate the percentage of MCs in the sample

5. For each tube, create a "Step 1 file" by acquiring and storing the information on a total of at least 1×10^5 events.
6. Be sure that CD117^{high} cells (gate R2) can be clearly discriminated from CD117^{dim+} cells.
7. For each tube, create a "Step 2 file" by acquiring and storing the information on the events falling inside the gate R1. Be sure to have a minimum of 100 CD117^{high} events (R2) for each tube.
8. Register the number of total processed events during the acquisition of "Step 2 files" for future calculation (usually this value is not saved by the acquisition software).

3.4 Data Analysis

Data analysis can be performed with any software capable of analyzing FCS 2.0 or FCS 3.0 files. In the examples shown, the Infinicyt software program (Cytognos) was used.

3.4.1 Mast Cell Identification and Enumeration

1. Create the following bivariate dot plots (*see* Fig. 3):
   (a) FSC-A versus SSC-A.
   (b) FSC-A versus FSC-W.
   (c) CD117 versus SSC.
   (d) CD45 versus CD117.

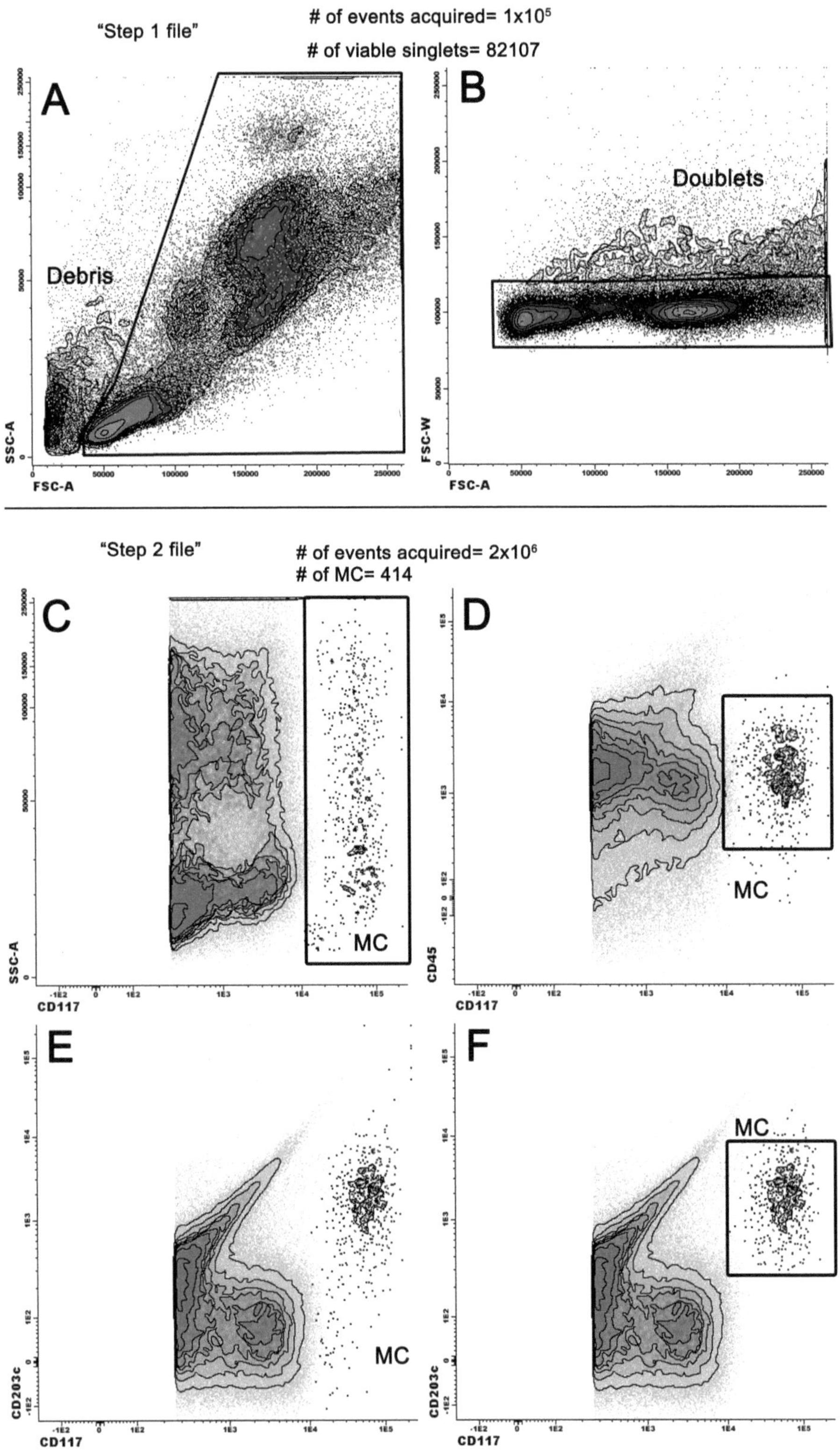

Fig. 3 Analysis procedure to identify MCs. In "Step 1 files," the number of viable singlets (events remaining after discarding doublets and debris) and the number of total events processed must be registered (panels

   - (e) If other "identification markers" are included, create CD117 versus "identification marker" dot plots to improve MC identification.
2. Analyze "Step 1 files":
   - (a) Exclude cell debris (*see* Fig. 3a) and doublets (*see* Fig. 3b and **Note 8**).
   - (b) Register the number of events that remain after excluding debris and doublets for each file (viable singlets).
3. Analyze "Step 2 files":
   - (a) Select and register the number of events fulfilling the phenotype of MC based on the identification markers used excluding cell debris and doublets (*see* **Note 9**).
4. Recall the following values registered during the process of acquisition and analysis of data from each tube:
   - (a) Number of total events processed in "Step 1 files."
   - (b) Number of viable singlets (events in "Step 1 files" after removing debris and doublets).
   - (c) Number of total events processed in "Step 2 files."
   - (d) Number of events fulfilling the phenotype of MC in "Step 2 files."
5. Calculate the percentage of MC (*see* **Note 10**):
   - (a) Apply the following formula using the values obtained from each tube:

$$\%\ \mathrm{MC} = \frac{D}{C \times \left(\frac{B}{A}\right)} \times 100.$$

   - (b) Calculate the mean value among the different tubes.

$$\%\mathrm{MC} = \frac{\text{number of MC} \times 100}{\text{processed events}} \times \frac{100}{\%\text{of cells of interest}}.$$

3.4.2 Mast Cell Phenotype Assessment

1. For each tube analyzed, register the information of fluorescence intensity for MCs in each channel including the fluorescence intensity values from the control tube to control the autofluorescence of MCs (*see* **Note 11**).
2. One may use the raw fluorescence intensity values to compare different samples or correct the values of each marker with the autofluorescence in the corresponding channel (*see* **Note 12**).

Fig. 3 (continued) a and **b**). In "Step 2 files," the number of events fulfilling the phenotype of MCs and the number of total events processed must be registered. The use of extra identification markers may help to correctly identify MCs (panels **c–f**)

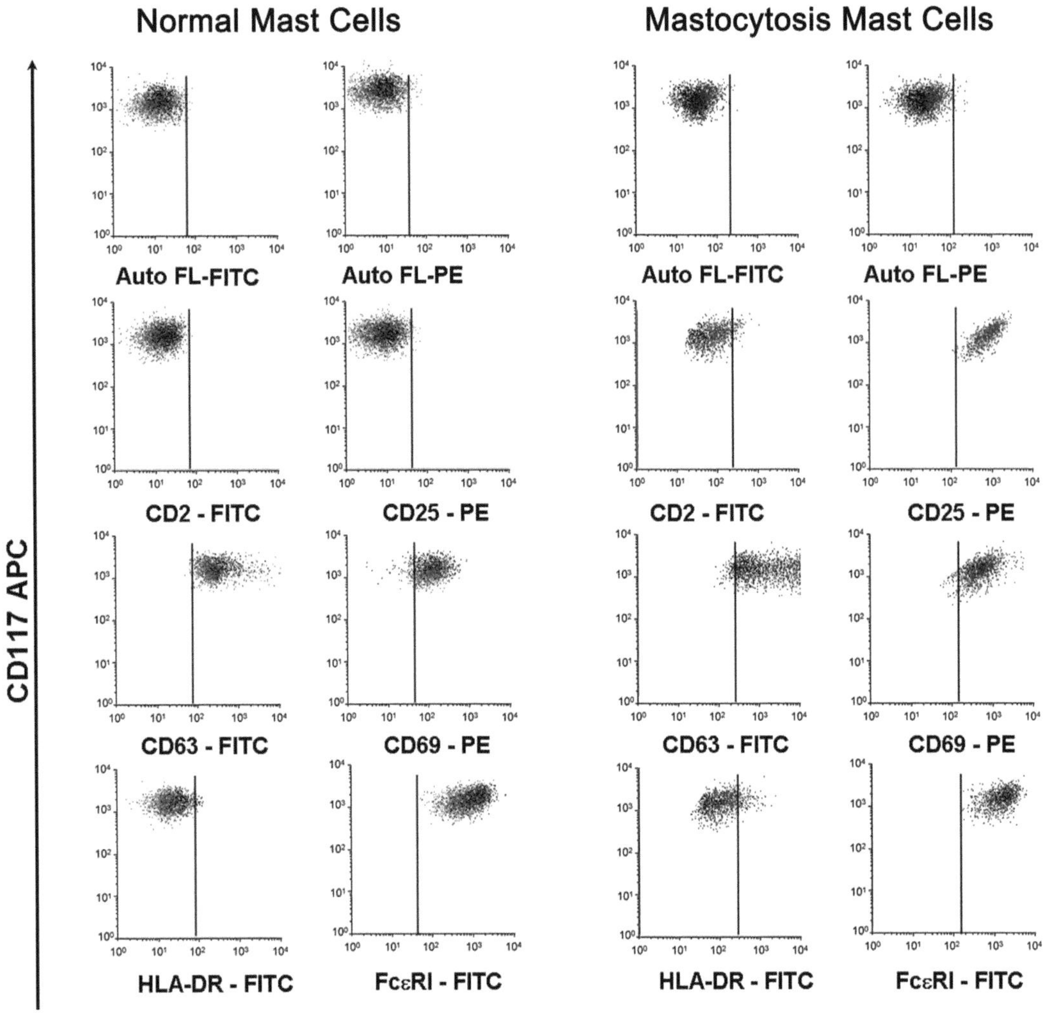

Fig. 4 MC phenotype assessment in two representative BM samples obtained from a nonmastocytosis case and from one patient with mastocytosis. The use of a "control" tube allows for the correction of autofluorescence of MCs which is variable among samples and is detected in the different channels. The autofluorescence levels in the corresponding channel must be used to determine the positivity and the relative intensity of expression for each antigen studied

3. Positivity must be established taking the autofluorescence in each channel (control tube) as a negative reference (*see* Fig. 4).
4. Record the phenotypic profile of MCs identified on the samples studied.

3.5 Expected Results

MCs display distinct immunophenotypes depending on their stage of maturation, tissue environment, and activation status [24]. For this reason, and for the correct interpretation of the distinct profiles detected in MC-related disorders, it is essential to know the normal immunophenotypic profile of MCs in each situation. Since the BM is the most common source of MCs to study by flow cytometry, mainly for diagnostic purposes, we present here the most common immunophenotypic features of MC in this tissue under normal and pathological situations (*see* Table 3).

3.5.1 Mast Cell-Committed Precursors

MCs mature from hematopoietic precursors which, based on in vitro experiments, was shown to be $CD34^+$, c-kit^+, $CD14^-$, $CD17^-$ [25, 26], $Fc\varepsilon RI^-$ [27], CD38 often positive, HLA-DR often negative [28], and $CD13^+$ [29]. Recent ex vivo studies, using BM from both healthy controls and patients with myelodysplastic syndrome (MDS), suggest that MC-committed precursors may be identified within the BM CD34+ HPC compartment as being $CD117^{hi}$/HLA-$DR^{-/int}$ [30]. Nevertheless, these precursors are extremely infrequent in normal BM [0 ± 0.005% (range 0–0.02%)] [30], and, consequently, our knowledge on the normal pattern of MC maturation is mainly based on in vitro differentiated MCs [31–34]. More recently, it has been reported that both immature MC-committed progenitors and mature MCs express CD117, CD58, CD63, CD147, CD151, CD203c, and CD172a independently of the growth factors used but that the expression of IL-3Rα (CD123) and of the granulocyte-macrophage colony-stimulating factor receptor (GM-CSFR) is restricted to early MC precursors [32]. There is a general consensus in the literature indicating that proteins typically associated with MCs, like the FcεRI, MC proteases (chymase and tryptase), or histamine, are only expressed in late stages of the maturation process and are absent in $CD34^+$ MC precursors [5, 27, 33, 35]. The exception seems to be tryptase which, based on unpublished observations of our group from ex vivo studies, may be dimly expressed in a small proportion of $CD34^+$ MC-committed precursors.

3.5.2 Mature Resting BMMC Phenotype Under Normal and Reactive Conditions

Normal MCs display high light scatter and autofluorescence and express a broad set of proteins, including myeloid-related antigens (e.g., CD33), proteins related with the initiation of the MC inflammatory response (e.g., FcεRI) [16, 36], or MC proteases (e.g., cytoplasmic carboxypeptidase and cytoplasmic total tryptase), displaying relatively high amounts of cytoplasmic immature tryptase [21]. Other proteins are described in the majority of normal/reactive MCs, like bcl2 (94%) or cytoplasmic mature tryptase [21], while others are only detected in a restricted number of them, like the IgG Fc receptors CD16 and CD64 [16, 17, 36, 37]. Furthermore, normal/reactive BMMCs constantly lack the expression of several proteins, some of them relevant from a SM diagnosis point of view, like CD2 or CD25 [16, 17, 36].

Table 3
Qualitative and semiquantitative patterns of expression of individual markers on normal, reactive, and systemic mastocytosis bone marrow mast cells

Functional group	Antigen	CD	Normal BM	Reactive BM	Overall SM	ISM	WDSM	ASM/MCL
Cytokine receptor	IL-2Rα	CD25	–	–	–/+ (93%)	+ (100%)	–/+dim (27%)	+ (100%)
	IL-3Rα	CD123	–	NR	–/+ (72%)	–/+ (84%)	–	–/++ (80%)
	c-Kit	CD117	+++ (100%)	+++ (100%)	++/+++ (100%)	++/+++ (100%)	+++ (100%)	++ (100%)
Adhesion-related molecules	LFA-2	CD2	–	–	–/+ (72%)	–/+ (82%)	–/+dim (9%)	–/+ (18%)
	Integrin alpha-L	CD11a	–/+ (20%)	NR	–	NR	NR	NR
	Integrin alpha-M	CD11b	–/+ (50%)	–/+ (50%)	–/+ (50%)	NR	NR	NR
	Integrin alpha-X	CD11c	–/+ (71%)	NR	+/++ (100%)	NR	NR	NR
	Integrin beta-2	CD18	+ (65%)	NR	–/+ (44%)	NR	NR	NR
	Siglec-2	CD22	–/+ (60%)	–/+ (50%)	–/+ (96%)	–/++ (95%)	+/++ (100%)	+/++ (100%)
	Integrin beta-1	CD29	++ (100%)	++ (100%)	+ (100%)	NR	NR	NR
	Siglec-3	CD33	++/+++ (100%)	++/+++ (100%)	+++ (100%)	NR	NR	NR
	HPCA-1	CD34	–	–	–	–	–	–
	Pgp-1	CD44	++ (100%)	++ (100%)	++ (100%)	NR	NR	NR
	Integrin alpha-4	CD49d	+/++ (100%)	NR	–/+ (80%)	NR	NR	NR
	Integrin alpha-5	CD49e	+ (100%)	NR	–/+ (30%)	NR	NR	NR
	Integrin alpha-V	CD51	+ (100%)	+	–/+ (45%)	NR	NR	NR
	ICAM-1	CD54	–/+ (75%)	NR	++ (100%)	NR	NR	NR
Complement-related molecules	CR1	CD35	–	+	+ (100%)	NR	NR	NR
	Complement decay-accelerating factor	CD55	+ (100%)	+ (100%)	+/++ (100%)	NR	NR	NR
	Membrane attack complex inhibition factor	CD59	+ (100%)	+ (100%)	+/++ (100%)	++ (100%)	+ (100%)	++ (100%)
	C5aR	CD88	–/+ (18%)	NR	–/+ (54%)	NR	NR	NR

Immunoglobulin receptors	FcεRI	NA	++/+++ (100%)	++/+++ (100%)	++/+++ (100%)	+/+++ (100%)	+++ (100%)	+/++ (100%)
	FcγRIIIB	CD16	−/+ (13%)	−/+ (13%)	−/+ (69%)	−/+ (75%)	−/+ (22%)	−/+ (18%)
	FcγRII	CD32	+ (100%)	+ (100%)	+/++ (100%)	++ (100%)	+ (100%)	++ (100%)
	FcγRI	CD64	−/+ (4%)	−/+ (4%)	−/+ (84%)	−/+ (87%)	−/+ (40%)	−/+ (18%)
MHC-related molecules	HLA-I	NA	++ (100%)	++ (100%)	+/+++ (100%)	++/+++ (100%)	++	+/++ (100%)
	HLA-DR	NA	−/+ (25%)	−/+ (25%)	−/+ (85%)	−/+ (90%)	−/+dim (50%)	−/+ (82%)
Tetraspanins	CD9 antigen	CD9	+++ (100%)	+++ (100%)	++/+++ (100%)	NR	NR	NR
	LAMP3	CD63	++ (100%)	++/+++[a] (100%)	++/+++ (100%)	++/+++ (100%)	++ (100%)	++/+++ (100%)
Mast cell proteases	Tryptase	NA	++ (100%)	++ (100%)	+/+++ (100%)	+/++ (100%)	+++ (100%)	+ (100%)
	Carboxypeptidase A	NA	++ (100%)	++ (100%)	+/+++ (100%)	+/++ (100%)	+++ (100%)	+ (100%)
Activation markers	Early-activation antigen CD69	CD69	+ (100%)	+ (100%)	+/++ (100%)	++ (100%)	+ (100%)	+/++ (100%)
	E-NPP 3	CD203c	+ (100%)	NR	+/++ (100%)	++ (100%)	+ (100%)	+/++ (100%)
Others	LCA	CD45	+ (100%)	+ (100%)	+/++ (100%)	++ (100%)	+ (100%)	+ (100%)
	Transferrin receptor protein 1	CD71	+ (100%)	NR	−/+ (38%)	NR	NR	NR
	Bcl-2	NA	−/+ (94%)	−/+ (94%)	−/+ (87%)	−/++ (81%)	+/++ (100%)	+/++ (100%)

Adapted from: [17–21, 36, 41, 49, 50, 52–54, 70]

Symbols, −/+: expressed in a subset of patients (percentage of positive cases); −, absent expression; +dim, dim positive expression; +, moderate positive expression; ++, positive expression; +++, strong positive expression. Mixed symbols indicate variable reactivity, either among cases or among cells from the same individual. Values within parentheses represent the percentage of positive cases

3.5.3 Activated Immunophenotype of Normal MCs

MCs undergo the final stages of their differentiation/maturation after the migration of their precursors into those vascularized tissues or serosal cavities in which they ultimately reside [38]. Mature tissue MCs are heterogeneous, and specific subpopulations with varied mediator profiles and different functional properties are observed in the different tissues [39] where they have an important role in the recognition of pathogens or other signs of infection [40]. A variety of stimuli can activate MCs to release a diverse array of biologically active products (reviewed in [38]), and MC activation by IgE-dependent stimuli or other triggers is associated with significant changes in antigen expression. These antigens include β2-integrins, cytokine receptors, complement receptors, and members of the tetraspan antigen family [41], like the lysosomal glycoprotein CD63 which is constitutively expressed on MCs and upregulated by IgE receptor cross-linking [41, 42]. Likewise, both the ectonucleotide pyrophosphatase/phosphodiesterase 3 (CD203c) and the CD69 activation-linked cell surface antigen are upregulated upon IgE activation [26, 41, 43, 44]. Other antigens, like MHC class II, are not expressed on resting MCs, but their expression on MCs isolated from pathogen-infected tissues and/or stimulated by tumor necrosis factor (TNF), INFγ, or bacterial lipopolysaccharide (LPS) reflects a role for MCs in adaptive immune responses since they may contribute to the presentation of antigens to T-cells [45–47].

Although MCs express multiple direct receptors for pathogens and their products, the indirect activation of MCs during infection allows them to respond to a wider range of organisms [39]. As previously referred, MCs constitutively express the high-affinity IgE receptor (FcεRI) and the IgG receptor (FcγRII, CD32). Nevertheless, expression for both FcγRI (CD64) and FcγRIII (CD16) is only induced after exposure to INFγ [48]. This Fc-receptor-mediated activation generally leads to production of lipid mediators, the generation of various cytokines and chemokines and degranulation, and a decrease in cytoplasmic contents of MC proteases such as tryptase [39].

3.5.4 MC Immunophenotype in Systemic Mastocytosis

Early studies on the immunophenotypic features of SM BMMC reported several aberrant phenotypes that allowed for a clear discrimination from normal and reactive MC [17, 49–55] (*see* Table 3). These studies reported the aberrant expression for CD2 and CD25 as a hallmark for SM, since these proteins were absent in normal/reactive BMMCs but expressed in the great majority of SM patients [17, 50, 51]. From these two markers, CD25 has been described to be the most sensitive, specific, and less variable, since positivity for CD2 has been described to vary depending on the fluorochrome used in the study [56].

Apart from the aberrant expression of CD2 and CD25, SM BMMCs display abnormally high expression of CD33 [52, 53] and of the CD2 ligand, CD58 [54]. Increased expression for activation markers such as CD69 [20] or CD63 [19] and complement-associated molecules, like CD11c, CD35 [19], CD59, or CD88, are also found in SM BMMC. In contrast, expression of CD117 [54], the CD71 transferrin receptor, and CD29 integrin is abnormally downregulated in these patients [17] (*see* Table 3).

Despite the fact that SM is a group of heterogeneous diseases regarding both clinical behavior and prognosis [57–64], the immunophenotypic profiles of the distinct subtypes of SM have only recently been studied individually [21]. Three different patterns were identified, which correlated with prognostic and molecular subtypes of SM and reflected the maturation status of the cells [21].

The most frequently detected immunophenotypic profile, an activated phenotype, is associated with the majority of ISM patients with (s+) or without (s−) skin lesions [59], usually associated with a good prognosis [61]. BMMCs from these patients showed a mature profile ($CD34^{-}$, CD117hi, $Fc\varepsilon RI^{hi}$) associated with the typical $CD2^{+}/CD25^{+}$ aberrant SM phenotype [21]. Furthermore, an activated phenotype with increased expression of CD63, CD69, and CD203c proteins, associated with expression of both CD64 and the MHC class II molecules –HLA-DR and HLA-DQ–, was detected in these patients. Such aberrant overexpression of some proteins could be related, at least in part, to the constitutive activation induced by the D816V KIT mutation, which virtually all these SM patients carry [21, 65] (*see* Fig. 2). Interestingly, BMMC burden in ISM patients can be extremely low [0.07% (range: 0.04–1.4%) on ISMs- and 0.06 (range: 0.00001–1.7%) on ISMs+ cases]. The coexistence of both normal and pathological MCs in the same patient is relatively frequent in these disease categories (33% of ISMs− patients without skin lesions and 18% of ISMs+ patients). Furthermore, aberrant MCs can be as few as 20% of the total BMMCs (personal observation), which raises the need of studying a large amount of cells so as to have enough sensitivity in order to avoid false-negative results.

A small group of patients, provisionally classified as well-differentiated SM (WDSM), do not have BMMC with the typical aberrant $CD2^{+}/CD25^{+}$ phenotype [21, 57, 58]. Most of these patients do not carry the D816V KIT mutation [65], and their BMMCs display a mature (CD34−, $CD117^{hi}$, $Fc\varepsilon RI^{hi}$) resting immunophenotype presenting only few immunophenotypic changes such as overexpression of bcl2, carboxypeptidase, or total tryptase [21]. The lack of an aberrant phenotype, as clear as the CD25 expression, on these patients, and the fact that their phenotype resembles that found in normal BMMC, leads to a lower sensitivity in the detection of WDSM cases by flow cytometry

[21]. However, we recently found that the BMMCs of these patients aberrantly express CD30, a receptor belonging to the TNF family, and present a typical $CD30^+/CD25^-$ phenotype; in contrast, ISM cases may present a $CD25^+/CD30^-$ or $CD25^+/CD30^+$ phenotype being found later in virtually all the aggressive forms of SM (results accepted for publication).

In contrast to the other groups of patients, poor-prognosis SM patients typically show a $CD2^-/CD25^+$ phenotype, which has been associated with a decreased expression of antigens acquired during MC maturation/differentiation, like CD117, FcεRI, or HLA-I; increased positivity for antigens present in early stages of MC differentiation, like CD123, HLA-DR, or HLA-DQ; and low levels of cytoplasmic tryptase and carboxypeptidase in association with low light scatter features [21]. This immunophenotypic profile reflects a more immature phenotype [21, 27, 30, 32, 37], which could be explained on the basis of the degree of clonal hematopoiesis, since, opposing to ISM patients, who usually carry the D816V KIT mutation restricted to MC compartment, these patients typically display a multilineage involvement [21, 65].

4 Notes

1. The combinations of antibodies to be used depend on the objective of the study and the equipment available. Despite this, some basic principles must be followed to identify and study MCs like the selection of identification markers and selection of sensitive fluorochromes. Tables 1 and 2 show basic combinations recommended for immunophenotypic analysis of MC in systemic mastocytosis studies and may serve as a basis for the establishment of other panels according to the goal of each investigator. Four-color (*see* Table 1) and 8-color (*see* Table 2) versions are shown.
2. Other red cell lysing solutions may be used if validated to study MC.
3. The fixation solution is usually formalin based, while the permeabilization solution is saponin based (commercial formulations unrevealed). The protocol here indicated assumes the usage of a commercial kit for fixation and permeabilization such as Fix&Perm (An der Grub, Bio Research GmbH, Vienna, Austria) or similar. However, other commercial or homemade formulations may be employed after validation to study MCs.
4. Perform BM aspiration in posterior iliac spine, using an 11–8 G biopsy needle in order to obtain enough BM particles. Since MCs are attached to the stroma and stromal cells, BM aspiration should be performed firmly and quickly. The harvest of higher volumes of sample in a single aspirate will not increase the number of MCs collected.

5. The red blood cells present in samples anticoagulated with heparin are more difficult to lyse. For this reason, the time of incubation may be adjusted according to the anticoagulant used: usually 10 min for samples containing EDTA and 15 min for those with heparin.
6. In this step, the clot formed may be difficult to dissolve. Complete disaggregation is critical for correct permeabilization of cell membranes, and particular attention must be paid to avoid false-negative results for cytoplasmic antigens.
7. The light that reaches each photomultiplier tube generates an electronic pulse, and up to three different measures of that pulse may be available: pulse width (*W*), pulse maximum intensity (height, *H*), and pulse area (*A*). The width of the pulse depends on the "time" that a particle takes to pass in front of the laser beam, while the amount of fluorescence is given by the maximum intensity (*H*) and reflects on the pulse area; these latter two measures should be proportional (if not consider revising the calibration of the equipment), and both can be used to quantify the fluorescence in each channel; usually pulse height is used in analogue equipment and pulse area is used in digital cytometers.
8. "Doublet discrimination" is a process whereby the area of the fluorescence light pulse is plotted against the width. Doublets will have greater pulse width than a single cell, as they take longer to pass through the laser beam, and therefore can be excluded from the analysis. The area of the fluorescence light pulse may also be plotted against the pulse height in order to exclude doublets.
9. MCs strongly express CD117, CD203c, and FcεRI. Nevertheless, none of these antigens is a specific marker for MCs; for instance, CD117 is also detected on a major fraction of hematopoietic precursor cells (HPC), CD56^{+}bright NK cells, and neoplastic cells from patients diagnosed with gammopathies, acute leukemia, and myelodysplastic syndromes, among other hematologic malignancies, and in cells derived from other non-hematopoietic tissues (reviewed in [66]); CD203c and FcεRI are also expressed on basophils [67–69]. The use of a CD117^{high}/CD45^{low} combination can accurately identify BMMC [21, 56], but further refinement may be done. The strong reactivity for CD117 in the absence of expression for CD38 and CD34 allows discriminating between MC and CD34 HPC and CD117+ plasma cells, while the expression of Fcε-RI and CD33 together with a strong expression for CD117 allows discriminating between MC and basophils [16].

10. MCs are present at low frequencies in most tissues. In both normal and SM BM samples, MCs usually represent less than 0.2% of nucleated cells [16, 17, 36]. Interestingly, despite the fact that in systemic MC disorders, like SM, these values are significantly increased, low and overlapping frequencies are still found on these patients, compared to normal BM (0.21 ± 0.27% [range: 0.001–1.7%] vs. 0.02 ± 0.02% [range: 0.001–0.09%], respectively) [16, 37]. Furthermore, it is also known that increased BMMCs can also be detected in reactive BM (0.087 ± 0.12% [range: 0.0021–0.54%]) [37] or in patients with other hematological disorders like Waldenström macroglobulinemia (0.095 ± 0.11% [range: 0.01–0.47%]) or myelodysplastic syndromes (0.099 ± 0.12% [range: 0.002–0.47%]) [37].
11. This is particularly important when studying MC since they present high autofluorescence and it may significantly vary between samples. Usually one measure of central tendency (i.e., mean, median, geometric mean) and a measure of dispersion (i.e., standard deviation, interquartile range, coefficient of variation) should be recorded.
12. The fluorescence intensity is reported in arbitrary units, and most of the fluorochrome-conjugated antibodies available do not have a stoichiometric amount of fluorochrome. Thus, the fluorescence intensity is directly proportional to the amount of protein, but in order to make comparisons, one must use the same reagents and the same procedures. Despite this, stoichiometric conjugates of fluorochrome antibodies are available which allow the quantification of the number of molecules identified using fluorescence standards.

References

1. Ishizuka T, Okajima F, Ishiwara M et al (2001) Sensitized mast cells migrate toward the antigen: a response regulated by p38 mitogen-activated protein kinase and Rho-associated coiled-coil-forming protein kinase. J Immunol 167:2298–2304
2. Kitaura J, Kinoshita T, Matsumoto M et al (2005) IgE- and IgE+Ag-mediated mast cell migration in an autocrine/paracrine fashion. Blood 105:3222–3229
3. Moon TC, St Laurent CD, Morris KE et al (2010) Advances in mast cell biology: new understanding of heterogeneity and function. Mucosal Immunol 3:111–128
4. Kitamura Y, Ito A (2005) Mast cell-committed progenitors. Proc Natl Acad Sci U S A 102:11129–11130
5. Shimizu Y, Sakai K, Miura T et al (2002) Characterization of "adult-type" mast cells derived from human bone marrow CD34(+) cells cultured in the presence of stem cell factor and interleukin-6. Interleukin-4 is not required for constitutive expression of CD54, Fc epsilon RI alpha and chymase, and CD13 expression is reduced during differentiation. Clin Exp Allergy 32:872–880
6. Austen KF, Boyce JA (2001) Mast cell lineage development and phenotypic regulation. Leuk Res 25:511–518
7. Jamur MC, Oliver C (2011) Origin, maturation and recruitment of mast cell precursors. Front Biosci (Schol Ed) 3:1390–1406
8. Gross HJ, Verwer B, Houck D, Recktenwald D (1993) Detection of rare cells at a frequency of

one per million by flow cytometry. Cytometry 14:519–526

9. Gross HJ, Verwer B, Houck D et al (1995) Model study detecting breast cancer cells in peripheral blood mononuclear cells at frequencies as low as 10(−7). Proc Natl Acad Sci U S A 92:537–541
10. Leary JF (1994) Strategies for rare cell detection and isolation. Methods Cell Biol 42 (Pt B):331–358
11. Leary JF, Schmidt DF, Gram JG et al (1994) High-speed flow cytometric analysis and sorting of human fetal cells from maternal blood for molecular characterization. Ann N Y Acad Sci 731:138–141
12. Sánchez ML, Almeida J, Vidriales B et al (2002) Incidence of phenotypic aberrations in a series of 467 patients with B chronic lymphoproliferative disorders: basis for the design of specific four-color stainings to be used for minimal residual disease investigation. Leukemia 16:1460–1469
13. Mayrhofer G, Gadd SJ, Spargo LD, Ashman LK (1987) Specificity of a mouse monoclonal antibody raised against acute myeloid leukaemia cells for mast cells in human mucosal and connective tissues. Immunol Cell Biol 65:241–250
14. Lerner NB, Nocka KH, Cole SR et al (1991) Monoclonal antibody YB5.B8 identifies the human c-kit protein product. Blood 77:1876–1883
15. Ashman LK, Cambareri AC, To, L.B et al (1991) Expression of the YB5.B8 antigen (c-kit proto-oncogene product) in normal human bone marrow. Blood 78:30–37
16. Orfao A, Escribano L, Villarrubia J et al (1996) Flow cytometric analysis of mast cells from normal and pathological human bone marrow samples: identification and enumeration. Am J Pathol 149:1493–1499
17. Escribano L, Orfao A, Díaz-Agustin B et al (1998) Indolent systemic mast cell disease in adults: immunophenotypic characterization of bone marrow mast cells and its diagnostic implications. Blood 91:2731–2736
18. Núñez-López R, Escribano L, Schernthaner G-H et al (2003) Overexpression of complement receptors and related antigens on the surface of bone marrow mast cells in patients with systemic mastocytosis. Br J Haematol 120:257–265
19. Escribano L, Orfao A, Díaz-Agustín B et al (1998) Human bone marrow mast cells from indolent systemic mast cell disease constitutively express increased amounts of the CD63 protein on their surface. Cytometry 34:223–228
20. Díaz-Agustín B, Escribano L, Bravo P et al (1999) The CD69 early activation molecule is overexpressed in human bone marrow mast cells from adults with indolent systemic mast cell disease. Br J Haematol 106:400–405
21. Teodosio C, García-Montero AC, Jara-Acevedo M et al (2010) Mast cells from different molecular and prognostic subtypes of systemic mastocytosis display distinct immunophenotypes. J Allergy Clin Immunol 125:719–726
22. Alvarez-Twose I, Matito A, Sanchez-Munoz L et al (2012) Contribution of highly sensitive diagnostic methods to the diagnosis of systemic mastocytosis in the absence of skin lesions. Allergy 67:1190–1191
23. Sánchez-Muñoz L, Morgado JMT, Alvarez-Twose I et al (2013) Flow cytometry criteria for systemic mastocytosis: bone marrow mast cell counts do not always count. Am J Clin Pathol 139:404–406
24. Schernthaner GH, Jordan JH, Ghannadan M et al (2001) Expression, epitope analysis, and functional role of the LFA-2 antigen detectable on neoplastic mast cells. Blood 98:3784–3792
25. Agis H, Willheim M, Sperr WR et al (1993) Monocytes do not make mast cells when cultured in the presence of SCF. Characterization of the circulating mast cell progenitor as a c-kit+, CD34+, Ly-, CD14-, CD17-, colony-forming cell. J Immunol 151:4221–4227
26. Agis H, Füreder W, Bankl HC et al (1996) Comparative immunophenotypic analysis of human mast cells, blood basophils and monocytes. Immunology 87:535–543
27. Rottem M, Okada T, Goff JP, Metcalfe DD (1994) Mast cells cultured from the peripheral blood of normal donors and patients with mastocytosis originate from a CD34+/Fc epsilon RI− cell population. Blood 84:2489–2496
28. Kempuraj D, Saito H, Kaneko A et al (1999) Characterization of mast cell-committed progenitors present in human umbilical cord blood. Blood 93:3338–3346
29. Kirshenbaum AS, Goff JP, Semere T et al (1999) Demonstration that human mast cells arise from a progenitor cell population that is CD34(+), c-kit(+), and expresses aminopeptidase N (CD13). Blood 94:2333–2342
30. Matarraz S, López A, Barrena S et al (2008) The immunophenotype of different immature, myeloid and B-cell lineage-committed CD34+ hematopoietic cells allows discrimination between normal/reactive and myelodysplastic

syndrome precursors. Leukemia 22:1175–1183
31. Dahl C, Hoffmann HJ, Saito H, Schiøtz PO (2004) Human mast cells express receptors for IL-3, IL-5 and GM-CSF; a partial map of receptors on human mast cells cultured in vitro. Allergy 59:1087–1096
32. Schernthaner G-H, Hauswirth AW, Baghestanian M et al (2005) Detection of differentiation- and activation-linked cell surface antigens on cultured mast cell progenitors. Allergy 60:1248–1255
33. Tedla N, Lee C-W, Borges L et al (2008) Differential expression of leukocyte immunoglobulin-like receptors on cord-blood-derived human mast cell progenitors and mature mast cells. J Leukoc Biol 83:334–343
34. Yokoi H, Myers A, Matsumoto K et al (2006) Alteration and acquisition of Siglecs during in vitro maturation of CD34+ progenitors into human mast cells. Allergy 61:769–776
35. Shimizu Y, Suga T, Maeno T et al (2004) Detection of tryptase-, chymase+ cells in human CD34 bone marrow progenitors. Clin Exp Allergy 34:1719–1724
36. Escribano L, Orfao A, Villarrubia J et al (1998) Immunophenotypic characterization of human bone marrow mast cells. A flow cytometric study of normal and pathological bone marrow samples. Anal Cell Pathol 16:151–159
37. Escribano L, Garcia-Montero AC, Núñez R, Orfao A (2006) Flow cytometric analysis of normal and neoplastic mast cells: role in diagnosis and follow-up of mast cell disease. Immunol Allergy Clin N Am 26:535–547
38. Galli SJ, Nakae S, Tsai M (2005) Mast cells in the development of adaptive immune responses. Nat Immunol 6:135–142
39. Marshall JS (2004) Mast-cell responses to pathogens. Nat Rev Immunol 4:787–799
40. Abraham SN, St John AL (2010) Mast cell-orchestrated immunity to pathogens. Nat Rev Immunol 10:440–452
41. Valent P, Schernthaner GH, Sperr WR et al (2001) Variable expression of activation-linked surface antigens on human mast cells in health and disease. Immunol Rev 179:74–81
42. Furuno T, Teshima R, Kitani S et al (1996) Surface expression of CD63 antigen (AD1 antigen) in P815 mastocytoma cells by transfected IgE receptors. Biochem Biophys Res Commun 219:740–744
43. Bühring HJ, Simmons PJ, Pudney M et al (1999) The monoclonal antibody 97A6 defines a novel surface antigen expressed on human basophils and their multipotent and unipotent progenitors. Blood 94:2343–2356
44. Ghannadan M, Baghestanian M, Wimazal F et al (1998) Phenotypic characterization of human skin mast cells by combined staining with toluidine blue and CD antibodies. J Invest Dermatol 111:689–695
45. Frandji P, Oskéritzian C, Cacaraci F et al (1993) Antigen-dependent stimulation by bone marrow-derived mast cells of MHC class II-restricted T cell hybridoma. J Immunol 151:6318–6328
46. Henz BM, Maurer M, Lippert U et al (2001) Mast cells as initiators of immunity and host defense. Exp Dermatol 10:1–10
47. Wong GH, Clark-Lewis I, McKimm-Breschkin JL, Schrader JW (1982) Interferon-gamma-like molecule induces Ia antigens on cultured mast cell progenitors. Proc Natl Acad Sci U S A 79:6989–6993
48. Woolhiser MR, Okayama Y, Gilfillan AM, Metcalfe DD (2001) IgG-dependent activation of human mast cells following up-regulation of FcgammaRI by IFN-gamma. Eur J Immunol 31:3298–3307
49. Bodni RA, Sapia S, Galeano A, Kaminsky A (2003) Indolent systemic mast cell disease: immunophenotypic characterization of bone marrow mast cells by flow cytometry. J Eur Acad Dermatol Venereol 17:160–166
50. Escribano L, Orfao A, Villarrubia J et al (1995) Expression of lymphoid-associated antigens in mast cells: report of a case of systemic mast cell disease. Br J Haematol 91:941–943
51. Escribano L, Orfao A, Villarrubia J et al (1997) Sequential immunophenotypic analysis of mast cells in a case of systemic mast cell disease evolving to a mast cell leukemia. Cytometry 30:98–102
52. Escribano L, Díaz-Agustín B, Bravo P et al (1999) Immunophenotype of bone marrow mast cells in indolent systemic mast cell disease in adults. Leuk Lymphoma 35:227–235
53. Escribano L, Díaz-Agustín B, Bellas C et al (2001) Utility of flow cytometric analysis of mast cells in the diagnosis and classification of adult mastocytosis. Leuk Res 25:563–570
54. Escribano L, Díaz-Agustín B, Núñez R et al (2002) Abnormal expression of CD antigens in mastocytosis. Int Arch Allergy Immunol 127:127–132
55. Pardanani A, Kim Linger T, Reeder T et al (2004) Bone marrow mast cell immunophenotyping in adults with mast cell disease: a prospective study of 33 patients. Leuk Res 28:777–783

56. Escribano L, Diaz-Agustin B, López A et al (2004) Immunophenotypic analysis of mast cells in mastocytosis: when and how to do it. Proposals of the Spanish Network on Mastocytosis (REMA). Cytometry B Clin Cytom 58:1–8
57. Akin C, Escribano L, Nuñez R et al (2004) Well-differentiated systemic mastocytosis: a new disease variant with mature mast cell phenotype and lack of codon 816 c-Kit mutations. J Allergy Clin Immunol 113:S327
58. Akin C, Fumo G, Yavuz AS et al (2004) A novel form of mastocytosis associated with a transmembrane c-kit mutation and response to imatinib. Blood 103:3222–3225
59. Alvarez-Twose I, González de Olano D, Sánchez-Muñoz L et al (2010) Clinical, biological, and molecular characteristics of clonal mast cell disorders presenting with systemic mast cell activation symptoms. J Allergy Clin Immunol 125:1269–1278
60. Bonadonna P, Perbellini O, Passalacqua G et al (2009) Clonal mast cell disorders in patients with systemic reactions to Hymenoptera stings and increased serum tryptase levels. J Allergy Clin Immunol 123:680–686
61. Escribano L, Alvarez-Twose I, Sánchez-Muñoz L et al (2009) Prognosis in adult indolent systemic mastocytosis: a long-term study of the Spanish Network on Mastocytosis in a series of 145 patients. J Allergy Clin Immunol 124:514–521
62. Lim K-H, Tefferi A, Lasho TL et al (2009) Systemic mastocytosis in 342 consecutive adults: survival studies and prognostic factors. Blood 113:5727–5736
63. Valent P, Horny HP, Escribano L et al (2001) Diagnostic criteria and classification of mastocytosis: a consensus proposal. Leuk Res 25:603–625
64. Valent P, Akin C, Escribano L et al (2007) Standards and standardization in mastocytosis: consensus statements on diagnostics, treatment recommendations and response criteria. Eur J Clin Investig 37:435–453
65. Garcia-Montero AC, Jara-Acevedo M, Teodosio C et al (2006) KIT mutation in mast cells and other bone marrow hematopoietic cell lineages in systemic mast cell disorders: a prospective study of the Spanish Network on Mastocytosis (REMA) in a series of 113 patients. Blood 108:2366–2372
66. Escribano L, Ocqueteau M, Almeida J et al (1998) Expression of the c-kit (CD117) molecule in normal and malignant hematopoiesis. Leuk Lymphoma 30:459–466
67. Gane P, Pecquet C, Crespeau H et al (1995) Flow cytometric monitoring of allergen induced basophil activation. Cytometry 19:361–365
68. Valent P, Majdic O, Maurer D et al (1990) Further characterization of surface membrane structures expressed on human basophils and mast cells. Int Arch Allergy Appl Immunol 91:198–203
69. Valent P, Bettelheim P (1992) Cell surface structures on human basophils and mast cells: biochemical and functional characterization. Adv Immunol 52:333–423
70. Valent P, Cerny-Reiterer S, Herrmann H et al (2010) Phenotypic heterogeneity, novel diagnostic markers, and target expression profiles in normal and neoplastic human mast cells. Best Pract Res Clin Haematol 23:369–378

Part VI

Whole Organism: Disease and Knockout Models

Chapter 28

Mast Cell-Mediated Reactions In Vivo

Stefan Frischbutter, Jörg Scheffel, and Frank Siebenhaar

Abstract

Mast cells are involved in many physiological reactions in which their functions can be very diverse. Models of allergic skin inflammation and systemic anaphylactic reactions in mice are validated methods in which the role of mast cells is well established. In this chapter, we will therefore present protocols for passive cutaneous anaphylaxis and contact hypersensitivity, i.e., models which can be used to identify and characterize the role of mast cells as well as mast cell mediators and receptors in allergic IgE-dependent and IgE-independent skin inflammation, and for passive systemic anaphylaxis, a model ideally suited to characterize the systemic effects of mast cell-derived mediators and mast cell receptors.

Key words Mast cells, Passive cutaneous anaphylaxis, Passive systemic anaphylaxis, Contact hypersensitivity

1 Introduction

Passive cutaneous anaphylaxis (PCA) and passive systemic anaphylaxis (PSA) in mice are classical models mimicking type I allergic responses. Here, mast cells are passively loaded with antigen-specific IgE, and mice are then subsequently challenged with the respective allergen. This leads to the cross-linking of IgE, activation of FcεRI receptors, the degranulation of mast cells, and the release of various mediators [1]. In PCA, this reaction occurs locally at the site of IgE injection in the skin and will lead to measurable extravasation and increase in skin thickness. In contrast, PSA reflects a systemic anaphylaxis resulting in cardiovascular and respiratory symptoms and in a pronounced drop in body temperature. Both models are ideally suited to characterize mediators and receptors involved in allergic type I responses and to identify and characterize effects of drugs aimed at affecting mast cell function [2–4].

It has to be noted, however, that other factors besides mast cells can potentially modulate some or all aspects of the observed reaction, for example, basophils, complement factors, or the strain and age of mice [5].

Bernhard F. Gibbs and Franco H. Falcone (eds.), *Basophils and Mast Cells: Methods and Protocols*, Methods in Molecular Biology, vol. 2163, https://doi.org/10.1007/978-1-0716-0696-4_28,

Contact hypersensitivity (CHS) is a classical type IV allergic reaction with a delayed skin reaction elicited by a hapten [6, 7]. This model is used to understand the pathophysiology of allergic contact dermatitis in humans and to identify and characterize interaction between immune cells in the skin and lymph nodes. The key players in the sensitization and elicitation phases of type IV reactions are T cells and dendritic cells. Therefore, this model is well suited to dissect the role of mast cells in the modulation of adaptive immune responses.

2 Materials

2.1 PCA

1. Mice between 6 and 12 weeks old (*see* **Note 1**).
2. Monoclonal IgE anti-DNP (Sigma-Aldrich) (*see* **Notes 2** and **3**).
3. Dinitrophenyl-human serum albumin (DNP-HSA) (Sigma-Aldrich).
4. Evans blue.
5. Dimethylformamide.
6. Caliper (e.g., Mitutoyo Thickness Gage).
7. Microfine syringes for insulin 0.3 mL (e.g., Becton Dickinson).
8. 27 and 30 G needles.
9. Micro-dissecting forceps (e.g., Braun, Harvard Apparatus) (×2).
10. Chirurgical scissors (e.g., Braun, Harvard Apparatus).
11. 96-well flat bottom plate for ELISA.

2.2 PSA

1. Mice between 6 and 12 weeks old (*see* **Note 1**).
2. Monoclonal IgE anti-DNP (Sigma-Aldrich) (*see* **Notes 2** and **3**).
3. Dinitrophenyl-human serum albumin (DNP-HSA) (Sigma-Aldrich).
4. Mouse restrainer (e.g., LABART).
5. Microfine syringes for insulin 0.3 mL (e.g., Becton Dickinson).
6. Needles: 27 and 30 G.
7. Microprobe thermometer with a rectal probe for mice (e.g., Physitemp Instruments) (*see* **Note 4**).

2.3 CHS Materials

1. Mice between 6 and 12 weeks old (*see* **Note 1**).
2. Hapten and appropriate solvent and vehicle (Table 1).
3. Caliper (Mitutoyo Thickness Gage).
4. 1–10 μL pipette.
5. Hair shaver.

Table 1
Haptens in CHS

Abbreviation	Vehicle	Type of reaction	Sensitization area	Challenge area	Sensitization dose	Challenge dose
OXA	Acetone	Th1	Belly	Ear	1–6% w/v	0.4–0.6% w/v
DNFB	Acetone	Th1	Belly	Ear	0.3–0.5% v/v	0.1–0.2% v/v
TNCB	Acetone-olive oil [3:1]	Th1	Belly	Ear	30 mg/mL	3 mg/mL
FITC isomer I	Acetone-DBP [1:1]	Th2	Belly	Ear	2% v/v	0.2% v/v
Substance name						
OXA: 4-ethoxymethylene-2-phenyl-2-oxazolin-5-one (oxazolone)[a]						
DNFB: 1-fluoro-2,4-dinitrobenzene						
TNCB: 2-chloro-1,3,5-trinitrobenzene						
FITC isomer I: fluorescein isothiocyanate isomer I						
DBP: dibutyl phthalate						

[a]Stored at −20 °C increases the time of use

3 Methods

3.1 PCA

3.1.1 Day 0 (Passive Sensitization)

1. Prepare and pre-warm the IgE anti-DNP working solution before injection: 20 μL of IgE anti-DNP [5 μg/mL] per animal is required. IgE anti-DNP concentration is ideally kept as a stock solution, i.e., at 2 mg/mL. To get a working solution, dilute the antibody 1:400 from the stock in saline (*see* **Note 3**).
2. Prepare antigen solution: 100 μL of dinitrophenyl-human serum albumin (DNP-HSA) (4 mg/mL) per animal is required. The concentration of DNP-HSA stock solution recommended is 100 mg/mL. To get working solution, dilute the stock in proportion 1:25 in saline or 1% Evans blue if the interest is edema quantification (*see* **Note 5**).
3. Perform an ear intradermal injection with a 30 G needle of 20 μL of IgE anti-DNP or vehicle (saline alone) in the ear pinna of each animal (*see* **Notes 6** and 7 and Fig. 1).

3.1.2 Day 1 (Cutaneous Anaphylaxis Elicitation)

1. Measure the ear thickness baseline (*see* **Notes** 7 and **8** and Fig. 2).
2. Perform a tail vein injection with a 27 G needle of 100 μL of DNP-HSA or saline. For simple ear swelling or histology, the DNP-HSA should be diluted in saline. For extravasation

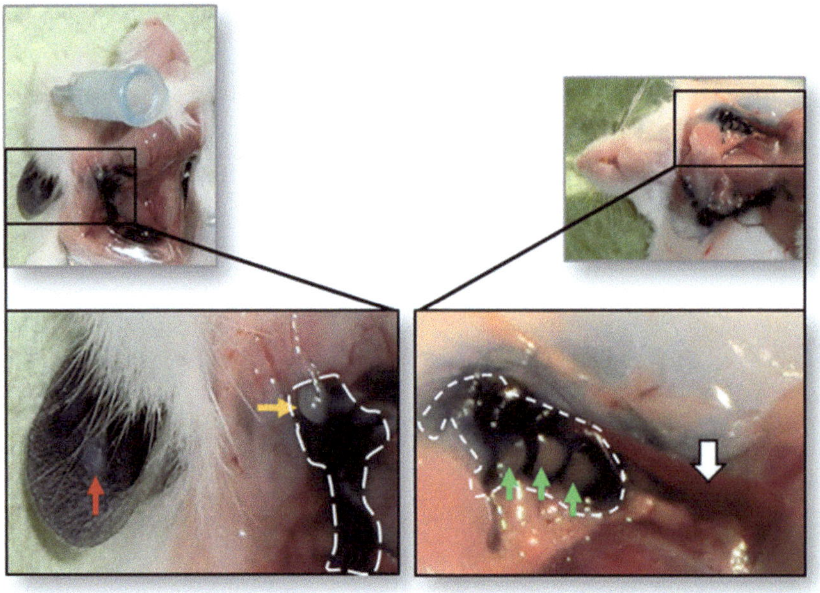

Fig. 1 Leakage of fluid injected in the ear pinnae. 50 μL of 10% Evans blue injected i.d. in the ear pinnae (red arrow) generates reflux trough extra peritoneal cavity (dotted white line) to the superficial cervical lymph nodes (yellow/green arrows). The dye was not delivered to the lymph nodes via blood flow; hence, neither the jugular vein (white arrow) nor the microvasculature is stained with the dye. That picture was taken immediately after injection

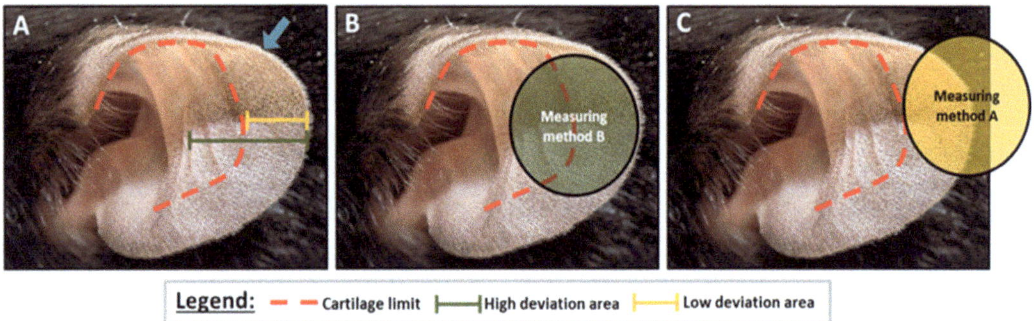

Fig. 2 Two different methods for ear swelling measurements. (**a**) Cartilage location in the mice ear, area of high and low deviation, and starting point to place the caliper spindle (blue arrow). (**b**) Including the cartilage (high deviation area) in the measured area may provoke high variability in the measurements (10–23%). (**c**) Excluding the cartilage enhances the sensitivity and diminishes the variability (10–3%)

assessment, the DNP-HSA should be diluted in Evans blue (*see* **Note 9**).

3.1.3 Assessment of Ear Swelling

1. Repeat the ear thickness measurement as described for the baseline (*see* **Note 7** and Fig. 2). The recommended time points are 30 min, 1, 2, 4, and 6 h after the challenge.

2. For analysis of ear swelling, ear thickness at baseline has to be subtracted from the ear thickness at each time point after elicitation (*see* **Note 7**).

3.1.4 Quantifying Evans Blue Dye Extravasation

This technique is used to assess the extent of extravasation which is mainly responsible for the swelling process.

1. Euthanize the animal at desired time point after elicitation (1–2 h).
2. Excise the ears from each animal.
3. Make a 6 μm punch biopsy.
4. Split the two halves.
5. Place each biopsy in a 2 mL vial filled with 0.5 mL of dimethylformamide (DMF).
6. Incubate 3–5 h at 55 °C and shake at 500 rpm.
7. Pass the whole ear exudate through a 40 μm cell strainer.
8. Pipette 150 μL of each ear exudate on the NUNC 96-well flat bottom plate.
9. Perform optical density measurement at 650 nm.
10. To perform the quantification of Evans blue, compare absolute values obtained within the different groups (*see* **Notes 10** and **11**).

3.2 PSA

3.2.1 Day 0 (Passive Sensitization)

1. Prepare monoclonal IgE anti-DNP working solution: 200 μL of IgE anti-DNP (200 μg/mL) per animal is required. The concentration of the IgE anti-DNP stock solution recommended is 2 mg/mL. To get working solution, dilute the antibody 1:10 from the stock in 0.9% NaCl. IgE anti-DNP could be obtained from hybridoma cells (*see* **Note 3**). The stock is normally stored at 2 mg/mL at −80 °C. Once diluted to working concentration, it may be stored at 4 °C and used within 2 weeks or discarded.
2. Perform an intraperitoneal injection of 200 μL of IgE anti-DNP or vehicle (saline alone).

3.2.2 Day 1 (Systemic Anaphylaxis Elicitation)

1. Prepare antigen working solution: 100 μL of dinitrophenyl-human serum albumin (DNP-HSA) (Sigma-Aldrich) (10 mg/mL) per animal is required. The concentration of DNP-HSA stock solution recommended is 100 mg/mL. To get the working solution, dilute the stock in proportion 1:10 in saline (*see* **Note 5**).
2. Baseline: Measure and note the initial body temperature by immobilizing the animal and register the rectal temperature.
3. Warming the animal: Place the animal under a warming lamp 5 min before the injection leading to the enhancement of the superficial blood flow. It may facilitate the injection.

4. Perform a tail vein injection with a 30 G needle of 100 μL of DNP-HSA or saline.

3.2.3 Body Temperature Measurement

1. Measure rectal temperature at 10-min intervals for the first hour and then at 90, 120, and 180 min following the challenge.

3.2.4 Assessment of Mast Cell Mediator Release

1. Sacrifice the mice after the first temperature measurement (10 min after induction of anaphylaxis).
2. Collect whole blood (e.g., by cardiac puncture) and peritoneal lavage fluid (PLF).
3. Leave blood sample for at least 1 h to clot.
4. Centrifuge the sample at 3300 × *g* for 20 min, and remove the serum from the clot by gently pipetting off into a clean tube.
5. To assess mast cell activation, measure mMCP-1 and/or histamine by ELISA in serum and/or PLF.

3.3 CHS (See Note 12)

3.3.1 Day 5 (Sensitization)

1. Hapten reconstitution: Perform according to Table 1. For FITC, the hapten must be stored at 4 °C after reconstitution. The recommended stock concentration is 16% w/v in acetone. Perform a dilution 1:4 in acetone thereafter; dilute 1:2 in DBP to obtain the sensitizing concentration. For TNCB, prepare 1 mL of vehicle mixture (proportion 1:3 olive oil/acetone). Weigh and reconstitute 30 mg of TNCB per 1 mL of vehicle to get the sensitizing dose. For DNFB and oxazolone, weigh 0.5 or 6 mg of hapten, respectively, per 1 mL of acetone to obtain the sensitizing dose (*see* Table 1 for additional concentrations and **Notes 13–15** to increase the performance of the hapten application).
2. Shave the abdominal region of the mice (around 1.5 cm^2).
3. Apply twice 10 μL of sensitizing solution over the shaved area by pipetting and smearing with the pipette tip (*see* **Note 16**).
4. Wait for 10 s to let the solution dry on the skin. Keep the animal under sedation during that time.

3.3.2 Day 0 (Challenge)

1. Hapten reconstitution: Perform according to Table 1. For FITC, perform a dilution 1:40 in acetone. Thereafter, dilute 1:2 in DBP to obtain the challenging concentration. For TNCB, prepare 1 mL of vehicle mixture (proportion 1:3 olive oil/acetone) and weigh and reconstitute 3 mg of TNCB per 1 mL of vehicle to get the challenging dose. For DNFB and oxazolone, reconstitute 0.1 or 0.6 mg of hapten, respectively, per 1 mL of acetone to obtain the challenging dose (*see* Table 1 for additional concentrations). To increase the performance of the hapten application, *see* **Notes 13–15**.

2. Place the animal under isoflurane sedation.
3. Measure and record the baseline of ear thickness baseline (*see* **Note** 7 and Fig. 2).
4. Apply 10 μL of challenging solution on the inner side of the ear and 10 μL on the dorsal side of the ear (*see* **Notes 16** and **17**).
5. Let dry for a minimum of 10 s. Keep the animal under sedation.
6. Perform ear measurement as needed for 2–5 days. The recommended time points are every 24 h for 5 days (*see* **Note 18**).

4 Notes

1. The number of mast cells increases along the life span of the mice, and different strains of mice have different mast cell numbers. This may result in altered responses in older mice or mice of different strain backgrounds. Furthermore, female mice are to be preferred, as male mice might have injured/inflamed ears due to fighting.
2. IgE specific for alternative antigens, e.g., TNP, may also be used but need close adjustment in terms of concentration.
3. Sigma-Aldrich distributes both the IgE anti-DNP (IgE clone: SPE-7) and the hybridoma cells (IgE clone: SP2). It is important to titrate the optimal concentration every time the lot is renewed. The aliquots of IgE should be stored in high concentrations and in small volume aliquots. In our lab, it is storage at 2 mg/mL in saline at −80 °C. Aliquots diluted to working concentration may be stored at 4 °C and used in the next 2 days and discarded afterward.
4. Alternatively, the use of subcutaneous transponder can be used to monitor the temperature over time (e.g., BMDSBio Medic Data Systems).
5. The stock solution should be stored at −20 °C (100 mg/mL) in DMEM-Pipes.
6. The site of injection should be approximately 2–4 mm under the distal end of the ear pinnae, and we highly recommended a volume of 20 μL per ear (*see* Fig. 1).
7. An established method to perform ear injection or measurement is to maintain the animal under isoflurane sedation (21% of O_2 and 2% isoflurane) using an isoflurane exsiccator.
8. Recommended guidelines for ear swelling:
   (a) Subtract the baseline ear thickness with the ear thickness at each time point after the elicitation of the reaction. Animals may differ to some extent in the baseline ear thickness.

(b) Measure always in the same place. Place the edge of the caliper spindle in the beginning of the upper ear folded area (*see* blue arrow, Fig. 1). If there is appearance of necrosis or other relevant signs outside of the measured area, it should be documented but not measured instead of the originally selected area.

(c) Methods of ear swelling: During the measurement process, the exclusion of the cartilage is preferred (*see* Fig. 2). Thus, excluding the cartilage by measuring the edges of the ear will reduce the variability within groups (Method A, Fig. 2).

9. The use of restrainers to perform intravenous tail injection is preferred over sedation with isoflurane since under sedation the blood pressure decreases and the tail vein diameter is reduced, making it more difficult to inject.

10. Collecting histological samples is an additional method to analyze mast cell morphology after the treatment (degranulation, localization, and numbers). We recommended semithin sections (1 μm) and staining with alkaline Giemsa.

11. Evans blue extravasation can be also applied to other skin areas, e.g., back skin, if PCA is elicited in those areas.

12. Reports of mast cell involvement in that model are controversial. That is mainly due to three reasons: (a) The contribution of mast cells in CHS may vary depending on the applied model. (b) Small variations in the protocol generate striking differences in the outcome (i.e., ear thickness). Such variations include type of hapten, dose, area of sensitization, and even measuring method. (c) The lot-dependent variation in the biological activity of the haptens. Table 1 summarizes the main experimental haptens used and other information.

13. Use always fresh reconstituted compounds. Haptens have a high tendency to be reduced. Use within 30 min after reconstitution.

14. Use tubes resistant to acetone degradation like polypropylene (PP) and high-density polyethylene (HDPE) tubes, for example, 1 and 2 mL Eppendorf tubes. Other plastics like current FACS tubes are not suitable for use with organic solvents such as acetone.

15. Some haptens like FITC create insoluble precipitates after reconstitution in the vehicle; therefore, it is necessary to filter the solution before application.

16. The grade of sensitization depends on the contact area. A larger contact area may result in a higher rate of sensitization.

17. Place the animal in dorsal position. Pipette following the exterior edge of the ear and let the liquid drop down toward the head. Avoid ear movements of the animal by keeping the

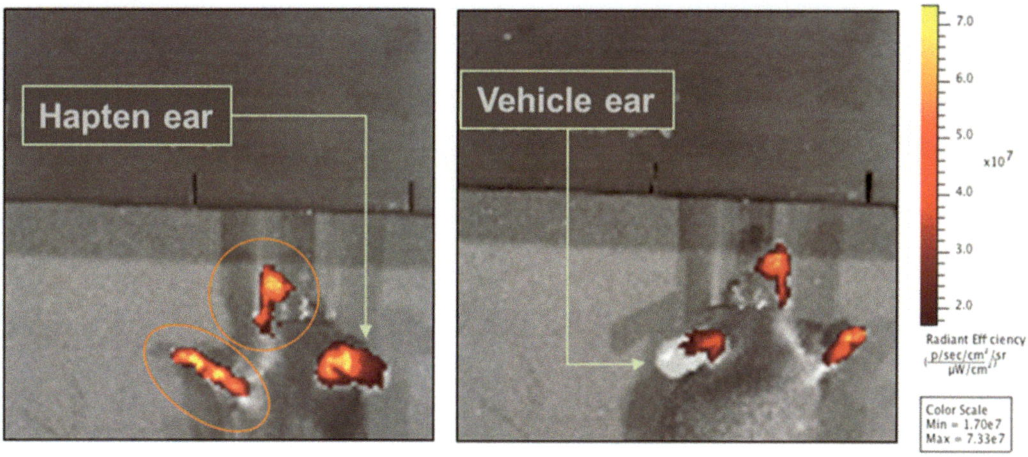

Fig. 3 Cross-contamination after hapten painting. The following day after hapten painting, the mice show inflammatory signal in forepaws, mouth (orange circles), and contralateral ear

animal in deep sedation with isoflurane. Sudden movements may reduce the amount of total hapten per unit area.

18. The topical application of substances in animals can have a disadvantage in the cross-contamination by the cleaning behavior of the mice. Normally, for measurement of ear swelling, it does not represent a major problem as little of the substances will be transferred from one ear to the other ear. However, it may have an effect on the systemic immunologic response. Figure 3 shows the inflammatory signal using a reporter system 24 h after application of a hapten in the left ear. Mouth, forepaws, and even the right ear are indeed showing an inflammatory signal.

References

1. Dearman RJ, Kimber I (2009) Animal models of protein allergenicity: potential benefits, pitfalls and challenges. Clin Exp Allergy 39:458–468
2. Siebenhaar F, Magerl M, Peters EM, Hendrix S, Metz M, Maurer M (2008) Mast cell-driven skin inflammation is impaired in the absence of sensory nerves. J Allergy Clin Immunol 121:955–961
3. Mori T, Kabashima K, Fukamachi S, Kuroda E, Sakabe J, Kobayashi M, Nakajima S, Nakano K, Tanaka Y, Matsushita S, Nakamura M, Tokura Y (2013) D1-like dopamine receptors antagonist inhibits cutaneous immune reactions mediated by Th2 and mast cells. J Dermatol Sci 71:37–44
4. Han SY, Bae JY, Park SH, Kim YH, Park JH, Kang YH (2013) Resveratrol inhibits IgE-mediated basophilic mast cell degranulation and passive cutaneous anaphylaxis in mice. J Nutr 143:632–639
5. Miyajima H, Watanabe N, Ovary Z, Okumura K, Hirano T (2002) Rat monoclonal anti-murine IgE antibody removes IgE molecules already bound to mast cells or basophilic leukemia cells, resulting in the inhibition of systemic anaphylaxis and passive cutaneous anaphylaxis. Int Arch Allergy Immunol 128:24–32
6. Christensen AD, Haase C (2012) Immunological mechanisms of contact hypersensitivity in mice. APMIS 120:1–27
7. Popov A, Mirkov I, Kataranovski M (2012) Inflammatory and immune mechanisms in contact hypersensitivity (CHS) in rats. Immunol Res 52:127–132

Chapter 29

Identification of Murine Basophils by Flow Cytometry and Histology

Christian Schwartz and David Voehringer

Abstract

Here, we describe how murine basophils can be detected in vivo by flow cytometry and immunofluorescence staining. Basophils constitute a homogeneous population of $CD4^{-}CD19^{-}CD49b^{+}IgE^{+}$ cells in flow cytometric analysis. When IgE levels are low, one can also use anti-FcεRI or anti-CD200R3 antibodies instead of anti-IgE. For immunofluorescence staining, we use an anti-Mcpt8 antibody since Mcpt8 is a specific marker for murine basophils. We describe how to prepare the tissue to cut cryo-sections and how to perform the staining using a tyramide-based amplification kit.

Key words Basophils, Mcpt8, Flow cytometry, Immunofluorescence staining, Tyramide amplification

1 Introduction

Basophils are the rarest of granulocytes, comprising only 0.5% of blood leukocytes. Basophils were first identified 130 years ago by Paul Ehrlich using basic aniline dyes to stain the contents of their cytoplasmic granules [1]. In mice, basophils were not detected until 1981 [2]. Due to their rarity and the lack of appropriate staining techniques, murine basophils were thought to be a redundant cell type. Only 22 years later, new techniques allowed the detection of basophils in mice and led to a renaissance of basophil research [3–6]. Instead of hematologic stainings and electron microscopy, detection of basophils by fluorochrome-labeled antibodies directed to surface antigens came to the fore. Over the past years, new flow cytometric and immunofluorescence stainings of basophils in mice have been developed. In this chapter, we outline the routine identification of murine basophils using both methods.

Flow cytometric analysis and fluorescence microscopy, both using fluorochrome-conjugated antibodies against cell surface antigens, are two of the most commonly used techniques of today's immunologist and allow the researcher to detect small cell populations. Flow cytometry is primarily used to quickly analyze vast

Bernhard F. Gibbs and Franco H. Falcone (eds.), *Basophils and Mast Cells: Methods and Protocols*, Methods in Molecular Biology, vol. 2163, https://doi.org/10.1007/978-1-0716-0696-4_29, © Springer Science+Business Media, LLC, part of Springer Nature 2020

numbers of cells, distinguish different cell populations, and characterize a specific cell type. While flow cytometry analyzes the composition of different organs, immunofluorescence is often used to determine the localization of a distinct cell type within tissues.

In contrast to mice, four monoclonal antibodies directed against human basophils are available for flow cytometry and immunohistochemistry [7–13]. Unfortunately, they cannot be used for the detection of murine basophils. Furthermore, CD63 and CD203c, activation markers for human basophils, have not been described for mouse basophils (reviewed in [14, 15]). Therefore, combinations of other markers are needed to reliably identify basophils in mice.

Among other surface markers, murine basophils express the Fc receptor for immunoglobulin (Ig)E, IL-3Rα, CD49b, CD90.2, CD200R3, and CD244 [16]. Although they are not exclusively expressed on basophils, a combination of the integrin CD49b and Fc receptor-bound IgE is routinely used to detect basophils [17–20]. In the first part of Subheading 2, we want to explain the routine staining procedure for the flow cytometric detection of murine splenic basophils using this combination of anti-CD49b and anti-IgE antibodies. Since basophils do not express markers specific for other major cell populations, like CD4 and CD19, we use these markers to gate out T and B cells and therefore get a more distinct $CD4^-CD19^-CD49b^+IgE^+$ basophil population.

In the second part of the Methods section, we want to describe the immunofluorescence staining of murine basophils with an antibody directed against the basophil-specific mast cell protease 8. Despite its name, this protease is exclusively expressed in mouse basophils and has been used for the generation of basophil-deficient and reporter mice [16, 18, 21–24]. Here, we describe how to prepare the tissue for cryo-sections and how we perform the staining using a tyramide-based amplification kit. The cryo-sections are stained with the anti-Mcpt8 antibody, followed by a peroxidase-conjugated secondary antibody. The peroxidase activity catalyzes covalent deposition of fluorophores in the immediate environment of the enzyme, thereby amplifying weak signals and allowing the detection of murine basophils.

2 Materials

2.1 Flow Cytometry Components

1. FACS buffer: 2 L Dulbecco's PBS (phosphate-buffered saline; 9.55 g/L), 50 mL FCS (fetal calf serum), and 0.2 g $NaNO_3$. Prepare 2 L and store at 4 °C (*see* **Note 1**).
2. ACK (ammonium-chloride-potassium) lysing buffer: 1 L H_2O, 8.29 g NH_4Cl, 1 g $KHCO_3$, and 0.037 g EDTA. Store at 4 °C.

3. Consumables: Plastic petri dish (9 cm), plastic syringe (5 mL), cell strainer or nylon mesh (70 μm), FACS tubes (5 mL), and conical tubes (15 mL).
4. Counting chamber, trypan blue.
5. Antibodies: anti-CD16/32, clone 2.4G2 (BioXCell); Alexa647-conjugated anti-CD49b, clone HMα2 (Biolegend); PE-conjugated anti-IgE, clone 23G3 (eBioscience); PerCP/Cy5.5-conjugated anti-CD4, clone RM4–5 (eBioscience); PerCP/Cy5.5-conjugated anti-CD19, clone ebio1D3 (eBioscience).
6. DAPI (4′,6-diamidino-2-phenylindole).

2.2 Histology Components

Components for tissue and slide preparation.

1. 4% PFA (paraformaldehyde): Dissolve 4 g paraformaldehyde (Roth) at 65 °C in 50 mL ddH_2O and 1 mL 1 M NaOH. Add 10 mL 10× PBS (95.5 g/L). Cool to room temperature. Adjust pH 7.4 with 1 M HCl. Add H_2O to 100 mL. Filter through 0.45 μm filter. Store at −20 °C.
2. Tissue freezing medium.
3. Disposable Vinyl Specimen Molds (15 × 15 × 5 mm) (Sakura Finetec).
4. Slides: SuperFrost Plus Gold (Thermo Scientific).
5. Cover slides.
6. Liquid barrier marker or Pap pen.
7. Fluoromount-G (Southern Biotech).

Components for Staining Procedure

8. Tyramide Signal Amplification Cyanine 3 Tyramide Reagent Pack (PerkinElmer).
9. Blocking buffer: 0.1 M Tris-Cl, pH 7.5, 0.15 mL NaCl, 0.5% blocking reagent (PerkinElmer included in Reagent Pack). Dissolve at 60 °C. Aliquot at 3 mL and store at −20 °C. Do not freeze after thawing.
10. PBS, pH 7.4.
11. H_2O_2.
12. Humid chamber.
13. Slide staining rack.
14. Antibodies: anti-CD16/32, clone 2.4G2 (BioXCell); rat serum (Sigma); mouse serum (Sigma); anti-Mcpt8, clone TUG8 (BioLegend); Horseradish-peroxidase-conjugated goat-anti-rat-IgG (Jackson ImmunoResearch).
15. DAPI.

3 Methods

3.1 Flow Cytometry

Always work on ice unless stated otherwise.

1. Prepare a single cell suspension from a freshly isolated spleen in 10 mL FACS buffer. Use the plunger of a 5 mL syringe to grind the organ through a 70 μm cell strainer into a petri dish. Collect the cell suspension in a 15 mL Falcon tube.
2. Pellet the cells by centrifugation for 5 min at 400 × *g* at 4 °C. Discard the supernatant and loosen the pellet.
3. Add 1 mL ACK buffer to lyse the erythrocytes for 2 min at room temperature. Add 9 mL FACS buffer to stop the reaction and centrifuge (400 × *g*, 5 min, and 4 °C). Resuspend the pellet in FACS buffer.
4. Count the cells using a counting chamber by mixing 20 μL of the cell suspension with 180 μL trypan blue.
5. Adjust cells to 3×10^7 cells/mL.
6. Transfer 100 μL (3×10^6 cells) into a separate 5 mL FACS tube (*see* **Notes 2** and **3**).
7. Add 2 μg anti-CD16/32 antibody to the cells for 5 min at room temperature to block unspecific Fc receptor binding sites.
8. Prepare antibody master mix in FACS buffer (4 °C) and keep on ice in the dark:
   (a) For every sample, use 100 μL FACS buffer.
   (b) Add anti-CD4-PerCP/Cy5.5 (1:800 or 0.25 μg/mL), anti-CD19-PerCP/Cy5.5 (1:400 or 0.5 μg/mL), anti-CD49b-Alexa647 (1:200 or 2.5 μg/mL), and anti-IgE-PE (1:200 or 0.5 μg/mL) (*see* **Notes 4** and **5**).
   (c) Vortex.
   (d) Add 100 μL to each sample.
9. Incubate for 20 min on ice in the dark.
10. Fill FACS tube with FACS buffer and centrifuge (400 × *g*, 5 min, 4 °C). Discard the supernatant and loosen the pellet (*see* **Note 6**).
11. Just before analysis on the flow cytometer, add 0.5 μg DAPI to the sample to discriminate live and dead cells (*see* **Note 7**).
12. At the cytometer, gate for all leukocytes, exclude doublets, and gate on DAPI negative, non-T/non-B cells ($CD4^-CD19^-$) (Fig. 1). The basophils constitute the $CD49b^+IgE^+$ population (*see* **Notes 8** and **9**).

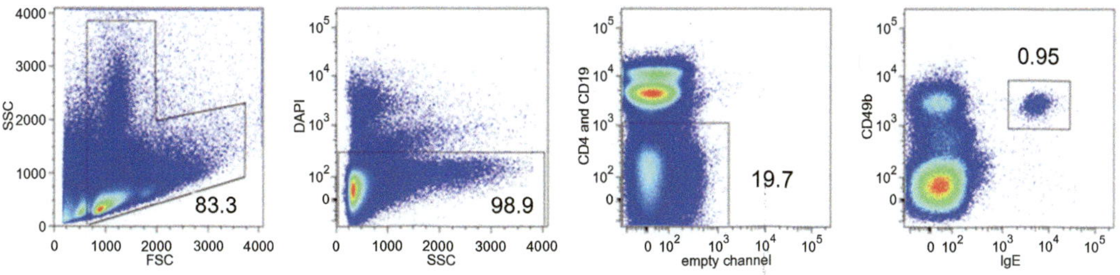

Fig. 1 Flow cytometric staining of basophils. From left to right: cells gated on leukocytes (FSC/SSC), DAPI-negative, non-B/non-T cells ($CD4^-CD19^-$), and $CD49b^+IgE^+$ basophils. Cells were measured on a FACS Canto II instrument (BD Biosciences, CA) and analyzed with FlowJo software (Treestar, OR)

3.2 Histology

1. Put freshly isolated organ (e.g., spleen) directly into 4% PFA and incubate for 2 h at 4 °C (*see* **Note 10**).
2. Transfer fixed tissue in 50 mL PBS and incubate overnight at 4 °C.
3. Embed tissue in cryomolds in tissue freezing medium, slowly freeze in liquid nitrogen, and store at −80 °C (*see* **Note 11**).
4. Cut sections at 6 μm, circle with liquid barrier marker, and let dry slides at least 3 h or overnight at room temperature.
5. Wash slides 2 × 5 min in PBS using a slide staining rack (*see* **Note 12**).
6. Incubate in 1% H_2O_2 in PBS for 1 h at room temperature. Carefully flick off gas bubbles from sections.
7. Wash slides 2 × 5 min in PBS (*see* **Note 13**).
8. Add 100 μL blocking buffer containing 1% mouse serum onto each section and incubate for 30 min at room temperature (*see* **Note 14**).
9. Wash slides 2 × 5 min in PBS.
10. Add 100 μL of 1:100 diluted unlabeled anti-Mcpt8 antibody (0.5 μg) to each section. Incubate for 2 h at room temperature (*see* **Notes 15** and **16**).
11. Wash 2 × 5 min in PBS.
12. Add 100 μL of 1:500 diluted horseradish-peroxidase-conjugated goat-anti-rat antibody (0.16 μg) to each section. Incubate for 45 min at room temperature in the dark.
13. Wash 2 × 5 min in PBS.
14. Prepare 1:100 dilution of tyramide-Cy3 in amplification buffer (supplied in the kit). Add 100 μl to each section and incubate for 7 min at room temperature in the dark (*see* **Note 17**).
15. Wash 2 × 5 min in PBS.
16. Add 100 μL of DAPI (2 μg) to each section and incubate for 20 min in the dark (*see* **Notes 18** and **19**).

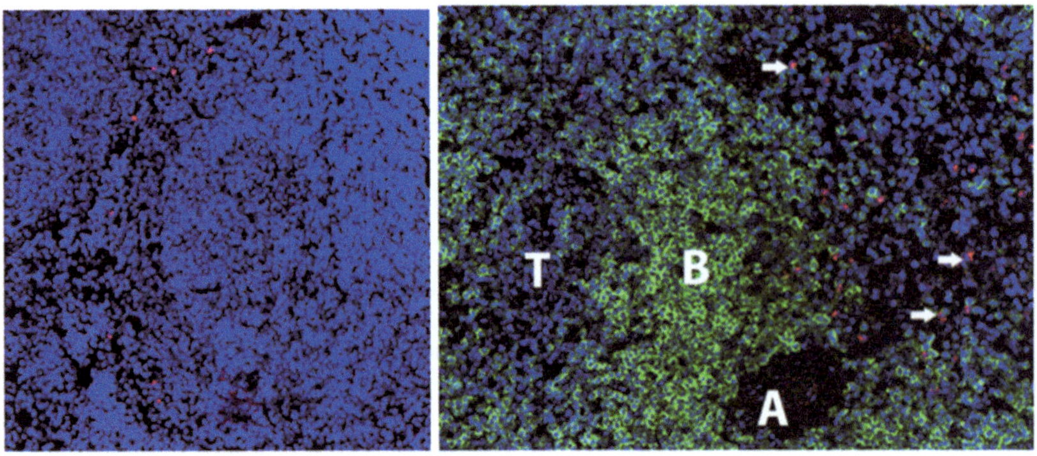

Fig. 2 Immunofluorescence staining of basophils (anti-Mcpt8, red) in spleens from C57BL/6 mice. Left: Nuclei were counterstained with DAPI (blue). Right: B cells were counterstained with anti-B220 (green). White arrows indicate basophils encircling the white pulp. *T* T cell zone, *B* B cell zone, *A* artery. Photomicrographs were taken on a Laser Scanning Microscope LSM 700 (Carl Zeiss, Jena, Germany) and analyzed with ZEN lite 2011 software (Carl Zeiss)

17. Wash 2 × 5 min in PBS.
18. Take slides and briefly put the edge of each slide onto a paper towel to remove PBS. Add one drop of Fluoromount-G onto each section and cover with cover slide (*see* **Note 20**).
19. Store slides at 4 °C in the dark and let them dry for at least 3 h before analysis on a fluorescence microscope (*see* **Notes 21** and **22**; *see* Fig. 2).

4 Notes

1. For the staining of blood basophils, heparin (2000 U/L) should be added to the buffer until erythrocytes have been lysed.
2. Dependent on which organ is used for analysis, basophil numbers can be very low. Therefore, more cells should be used for the staining procedure.
3. When analyzing many samples, we find it useful to stain in a 96-well plate, reducing antibody cocktails to 50 μL/well and an additional washing step with 200 μL FACS buffer after the antibody staining.
4. There are some situations when Fc receptor-bound IgE might not be detectable. This is dependent on the age of the mouse and the hygiene status of the animal facility, and some knockout mouse strains are not able to produce IgE (Stat6-KO,

Table 1
Other markers suitable for detection of basophils in flow cytometry

Surface marker	Expression on basophils	Also expressed on	Clone	Company
FcεRIα	++	Mast cells, subset of DCs	MAR-1	BioLegend
CD49b	++	NK, NKT, subset of T cells	HMα2	BioLegend
CD200R3	++	Mast cells	Ba13	Biolegend
CD90.2 (depends on mouse strain)	+	T cells, NKT, thymocytes, stem cells	53–2.1	eBioscience
IgE	– to +++	B cells, mast cells	23G3	Southern Biotech
CD16/32	++	Macrophages, NK cells, B cells	2.4G2	eBioscience
CD244	++	NK cells, T cells, mast cells	2B4	eBioscience

IL-4/IL-13-KO) or the receptor (FcR-γ-chain-KO, FcεRI-KO). To stain basophils in these situations, other antibodies must be used. Suitable antibodies are listed in Table 1.

5. There are currently two clones for the anti-CD49b antibody available from commercial sources. In our hands, the HMα2 clone gives a brighter CD49b$^+$ population compared to the DX-5 clone.
6. Cells can be fixed for later analysis or intracellular staining in 4% PFA for 30 min at 4 °C followed by two washing steps in PBS and the appropriate intracellular staining. For live/dead discrimination, use fixable viability dyes.
7. We find that stained samples should be analyzed quickly, to reduce apoptosis of granulocytes.
8. During flow cytometric analysis, always acquire enough cells to gain a distinct basophil population. A good way to do so is to set the stopping gate to the basophil population and acquire at least 500 cells in this gate.
9. Basophils are somewhat autofluorescent; therefore, when analyzing cell surface markers, an isotype control is essential.
10. Be careful not to fix the tissue for too long.
11. The embedded tissue can also be frozen on dry ice.
12. To reduce background of some tissues, pretreat the slides with sodium citrate buffer (2.94 g trisodium citrate, 1 L distilled water; adjust pH to 6.0 with 1 M HCl, add 0.5 mL Tween-20, and store at 4 °C). Heat the buffer to 95 °C and incubate the slides for 10 min. This also de-masks some antigens.

13. Never let the slides run dry. Work in a moist chamber. Always use blocking solution to dilute antibodies and DAPI. As a humid chamber, we use a box with a lid, line it with wet paper towels, and put the slides on two plastic pipettes, so they do not touch the wet paper towels.
14. When adding any solution to the sections, be careful not to wash away the section. Best pipette it next to the tissue and always use the same spot to add the solution.
15. Instead of staining at room temperature for 2 h, staining overnight at 4 °C is also applicable.
16. Instead of using the anti-Mcpt8 staining, basophils can be identified as c-kit^{-}IgE^{+}, c-kit^{-}FcεRI^{+}, or IL-4^{+}c-kit^{-}MBP^{-}CD4^{-} cells in immunofluorescence microscopy.
17. Other companies also offer signal amplification kits (Vector, Invitrogen). In addition, there are also other conjugates for the tyramide available. When a biotin-conjugated antibody is involved, it is mandatory to perform a block of endogenous biotin-binding sites. Therefore, we use the Streptavidin/Biotin Blocking Kit (Vector Laboratories). After the first block with serum, wash and add one drop of streptavidin-blocking solution to each section. Incubate for 15 min at room temperature. Rinse briefly in PBS. Add one drop of biotin-blocking solution onto each section and incubate for 15 min. Wash 2 × 5 min and continue with the appropriate antibody.
18. At this point, a second staining can be integrated. Do not forget to block free antibody binding sites.
19. If there is no violet laser installed in the fluorescence microscope, other viability dyes can be used. Propidium iodide gives a bright signal but cannot be used together with Cy3-conjugated antibodies.
20. To apply one drop, we normally cut off the tip of a blue 1000 μL pipette tip.
21. Control the immunofluorescence staining by using an isotype control and a biological control. As biological control, basophil-deficient mice (Mcpt8Cre, Mcpt8DTR) or basophil-depleted mice (MAR-1, Ba103) can be used.
22. Fluorescence fades over time, so timely analysis is recommended.

Acknowledgments

This work was supported in part by an ERC starting grant (PAS_241506) from the European Union.

References

1. Ehrlich P (1879) Beiträge zur Kenntnis der granulierten Bindegewebszellen und der eosinophilen Leukozyten. Arch Anat Physiol 3:166
2. Urbina C, Ortiz C, Hurtado I (1981) A new look at basophils in mice. Int Arch Allergy Appl Immunol 66:158–160
3. Gessner A, Mohrs K, Mohrs M (2005) Mast cells, basophils, and eosinophils acquire constitutive IL-4 and IL-13 transcripts during lineage differentiation that are sufficient for rapid cytokine production. J Immunol 174:1063–1072
4. Min B et al (2004) Basophils produce IL-4 and accumulate in tissues after infection with a Th2-inducing parasite. J Exp Med 200:507–517
5. Mukai K et al (2005) Basophils play a critical role in the development of IgE-mediated chronic allergic inflammation independently of T cells and mast cells. Immunity 23:191–202
6. Voehringer D, Shinkai K, Locksley RM (2004) Type 2 immunity reflects orchestrated recruitment of cells committed to IL-4 production. Immunity 20:267–277
7. Bodger MP et al (1987) A monoclonal antibody reacting with human basophils. Blood 69:1414–1418
8. Gane P et al (1993) Flow cytometric evaluation of human basophils. Cytometry 14:344–348
9. Han X et al (2008) Immunophenotypic study of basophils by multiparameter flow cytometry. Arch Pathol Lab Med 132:813–819
10. Irani AM et al (1998) Immunohistochemical detection of human basophils in late-phase skin reactions. J Allergy Clin Immunol 101:354–362
11. Kepley CL, Craig SS, Schwartz LB (1995) Identification and partial characterization of a unique marker for human basophils. J Immunol 154:6548–6555
12. McEuen AR et al (1999) Development and characterization of a monoclonal antibody specific for human basophils and the identification of a unique secretory product of basophil activation. Lab Investig 79:27–38
13. McEuen AR et al (2001) Mass, charge, and subcellular localization of a unique secretory product identified by the basophil-specific antibody BB1. J Allergy Clin Immunol 107:842–848
14. Buhring HJ, Streble A, Valent P (2004) The basophil-specific ectoenzyme E-NPP3 (CD203c) as a marker for cell activation and allergy diagnosis. Int Arch Allergy Immunol 133:317–329
15. Kleine-Tebbe J et al (2006) Diagnostic tests based on human basophils: potentials, pitfalls and perspectives. Int Arch Allergy Immunol 141:79–90
16. Ohnmacht C, Voehringer D (2009) Basophil effector function and homeostasis during helminth infection. Blood 113:2816–2825
17. Liu AY et al (2013) Mast cells recruited to mesenteric lymph nodes during helminth infection remain hypogranular and produce IL-4 and IL-6. J Immunol 190:1758–1766
18. Ohnmacht C et al (2010) Basophils orchestrate chronic allergic dermatitis and protective immunity against helminths. Immunity 33:364–374
19. Nabe T et al (2013) Roles of basophils and mast cells infiltrating the lung by multiple antigen challenges in asthmatic responses of mice. Br J Pharmacol 169:462–476
20. Torrero MN et al (2009) CD200R surface expression as a marker of murine basophil activation. Clin Exp Allergy 39:361–369
21. Lunderius C, Hellman L (2001) Characterization of the gene encoding mouse mast cell protease 8 (mMCP-8), and a comparative analysis of hematopoietic serine protease genes. Immunogenetics 53:225–322
22. Lutzelschwab C et al (1998) Characterization of mouse mast cell protease-8, the first member of a novel subfamily of mouse mast cell serine proteases, distinct from both the classical chymases and tryptases. Eur J Immunol 28:1022–1033
23. Wada T et al (2010) Selective ablation of basophils in mice reveals their nonredundant role in acquired immunity against ticks. J Clin Invest 120:2867–2775
24. Sullivan BM et al (2011) Genetic analysis of basophil function in vivo. Nat Immunol 12:527–535

Correction to: Measuring Histamine and Cytokine Release from Basophils and Mast Cells

Bettina M. Jensen, Ewa A. Bartko, Katrine Baumann, and Per S. Skov

Correction to:
Chapter 21 in: Bernhard F. Gibbs, Franco H. Falcone (eds), *Basophils and Mast Cells: Methods and Protocols*, Methods in Molecular Biology, vol. 2163,
https://doi.org/10.1007/978-1-0716-0696-4_21

The original version of the book was inadvertently published without incorporating the author's proof corrections in chapter 21 wherein the equation ng/mL histamine in sample = (Sample FU − Blank FU)/(50 ng FU −Blank FU) has been updated as ng/mL histamine in sample = (Sample FU − Blank FU) × 50/(50 ng FU −Blank FU).

The updated original version of this chapter can be found at
https://doi.org/10.1007/978-1-0716-0696-4_21

Bernhard F. Gibbs and Franco H. Falcone (eds.), *Basophils and Mast Cells: Methods and Protocols*, Methods in Molecular Biology, vol. 2163, https://doi.org/10.1007/978-1-0716-0696-4_30, © Springer Science+Business Media, LLC, part of Springer Nature 2023

INDEX

Bernhard F. Gibbs and Franco H. Falcone (eds.), *Basophils and Mast Cells: Methods and Protocols*, Methods in Molecular Biology, vol. 2163, https://doi.org/10.1007/978-1-0716-0696-4, © Springer Science+Business Media, LLC, part of Springer Nature 2020

MIX
Papier aus verantwortungsvollen Quellen
Paper from responsible sources
FSC® C105338

If you have any concerns about our products,
you can contact us on
ProductSafety@springernature.com

In case Publisher is established outside the EU,
the EU authorized representative is:
Springer Nature Customer Service Center GmbH
Europaplatz 3, 69115 Heidelberg, Germany

Printed by Libri Plureos GmbH
in Hamburg, Germany